Anatomy & Physiology Coloring Book

Senior Vice President, STEM and Original Custom: Daryl Fox
Executive Program Director, Life Sciences and Original Custom: Sandy Lindelof
Program Director, Original Custom: Christian Perlee
Senior Teaching & Learning Strategist: Jim Zubricky
Marketing Manager, Original Custom and Labs: Lauren Harp
Senior Content Production Manager: Heather Christian
Senior Pre-Production Manager: Kathy Lopez
Production Manager: Carolyn Coole
Associate Editor: Samantha Hunt
Editorial Assistant: Megan Johannessen
Senior Scientific Illustrator: Katie Nealis
Scientific Illustrator: Archana Venkataramani
Typesetter: Kimberly Ritter
Proofreader: Lori Emelander

Cover image: chalongrat/stock.adobe.com

Printed in the United States of America

10 9 8 7 6 5 4 3 2 1

ISBN 978-1-5339-6549-3

Macmillan Learning
120 Broadway
New York, NY 10271
www.macmillanlearning.com

COLORBOOK 6549-3 F24

Table of Contents

Acknowledgments ix
Anatomy Coloring Book Instructions xi
Chapter 1: Introduction 1
Body planes
Body directions
Anatomical areas
Body cavities
Arm, leg, and trunk movements
Shoulder, neck, foot, and hand movements
Chapter 2: Homeostasis, Chemistry, and Cells 13
Homeostasis
pH balance
Atom
Multicellular organization
Generalized animal cell
Cell membrane
Cellular transport
Mitochondrion
Metabolic physiology
Microscope
Chapter 3: Tissues 27
Epithelial tissues
Connective tissues
Muscle tissues
Cartilage tissues
Chapter 4: The Integumentary System 35
Integument (skin)
Skin sensory
Epidermis
Nail structure

Chapter 5: The Skeletal System 43

Bone tissue and long bone anatomy

Bone shapes

Skeletal divisions

The complete skeleton

Bones of the skull

Additional views of the skull

Interior views of the skull

Bones of the spine

Typical vertebrae

Sacrum and coccyx

Rib cage and sternum

Bones of the upper limb

Scapula and clavicle

Long bones of the upper limb

Bones of the wrist and hand

Bones of the pelvis

Bones of the lower limb

Bones of the thigh and leg

Bones of the foot

Chapter 6: Joints 79

Joint classification

Joint movement

Spinal ligaments

Shoulder joint

Elbow joint

Wrist and finger joints

Hip joint

Knee joint

Ankle and foot joints

Chapter 7: The Muscular System 97

Neuromuscular junction

Skeletal muscle shapes

Sliding filament theory of muscle contraction

Brain meninges
Precentral and postcentral regions
Cranial nerves
Spinal cord
Spinal cord tracts
Spinal cord and meninges
Neural pathways: Ascending tracts
Neural pathways: Descending tracts
Cervical plexus
Nerves of the upper limb
Brachial plexus
Lumbosacral plexus
Nerves of the lower limb
Dermatomes
Autonomic nervous system
Chapter 9: The Special Senses 183
Eye
Retina
Eye muscles
Optic pathway
Ear
Inner ear equilibrium structures
Olfactory reception
Taste reception
Taste buds
Chapter 10: The Endocrine System 199
Endocrine glands
Pituitary gland
Thyroid gland
Adrenal gland
Hormone secretions
Pancreas
Islet of Langerhans

Chapter 14: The Digestive System 271

Digestive system overview

Mouth

Tooth anatomy

Teeth

Swallowing

Mesentery and peritoneal spaces

Stomach

Small intestine

Large intestine

Layers of intestinal wall

Rectum and anus

Liver

Accessory digestive organs

Digestion

Chapter 15: The Urinary System 295

Urinary system

Kidney

Renal pyramid with juxtamedullary and cortical nephrons

Nephron function and formation of urine

Chapter 16: The Reproductive System 305

Male reproductive system

Spermatic cord

Penis and testes

Spermatogenesis

Female reproductive system

Uterus and ovaries

Follicle maturation

Reproductive cycle

Breast

Index 321

Acknowledgments

We would like to thank our team at Hayden-McNeil for all their hard work and efforts that went into this project: Chris Perlee, Program Director for Original Custom; Jim Zubricky, Senior Teaching & Learning Strategist; Samantha Hunt, Associate Editor; Katie Nealis, Senior Scientific Illustrator; Archana Venkataramani, Scientific Illustrator; Megan Johanessen, Editorial Assistant; Kimberly Ritter, Typesetter; Lori Emelander, Proofreader; Heather Christian, Senior Content Production Manager; Kathy Lopez, Senior Pre-Production Manager; Carolyn Coole, Production Manager; and Lauren Harp, Marketing Manager for Original Custom and Labs. We would also like to thank Sandy Blood from SB Indexing Services for creating the index for this manuscript.

This project would not have been possible without four individuals who laid the groundwork for this project in different forms and iterations over the past decade: Cheri Bowman, Proofreading Supervisor; Jonathan Higgins, Medical Illustrator; Amanda Humphrey, Illustration Supervisor; and Rachel Huhta, Scientific Illustrator.

We are indebted to the many educators who helped shape this project. With their guidance, we have created a text to better meet the needs of students everywhere. To all of them, we are deeply grateful:

- Mercedes Alba, San Antonio College
- Robin Altman, California State University, Sacramento
- Matt Daniels, Elyria High School and Lorain County Community College
- Haley Eaton, Northwood High School
- Dr. Mariem Hathout, MD, MPH, Aurora University
- Alex Imholtz, Prince George's Community College
- Kebret T. Kebede, MD, Nevada State University
- Dana Kurpius, Elgin Community College
- Zak Linczeski, Northern Michigan University
- Shari Litch Gray, PhD, Regis College
- Margaret Long, Gwinnett Technical College
- Cristy Tower-Gilchrist, Emory University

Anatomy Coloring Book Instructions

- Note any specific coloring instructions on each activity first. These may dictate certain colors for certain structures or specify leaving certain structures uncolored.
- Next, color in the circle next to a term and then use the same color for the corresponding structure on the illustration.
- If the activity includes a key, the term and corresponding structure will have matching letters. Unkeyed activities will have no letters, allowing you to test your knowledge of the proper location of the structures. Both of these kinds of activities will help you establish a mental connection between each component and its name to aid in memory retention and recall.
- Using 20–24 unique colors is recommended. When there are more structures to color than you have unique colors for, you will need to duplicate colors.
- Structures to be colored are outlined in thick, black lines. Thinner black lines, gray lines, and hatching within these structures represent surface details and are to be colored over.
- Pay attention to any thick, gray dashed lines. These indicate an area to be colored but do not depict an actual edge.
- Use lighter colors for larger shapes.
- Do not use similar shades of the same color for abutting structures.
- Subhead structures (denoted by a letter with a subscript) should be the same color as the main structure.
- Note the areas that should not be colored (marked with *), and the areas to be colored gray (marked with •).
- Symmetrical structures won't always be labeled; however, they should be colored on both sides.

1

Introduction

Body planes
Body directions
Anatomical areas
Body cavities
Arm, leg, and trunk movements
Shoulder, neck, foot, and hand movements

In this first chapter of your prismatic journey, you will be filling in the full picture of human **movement and the bodily planes** that form the frame. You will be studying the three quintessential geometric planes that anatomists and physiologists often refer to when discussing our bodies as well as the myriad of movements our marvelous bodies are capable of. As you thoughtfully traverse through each section of this chapter, you will discover the astonishing array of angles our bodies have to explore. Let's bring our bodies to life!

Body planes

- (a) Sagittal
- (b) Frontal (coronal)
- (c) Transverse (horizontal)

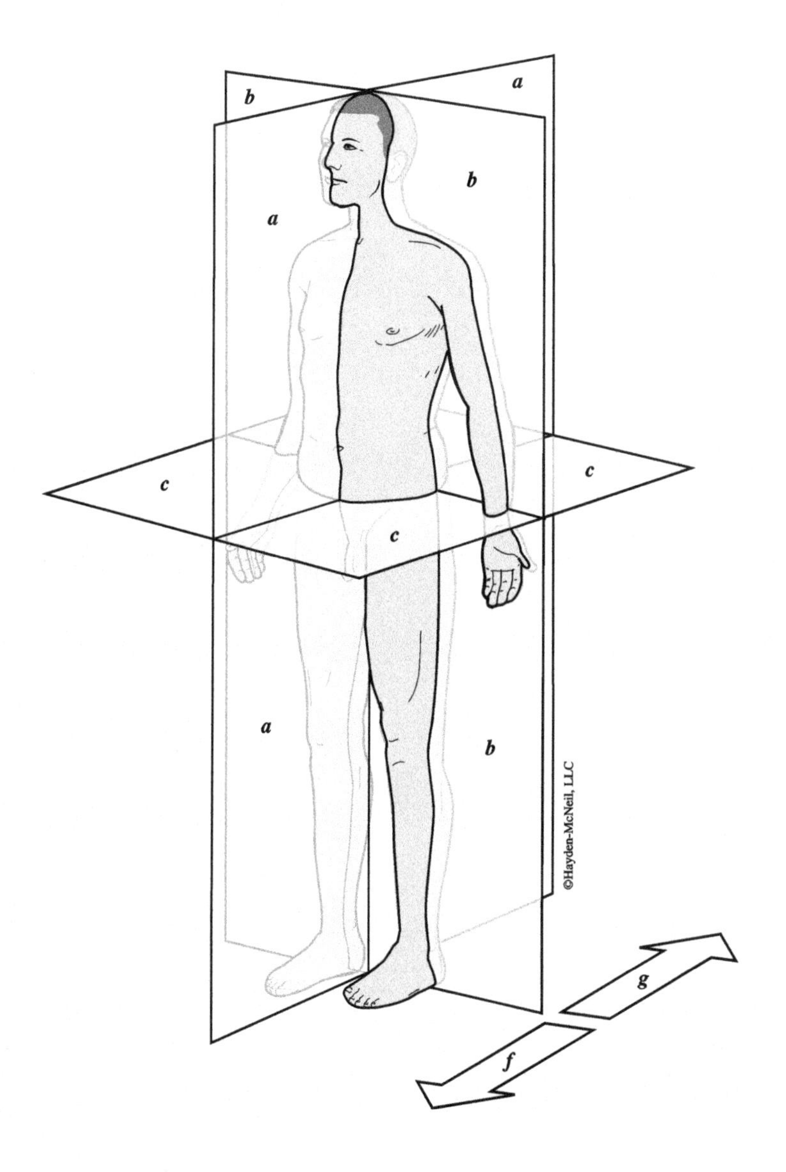

Body directions

- (d) Cranial, superior
- (e) Caudal, inferior
- (f) Anterior, ventral
- (g) Posterior, dorsal
- (h) Medial
- (i) Lateral
- (j) Proximal
- (k) Distal

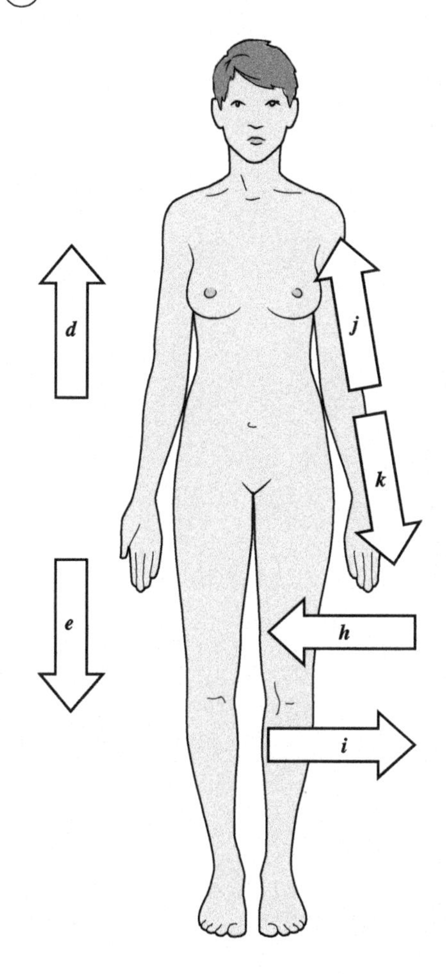

Anatomical areas

- (a) Head (cephalic)
- (b) Neck (cervical)
- (c) Chest (thoracic)
- (d) Abdomen (abdominal)
- (e) Pelvis (pelvic)
- (f) Upper limb
- (g) Lower limb

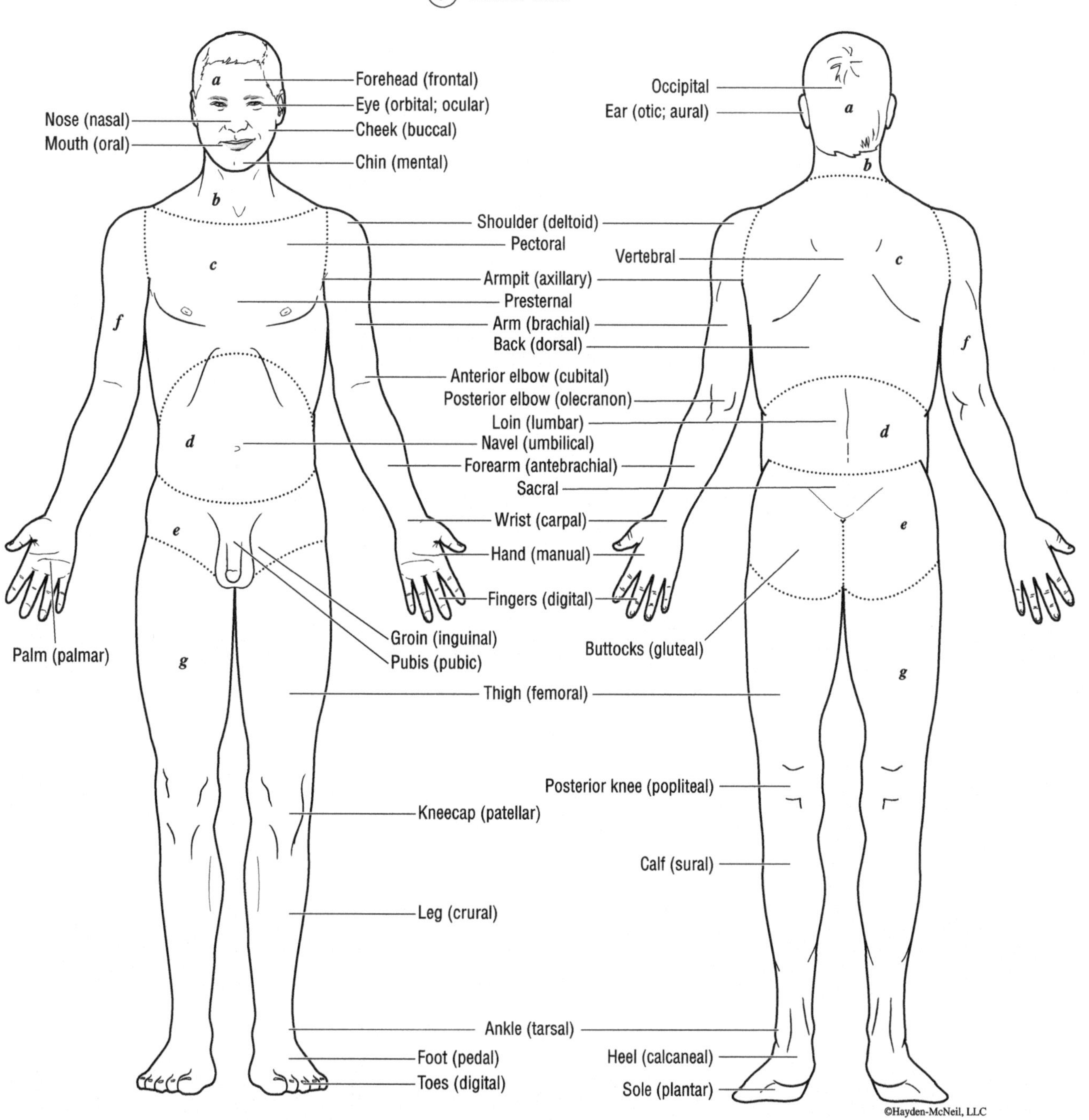

Body cavities

- (a) Cranial cavity
- (b) Vertebral cavity
- (c) Thoracic cavity
- (d) Abdominal cavity
- (e) Pelvic cavity
- (f) Pleural cavity
- (g) Pericardial cavity

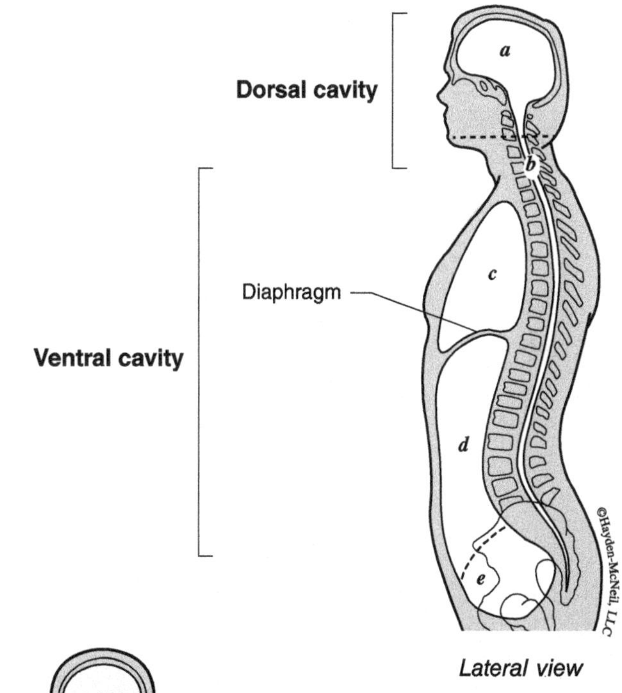

Lateral view

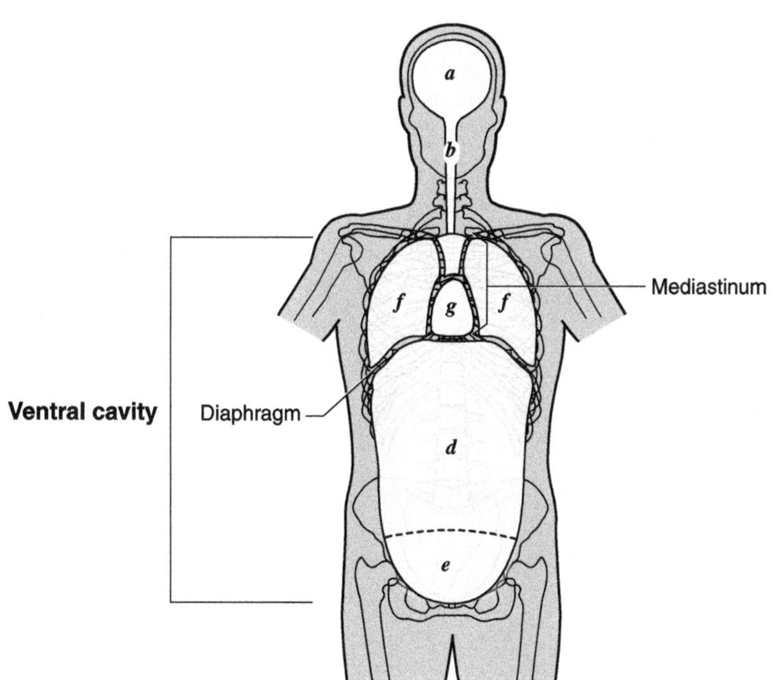

Anterior view

Arm, leg, and trunk movements

- (a) Flexion
- (b) Extension
- (c) Hyperextension
- (d) Lateral flexion
- (e) Rotation
- (f) Medial rotation
- (g) Lateral rotation
- (h) Circumduction
- (i) Abduction
- (j) Adduction
- (k) Supination
- (l) Pronation

Arm movements

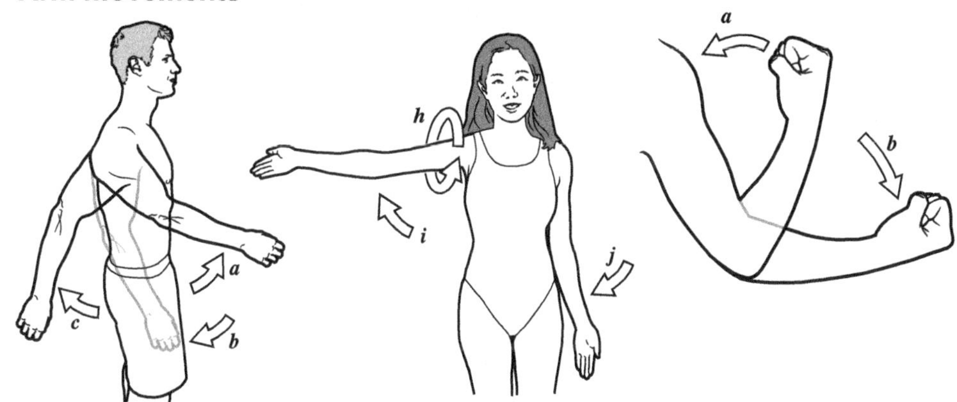

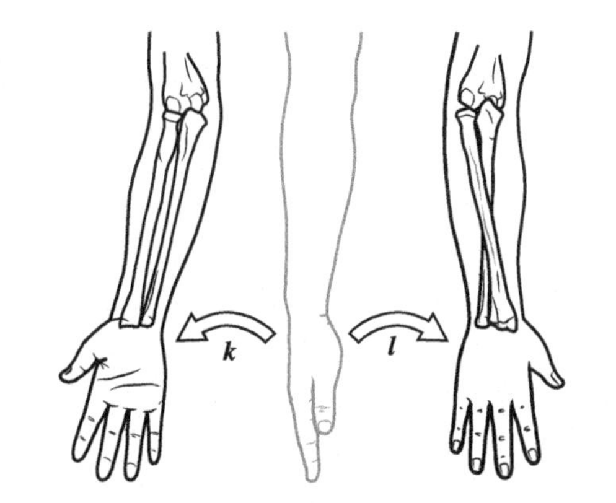

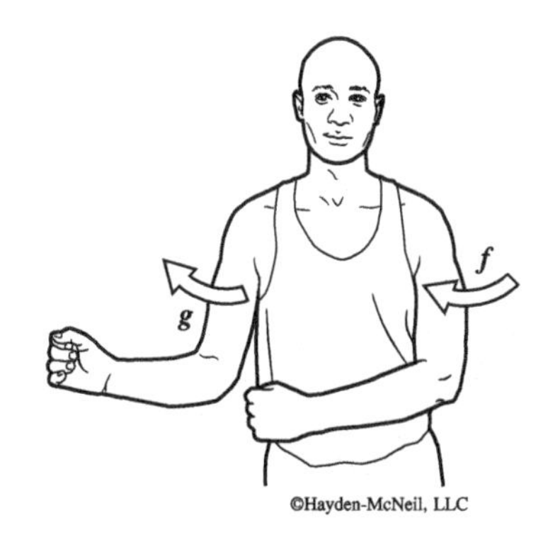

Trunk movements

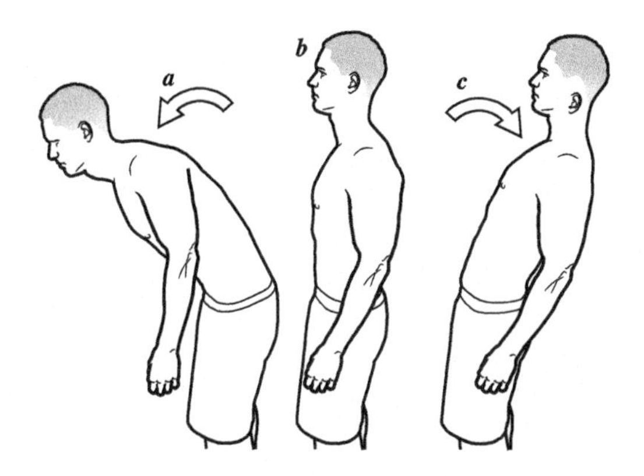

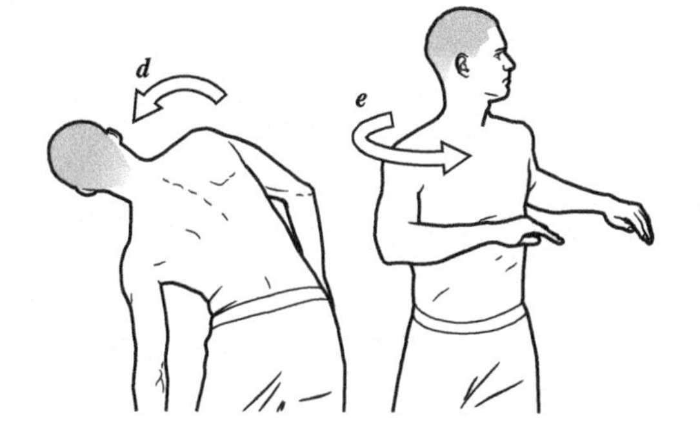

Leg movements

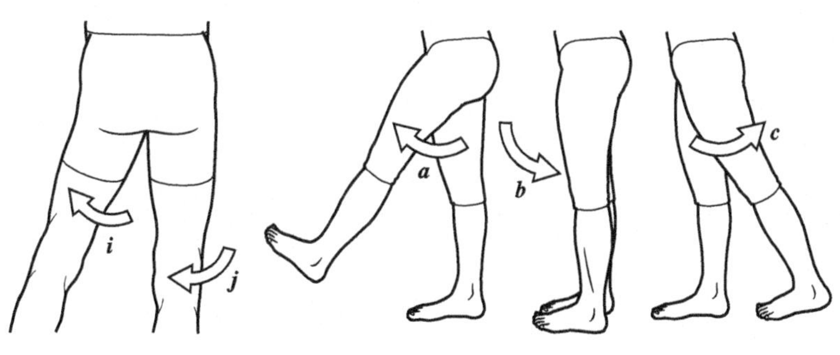

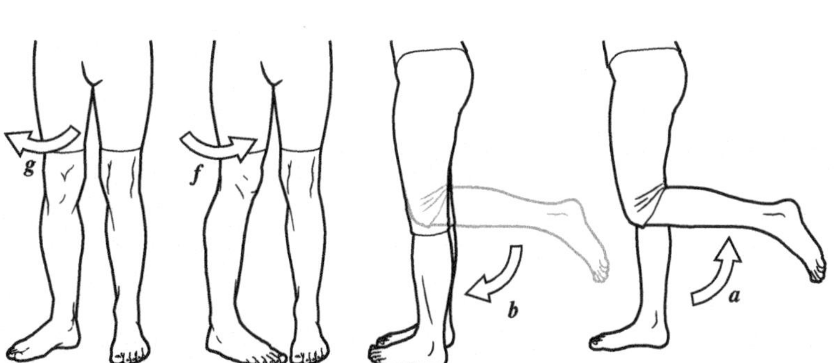

Shoulder, neck, foot, and hand movements

- (a) Flexion
- (b) Extension
- (c) Hyperextension
- (d) Lateral flexion
- (e) Rotation
- (f) Circumduction
- (g) Abduction
- (h) Adduction
- (i) Depression
- (j) Elevation
- (k) Protraction
- (l) Retraction
- (m) Dorsiflexion
- (n) Plantar flexion
- (o) Eversion
- (p) Inversion
- (q) Opposition of thumb

Shoulder movements

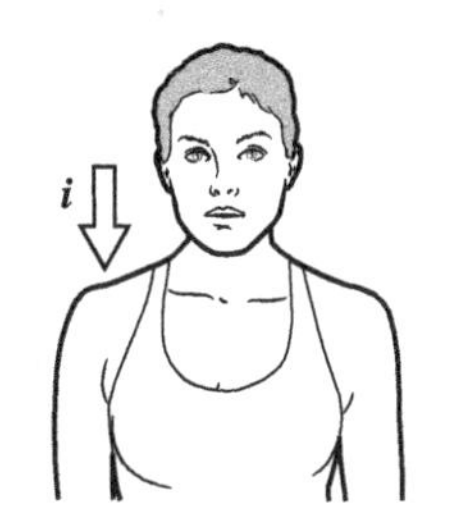

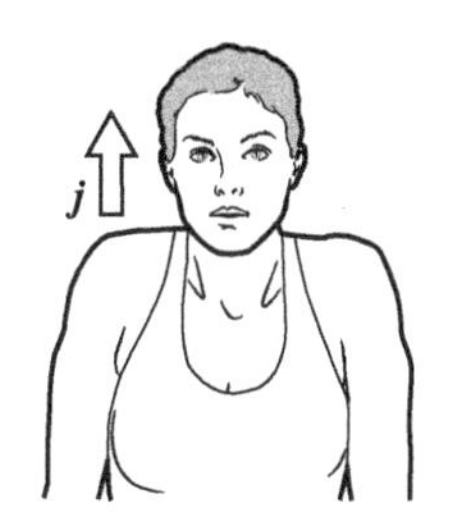

Neck movements

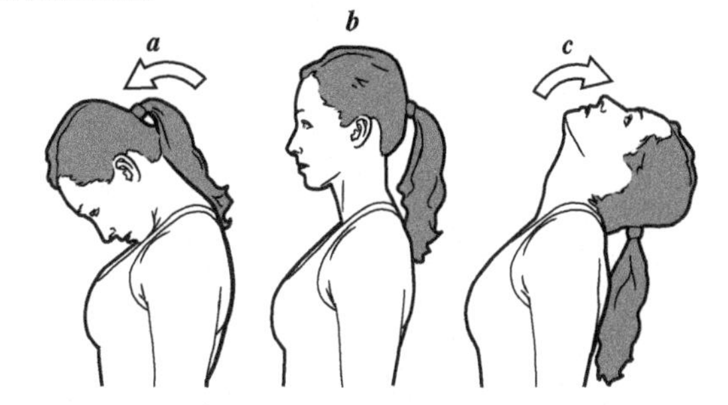

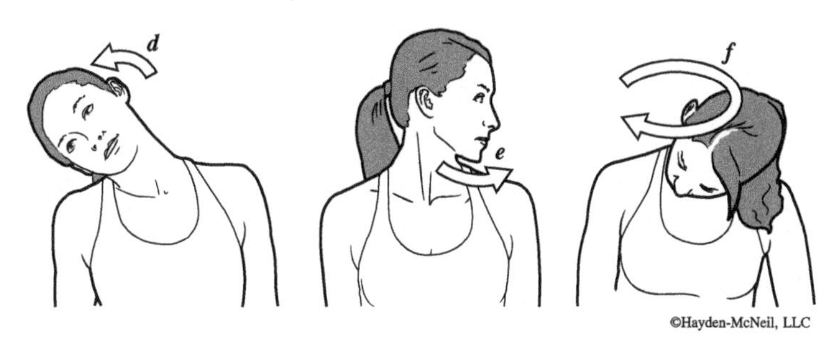

Foot movements

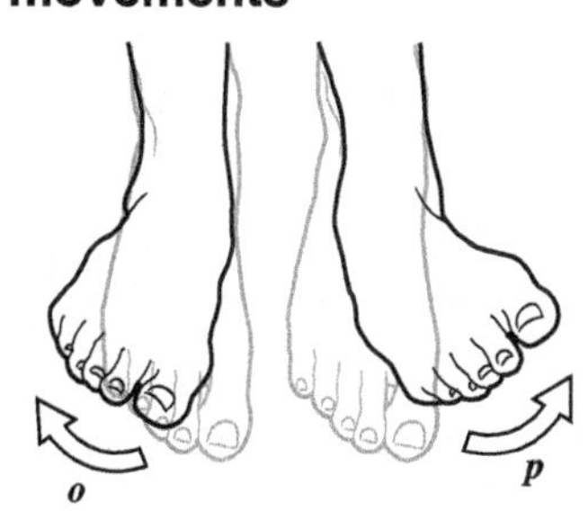

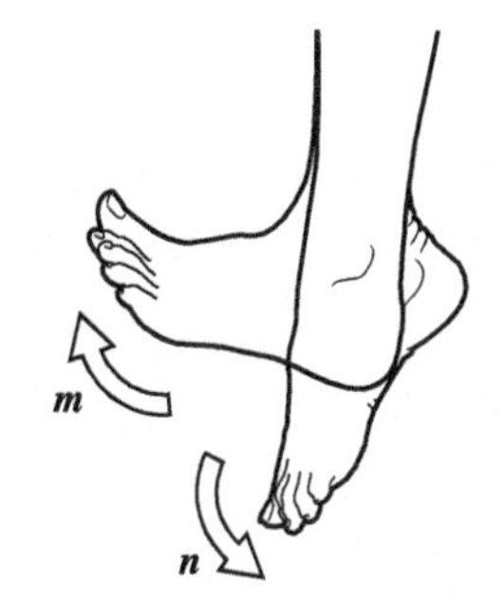

Hand movements

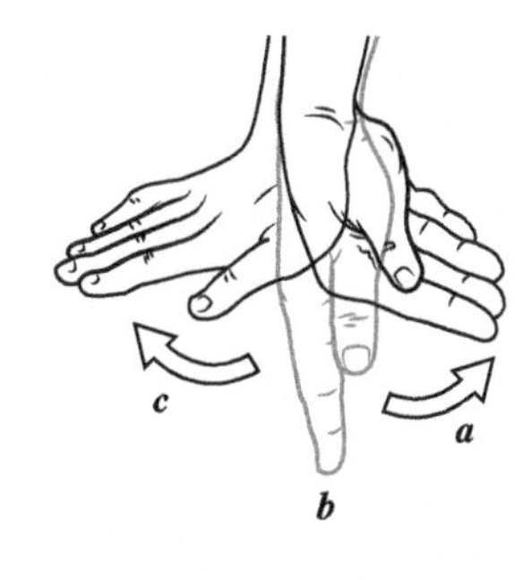

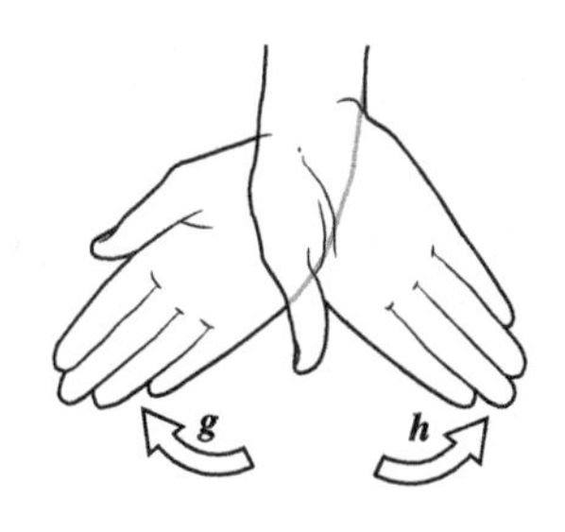

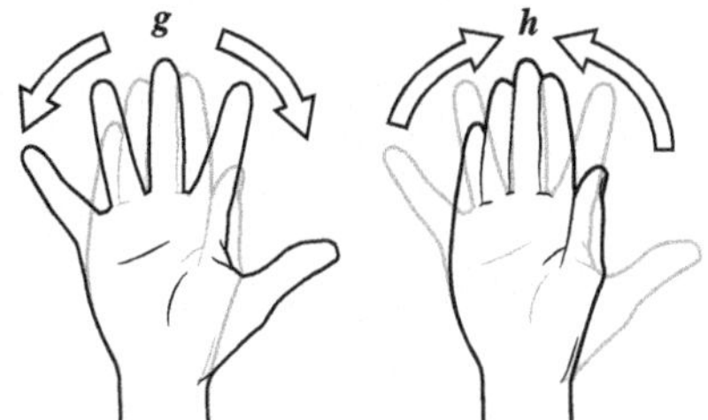

2

Homeostasis, Chemistry, and Cells

Homeostasis
pH balance
Atom
Multicellular organization
Generalized animal cell
Cell membrane
Cellular transport
Mitrochondrion
Metabolic physiology
Microscope

In this chapter, you will be pulling back the curtain on the molecular stage and witnessing the delicate dance of **homeostasis and multicellular organization**. You will be cataloging the awe-inspiring inner workings of our cells that ensure our body's equilibrium as well as piecing together the kaleidoscopic lenses of the remarkable microscopes we can see them with—you will have a backstage pass to it all. So, find your seat and prepare to experience the fascinating world of cellular biology in color!

Homeostasis

- (a) Stimulus
- (b) Receptor (sensor)
- (c) Control center
- (d) Effector

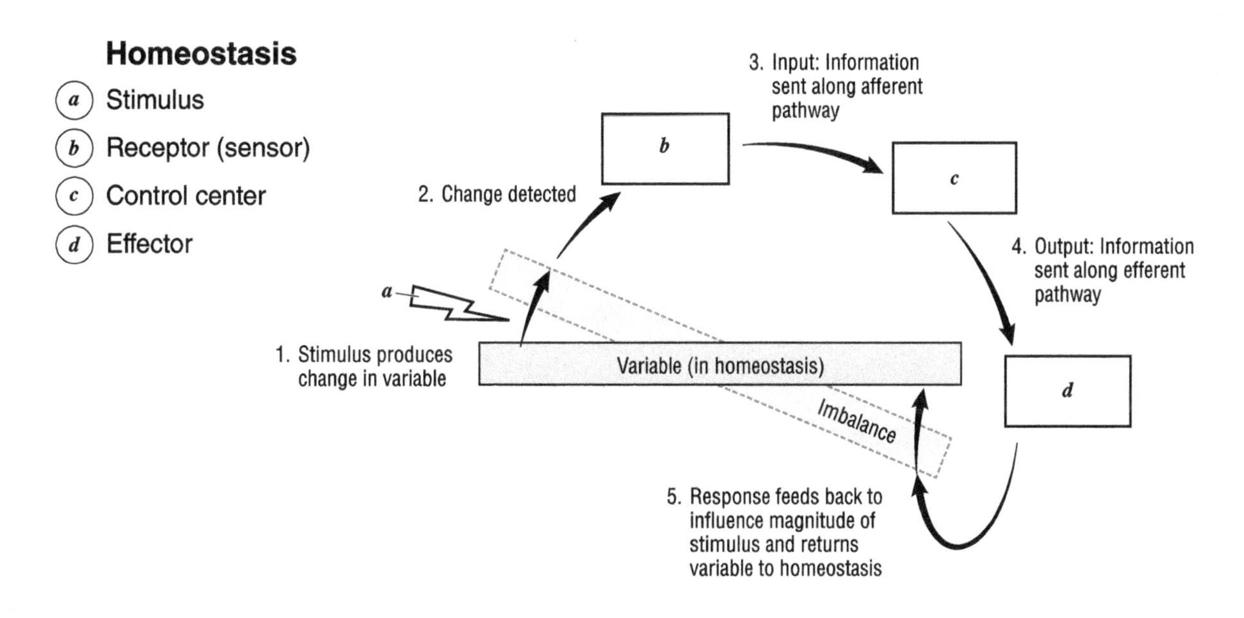

pH balance

- (e) Acidic
- (f) Basic

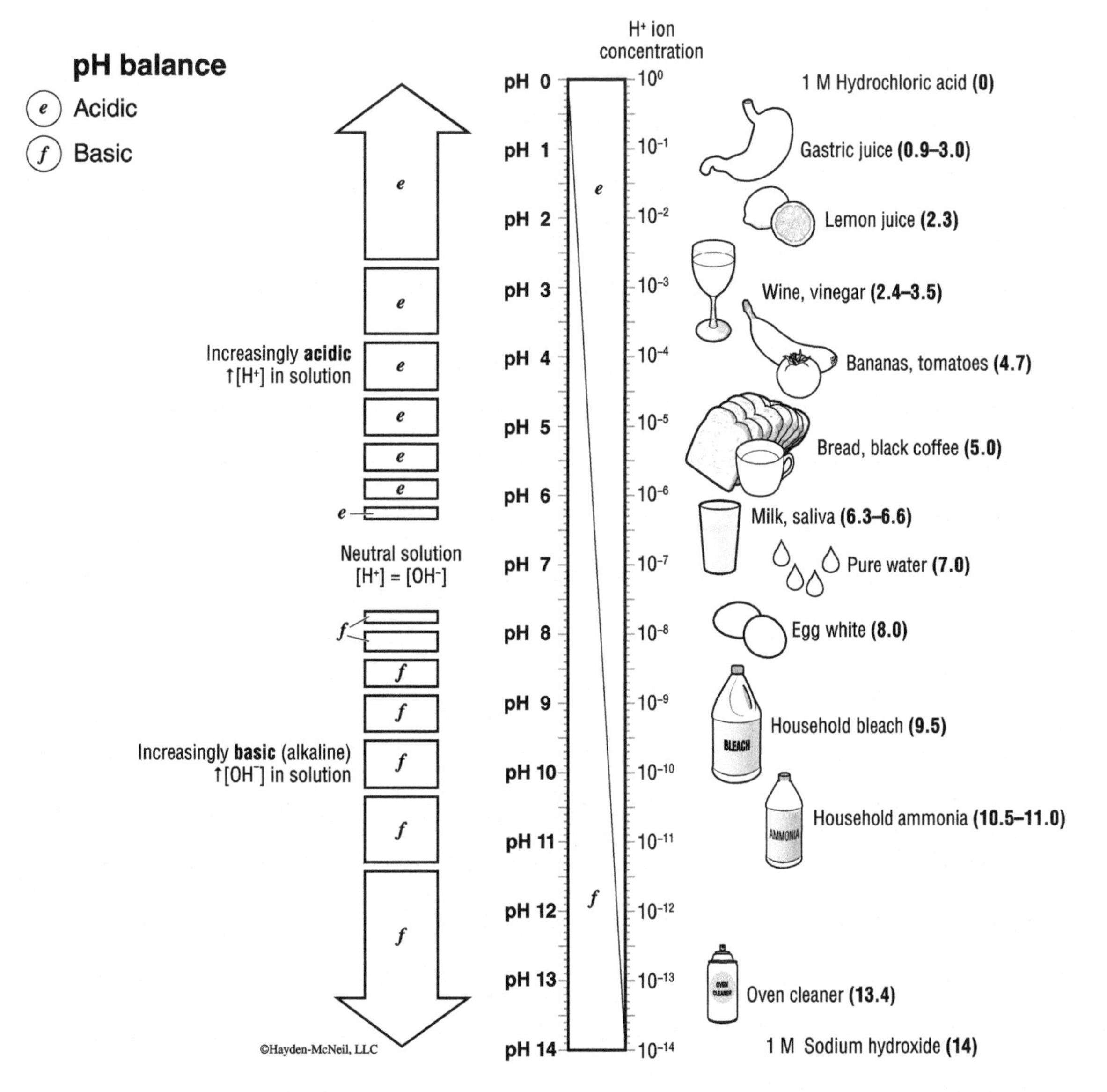

Atom

Note that electrons are found in a cloud surrounding the nucleus. Color the electron area accordingly, with a higher density of color closer to the nucleus.

(a) Proton, p^+ (positive charge)

(b) Neutron, n^0 (no charge)

(c) Electron, e^- (negative charge)

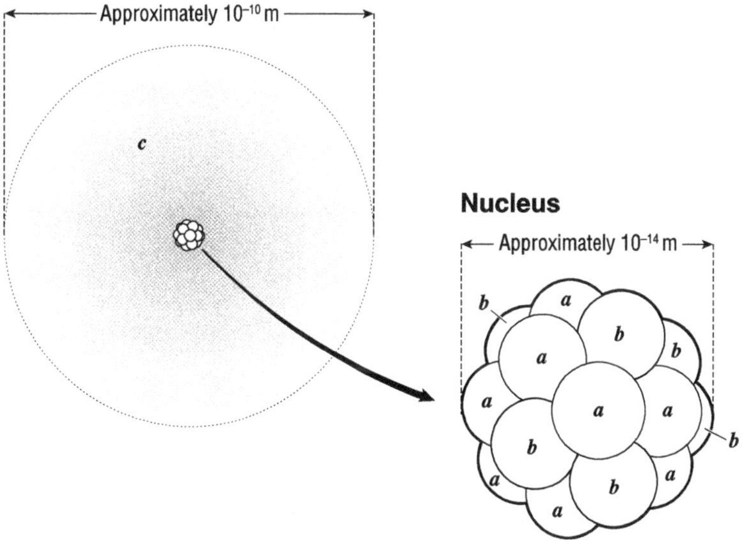

Multicellular organization

(d) Atomic level

(e) Molecular level

(f) Cellular level

(g) Tissue level

(h) Organ level

(i) System level

(j) Organism level

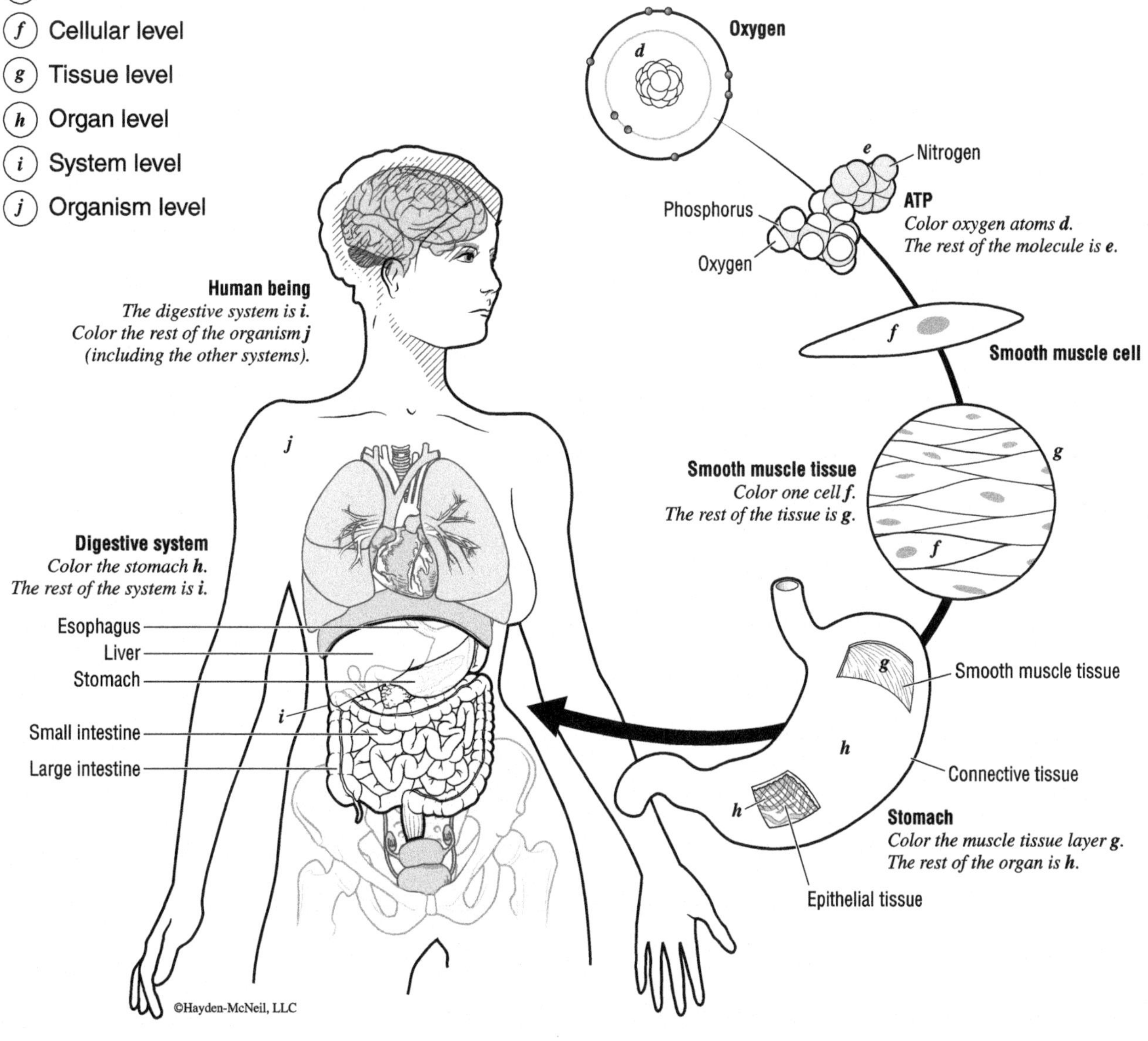

Generalized animal cell

- (a) Nucleus
  - a_1 Nucleolus
- (b) Smooth endoplasmic reticulum
- (c) Rough endoplasmic reticulum
- (d) Ribosomes
- (e) Centriole
- (f) Lysosome
- (g) Cytoplasm
- (h) Golgi apparatus
- (i) Mitochondrion
- (j) Plasma membrane
- (k) Cytoskeleton
  - k_1 Microfilament
  - k_2 Intermediate filament
  - k_3 Microtubule

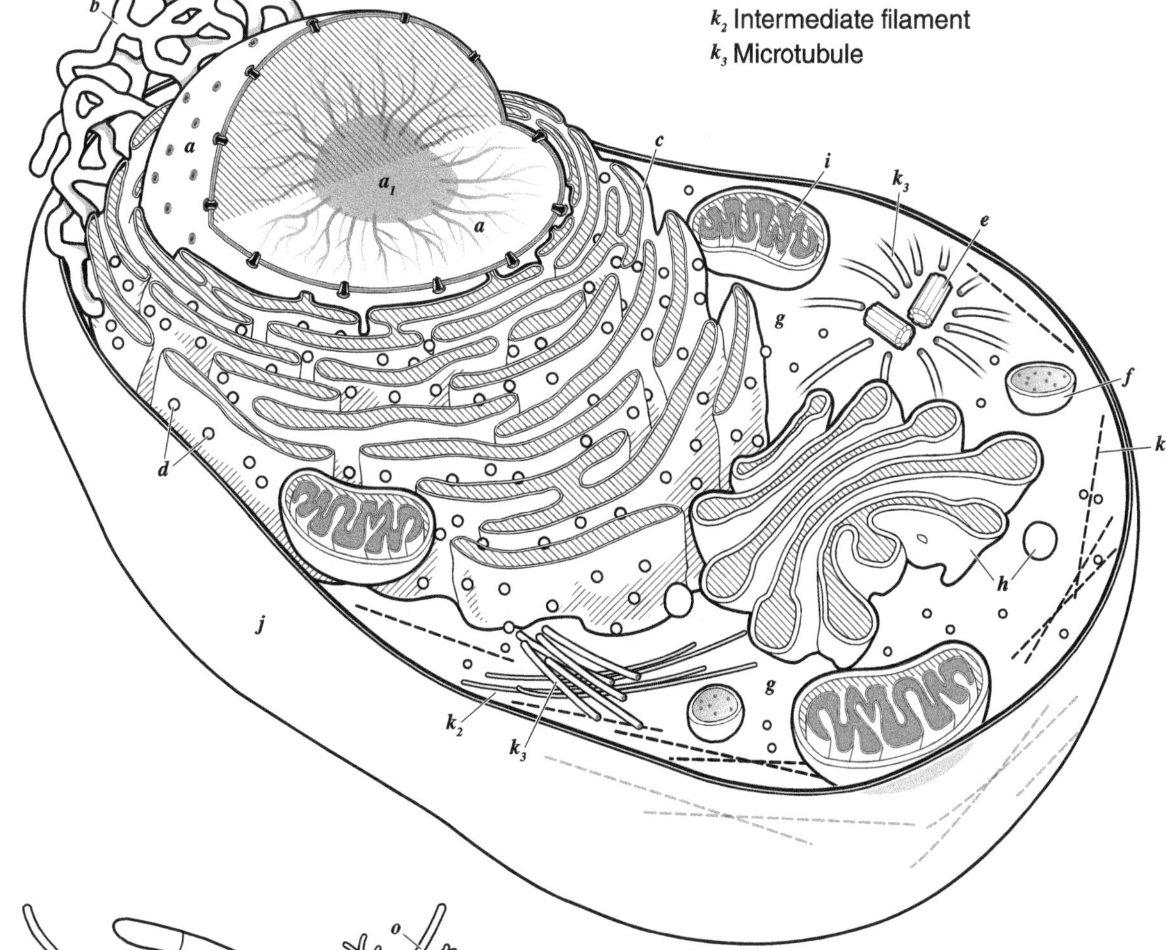

Outside cell

Inside cell

Cell membrane

- (l) Phospholipid bilayer
- (m) Glycolipid
- (n) Glycoprotein
- (o) Extracellular matrix protein
- (p) Integral (transmembrane) protein
- (q) Cholesterol
- (r) Peripheral protein

Cellular transport

- (a) Food particle
- (b) Plasma membrane; food vacuole
- (c) Golgi apparatus; lysosome
- (d) Endoplasmic reticulum; transport vesicle
- (e) Hydrolytic enzymes
- (f) Nucleus

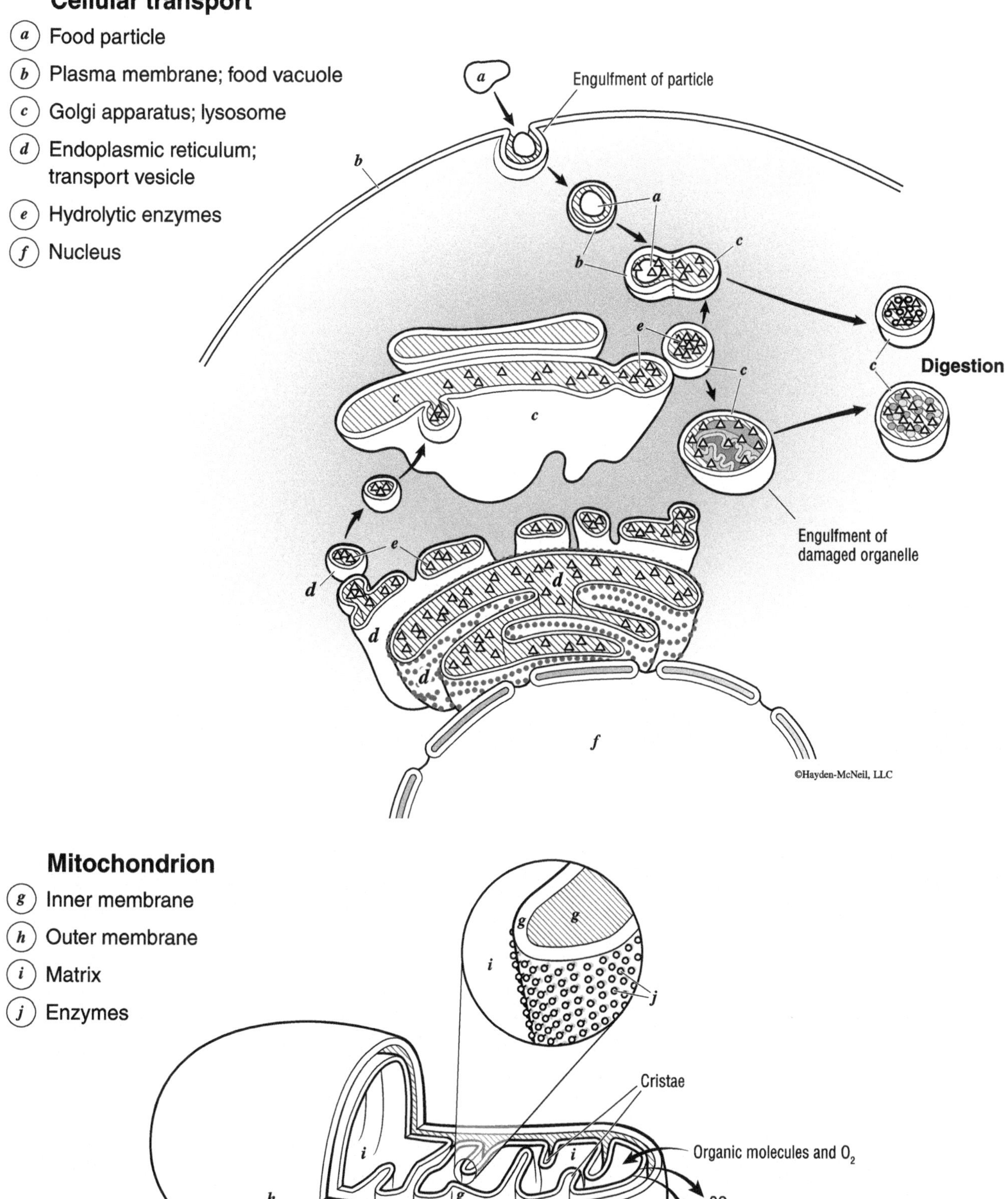

Mitochondrion

- (g) Inner membrane
- (h) Outer membrane
- (i) Matrix
- (j) Enzymes

Metabolic physiology

Color arrows the same in both.

Cellular respiration

- (a) Glycolysis
- (b) Acetyl CoA
- (c) Krebs cycle
- (d) Electron transport chain

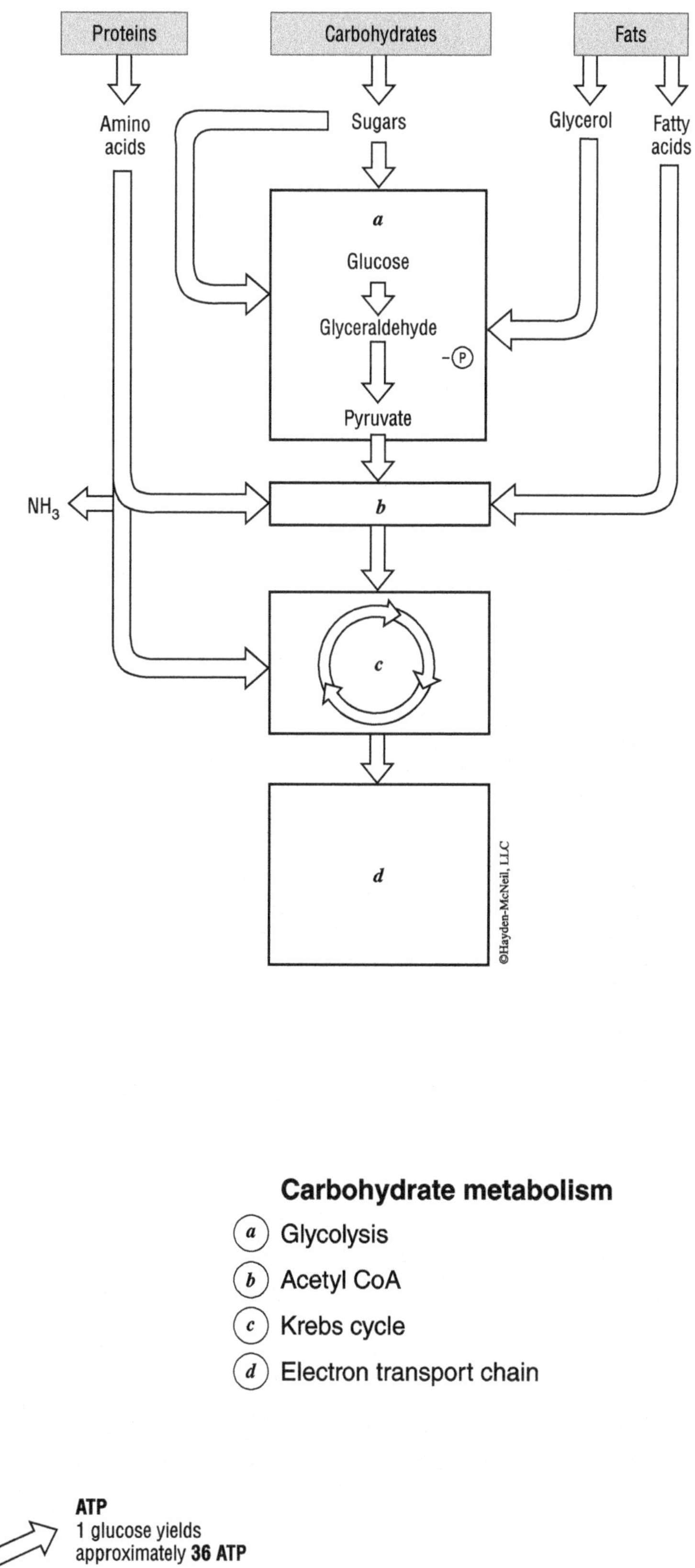

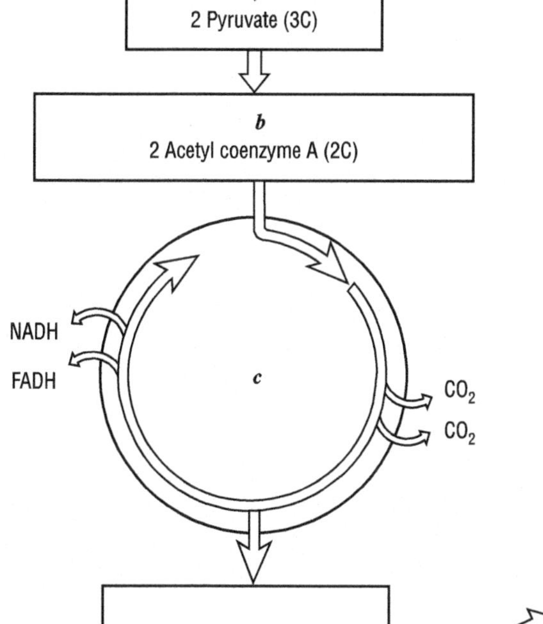

Carbohydrate metabolism

- (a) Glycolysis
- (b) Acetyl CoA
- (c) Krebs cycle
- (d) Electron transport chain

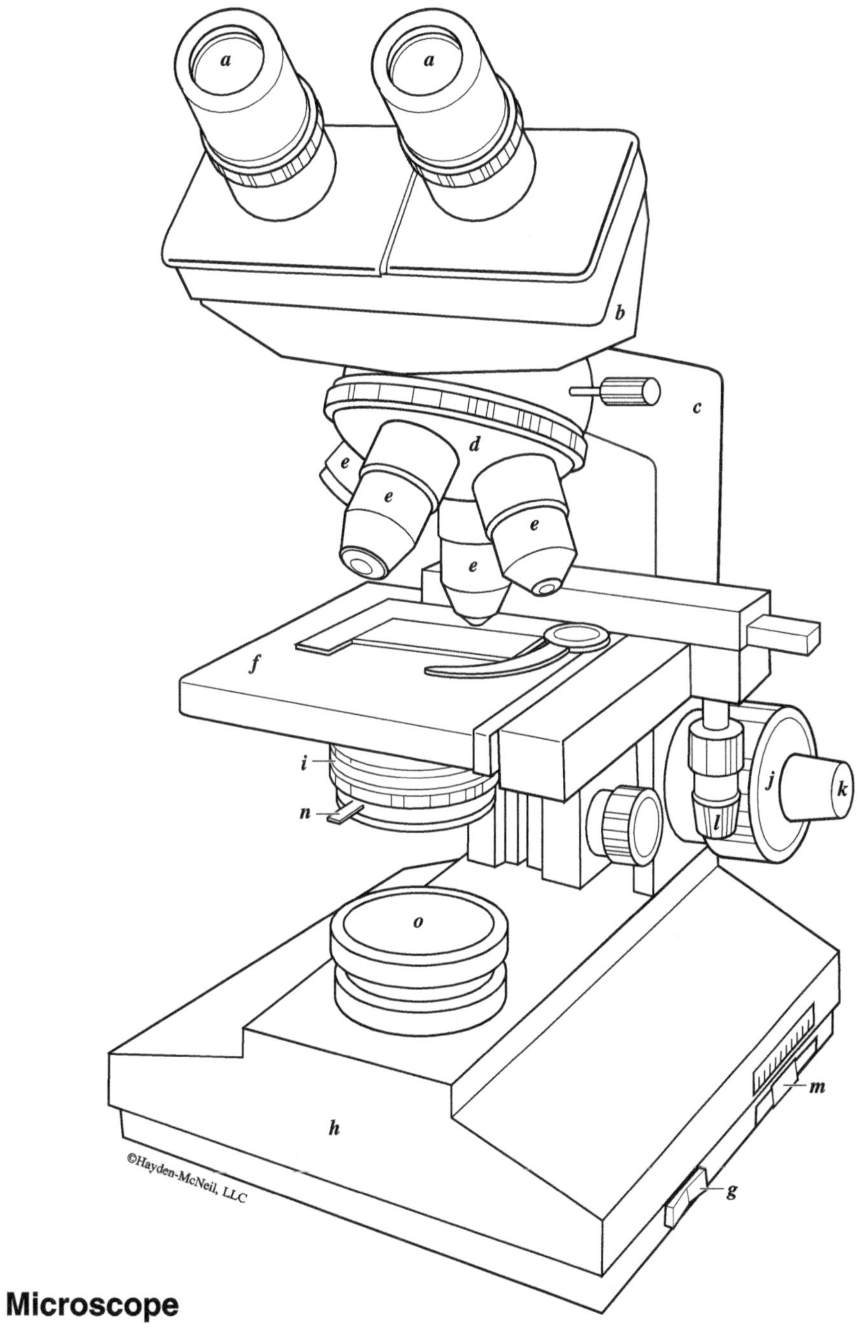

Microscope

- (a) Ocular lenses
- (b) Body tube
- (c) Arm
- (d) Revolving nosepiece
- (e) Objective lenses
- (f) Stage
- (g) Power switch
- (h) Base
- (i) Condenser
- (j) Coarse adjustment knob (coarse focus)
- (k) Fine adjustment knob (fine focus)
- (l) Stage adjustment knob (moves the slide)
- (m) Light intensity control
- (n) Iris diaphragm adjustment lever
- (o) Light source

3

Tissues

Epithelial tissues

Connective tissues

Muscle tissues

Cartilage tissues

In the following chapter, you will be weaving together the vibrant **tissues** that make up the human body's biological fabric. With every new thread, you will be discovering the cooperative connection between all four of the human body's diverse and differing tissues, each with their own unique strengths and capabilities. Join us in enriching the tapestry of our tissues and knitting together our expansive epithelial tissue, resilient muscle tissue, communicative nerve tissue, and, finally, the connective tissue that binds us all together.

Epithelial tissues

Color the blood vessel red.

Simple (one layer)

(a) Squamous
(b) Cuboidal
(c) Columnar
(d) Pseudostratified columnar

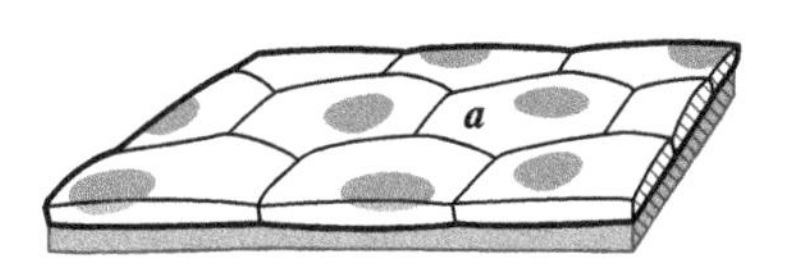

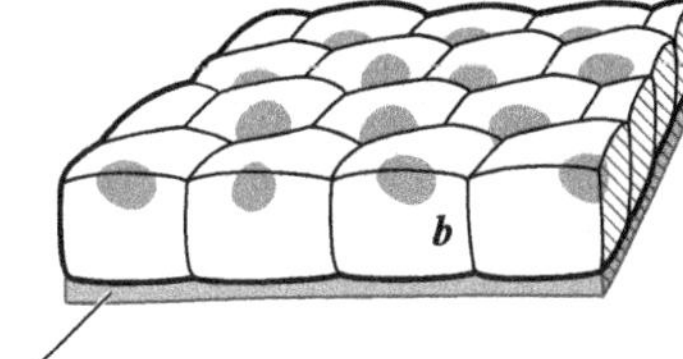

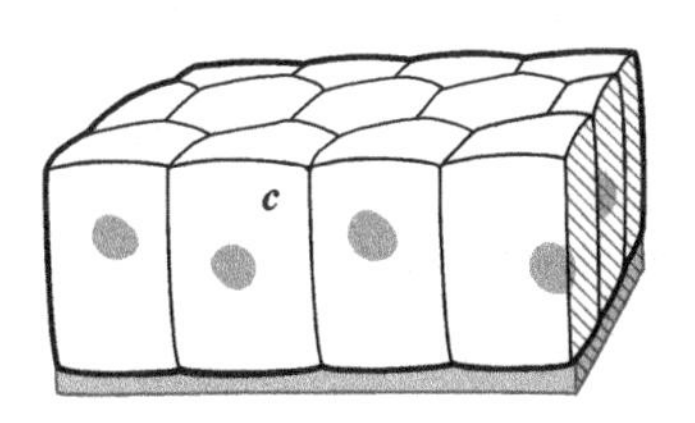

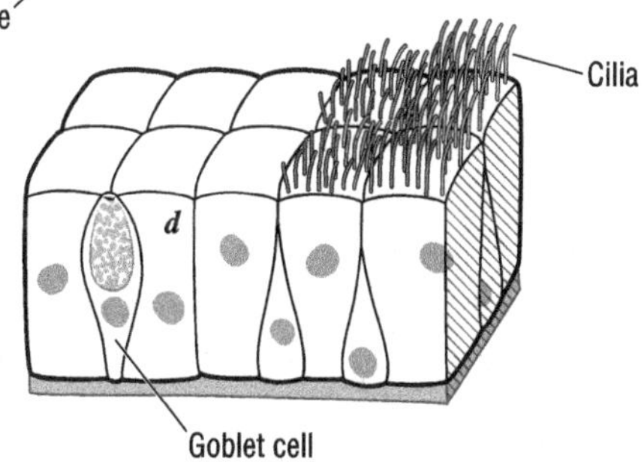

Stratified (several layers)

(a) Squamous
(c) Columnar
(e) Transitional

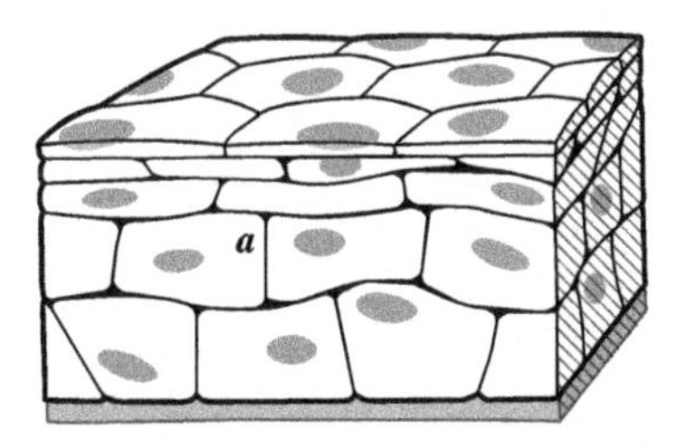

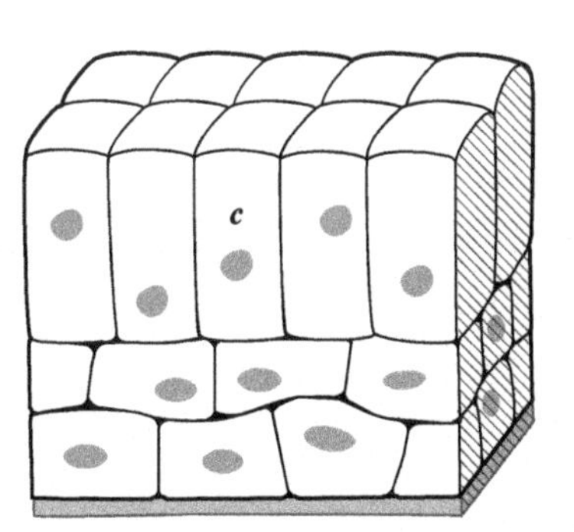

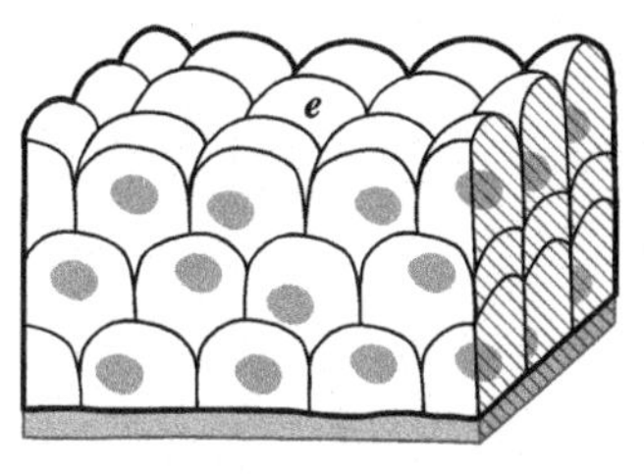

Contracted

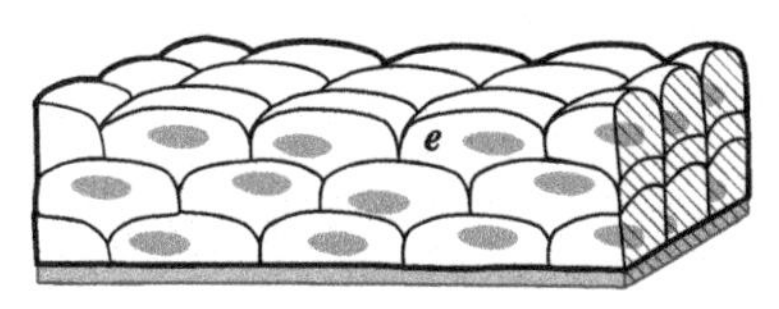

Distended

Glandular

(f) Exocrine gland
(g) Endocrine gland

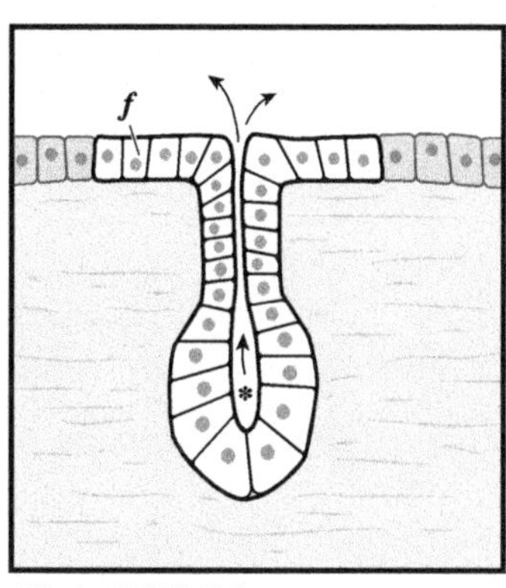

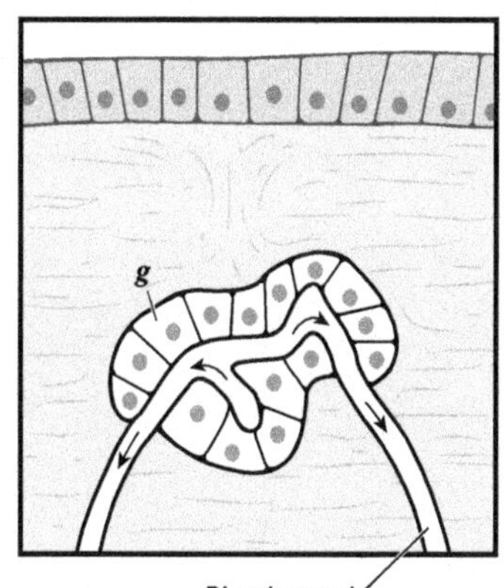

Connective tissues

Use a light color for collagen fibers and a dark color for elastic fibers.

Loose areolar

- (a) Fibroblasts
- (b) Collagen fibers
- (c) Elastic fibers

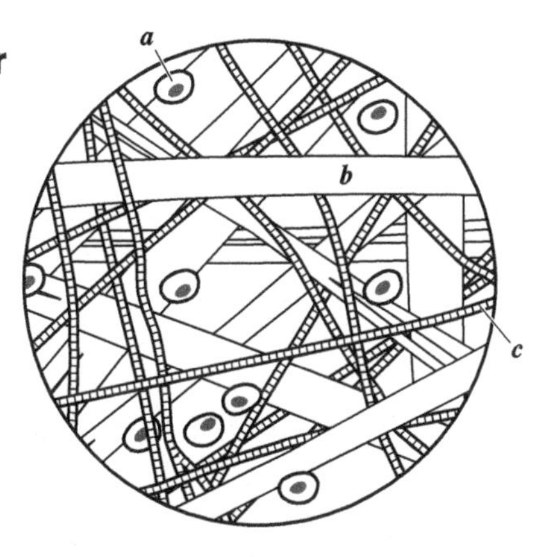

Adipose

- (d) Adipocyte
- (e) Lipid droplet

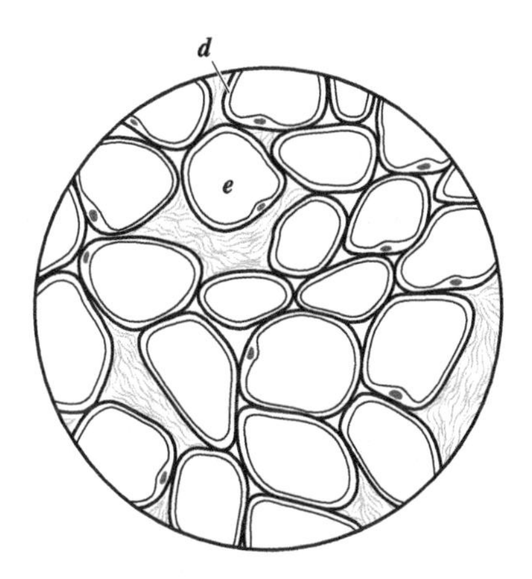

Dense regular

- (a) Fibroblasts
- (b) Collagen fibers

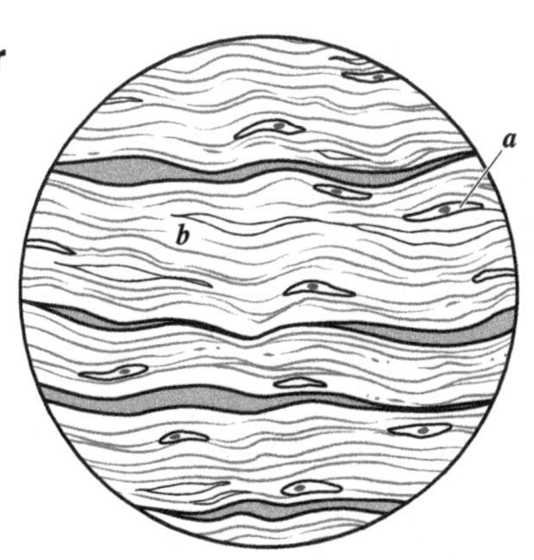

Dense irregular

- (a) Fibroblasts
- (b) Collagen fibers
- (c) Elastic fibers

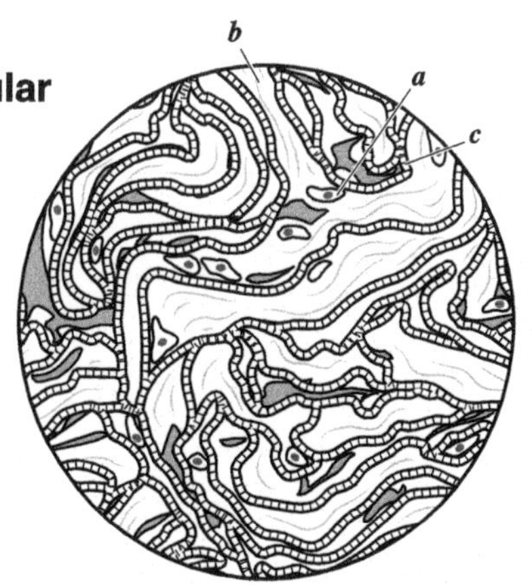

Reticular

- (f) Reticular fibers
- (g) Organ cells

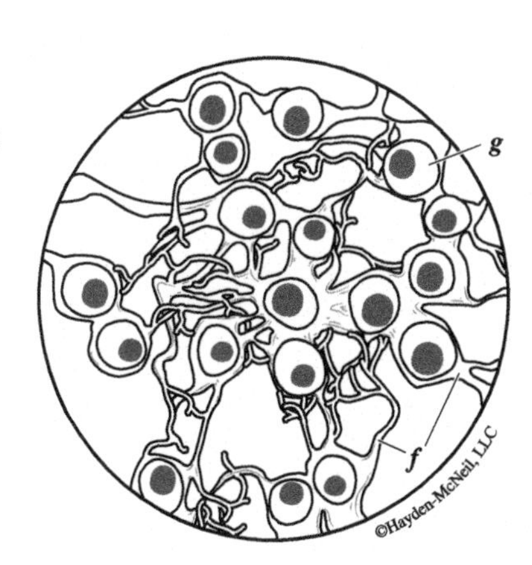

Elastic

- (a) Fibroblasts
- (c) Elastic fibers

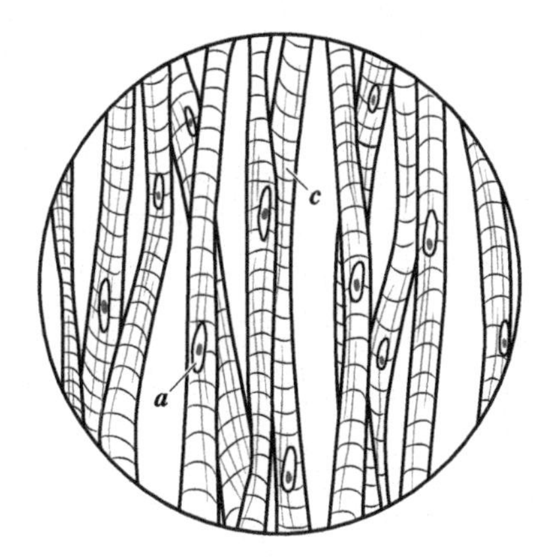

Muscle tissues

Smooth muscle

(a) Smooth muscle cells
(b) Nucleus

Striated muscle

(c) Striated muscle cells
(b) Nucleus

Cardiac muscle

(d) Cardiac muscle cells
(b) Nucleus
(e) Intercalated discs

Cartilage tissues

Hyaline cartilage

(f) Chondrocytes
(g) Matrix
(h) Lacunae

Elastic cartilage

(f) Chondrocytes
(g) Matrix
(h) Lacunae
(i) Elastic fibers

Fibrocartilage

(f) Chondrocytes
(g) Matrix
(h) Lacunae
(j) Collagen fibers

Nervous tissue

(k) Neurons
(b) Nucleus
(l) Glial cells

4

The Integumentary System

Integument (skin)

Skin sensory

Epidermis

Nail structure

This chapter covers the chromatic complexities of our body's outermost layer, the **integumentary system**! Well-versed in versatility and elasticity, our skin is exceedingly more elaborate than what first meets the eye. Throughout these pages, you will be examining the resilient role of our protective epidermis, getting to the root of how our hair grows, and uncovering the intricate layers beneath that give way to sensation. "Skin deep" has never had a deeper meaning!

Integument (skin)

Epidermis

(a) Stratum corneum

(b) Stratum lucidum *Leave white.*

(c) Stratum granulosum

(d) Stratum spinosum

(e) Stratum basale

Dermis

(f) Connective tissue *Use a light color.*
f_1 Papilla

(g) Nerve *Color yellow.*

(h) Vessels *Color red and blue.*

Hair

(i) Shaft

(j) Root

(k) Follicle
k_1 Bulb
k_2 Papilla

(l) Arrector pili m.

(m) Sebaceous gland

(n) Sweat gland

(o) Subcutaneous layer

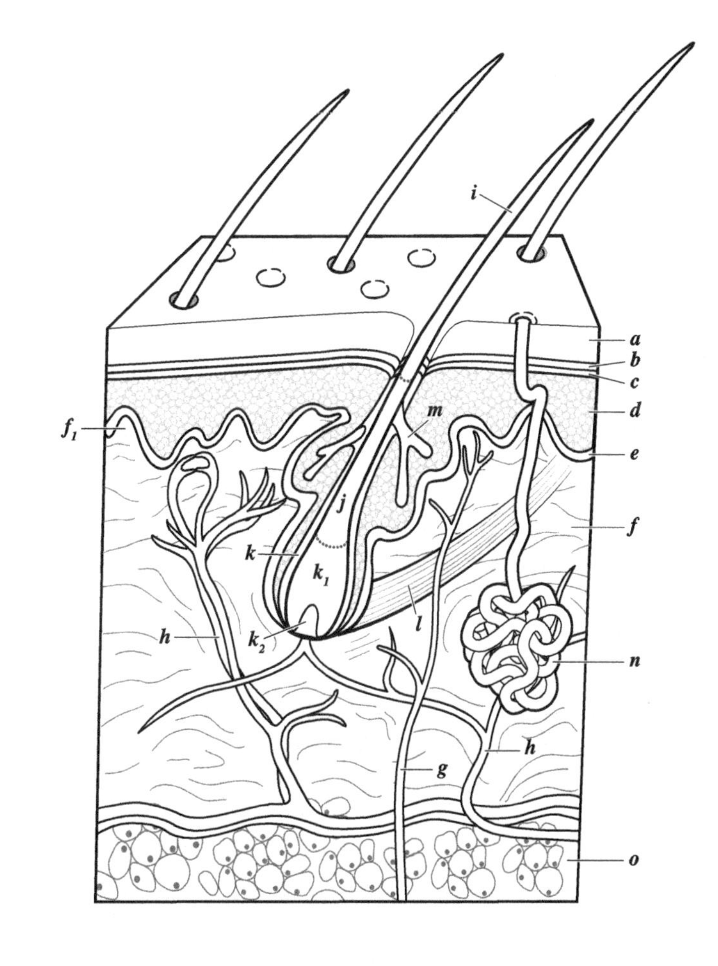

Skin sensory

(p) Tactile epithelial (Merkel) cells

(q) Tactile (Meissner's) corpuscle

(r) Thermoreceptor

(s) Nociceptor

(t) Bulbous (Ruffini) corpuscle

(u) Lamellar (Pacinian) corpuscle

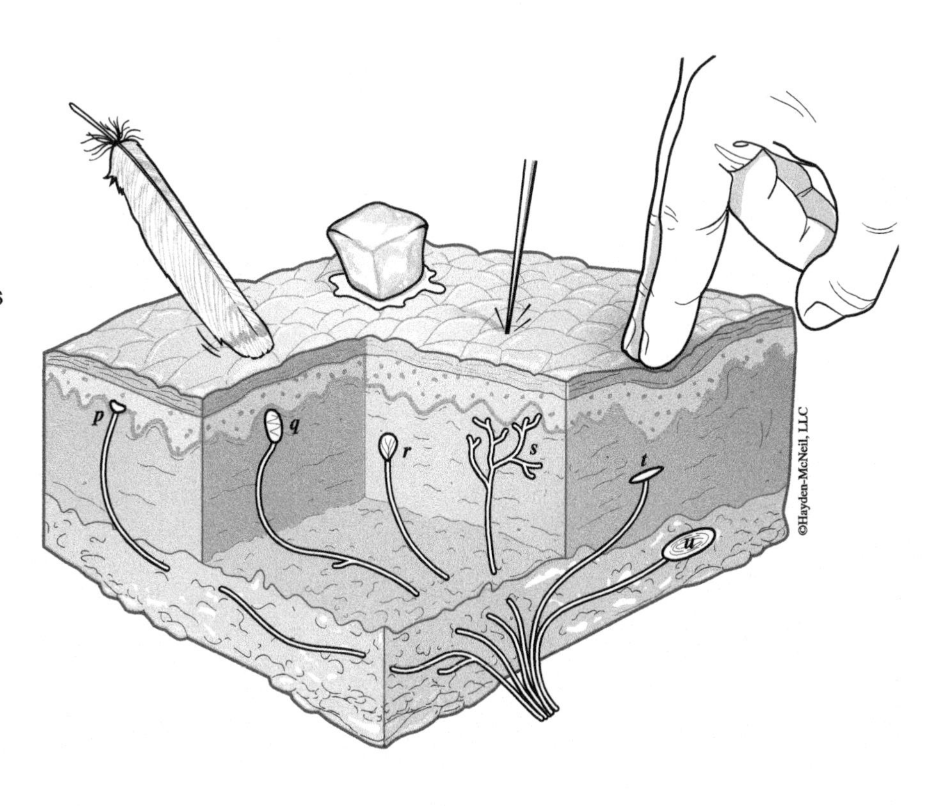

Epidermis

(a) Stratum corneum

(b) Stratum lucidum *Leave white.*

(c) Stratum granulosum

(d) Stratum spinosum

(e) Stratum basale

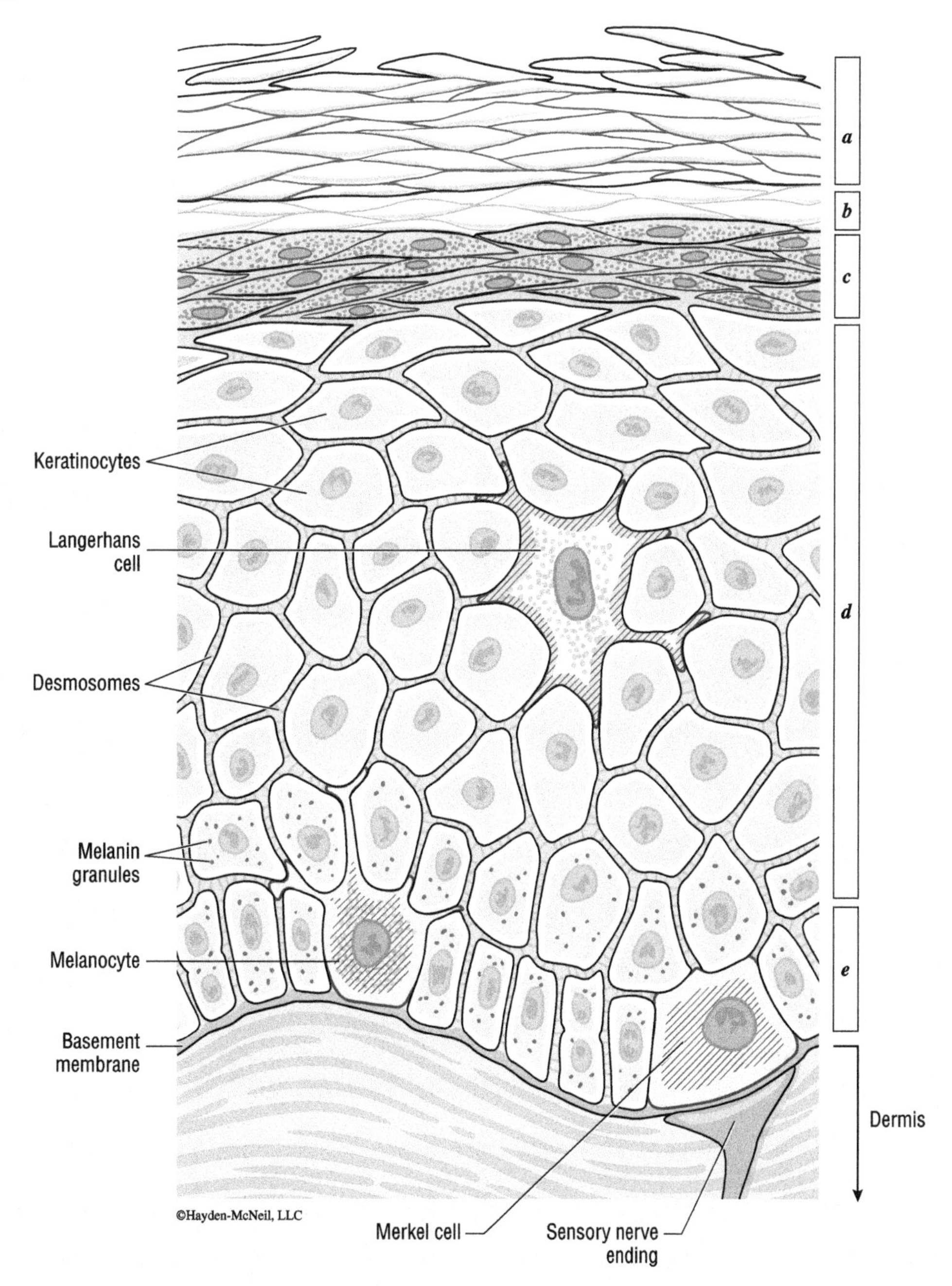

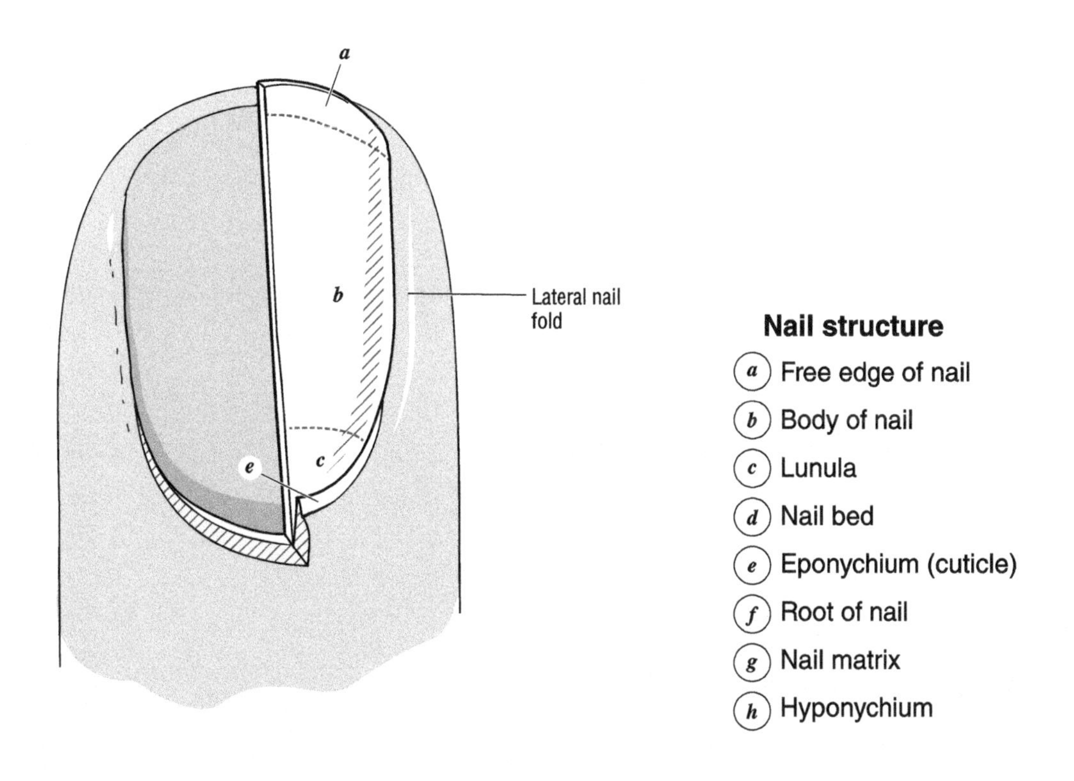
a
b
c
e
Lateral nail fold
Nail structure
a Free edge of nail
b Body of nail
c Lunula
d Nail bed
e Eponychium (cuticle)
f Root of nail
g Nail matrix
h Hyponychium

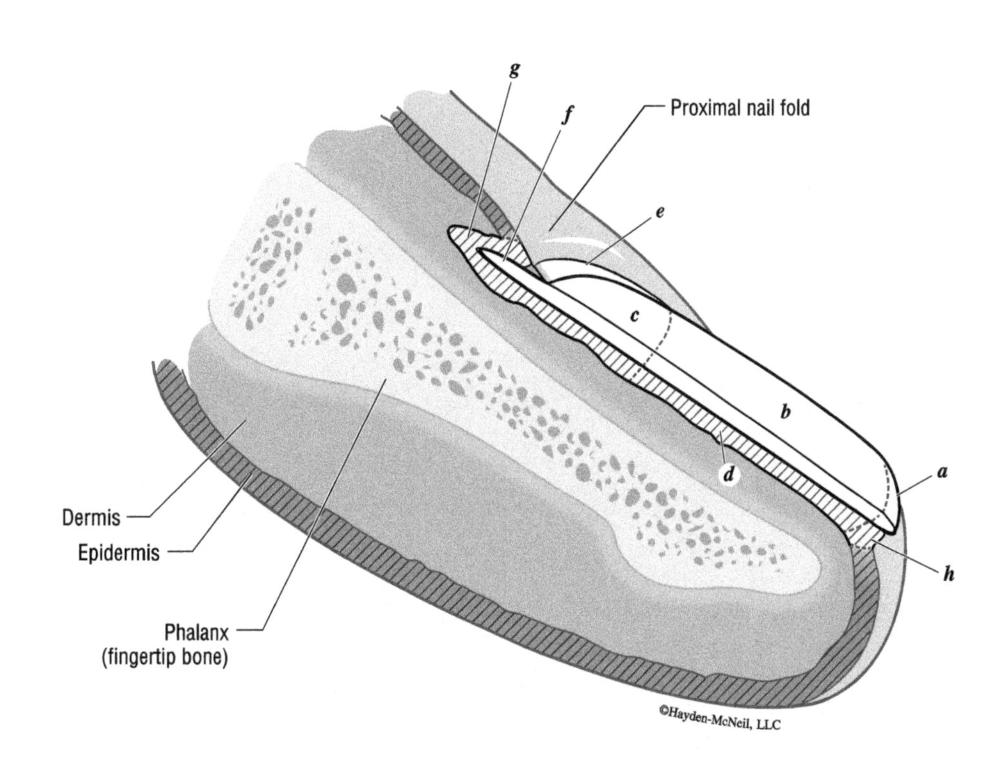
g
f
Proximal nail fold
e
c
b
d
a
h
Dermis
Epidermis
Phalanx
(fingertip bone)
©Hayden-McNeil, LLC

5

The Skeletal System

Bone tissue and long bone anatomy
Bone shapes
Skeletal divisions
The complete skeleton
Bones of the skull
Additional views of the skull
Interior views of the skull
Bones of the spine
Typical vertebrae
Sacrum and coccyx
Rib cage and sternum
Bones of the upper limb
Scapula and clavicle
Long bones of the upper limb
Bones of the wrist and hand
Bones of the pelvis
Bones of the lower limb
Bones of the thigh and leg
Bones of the foot

In this leg of your journey, you will be playing the role of an inspired architect, constructing an understanding of our **skeletal system's** framework that can stand all on its own. As you skillfully piece together our sturdy infrastructure from the ground up, you will discover the intelligent diversity of our skeleton in full color and its fundamental role in our functionality. Within these pages, the unassuming stapes—the smallest bone—will be lauded for its indispensability, and the strong, crucial femur boasts its status as our largest bone. Every fragment has its purpose, and every piece has its place, all actively contributing to what makes us whole.

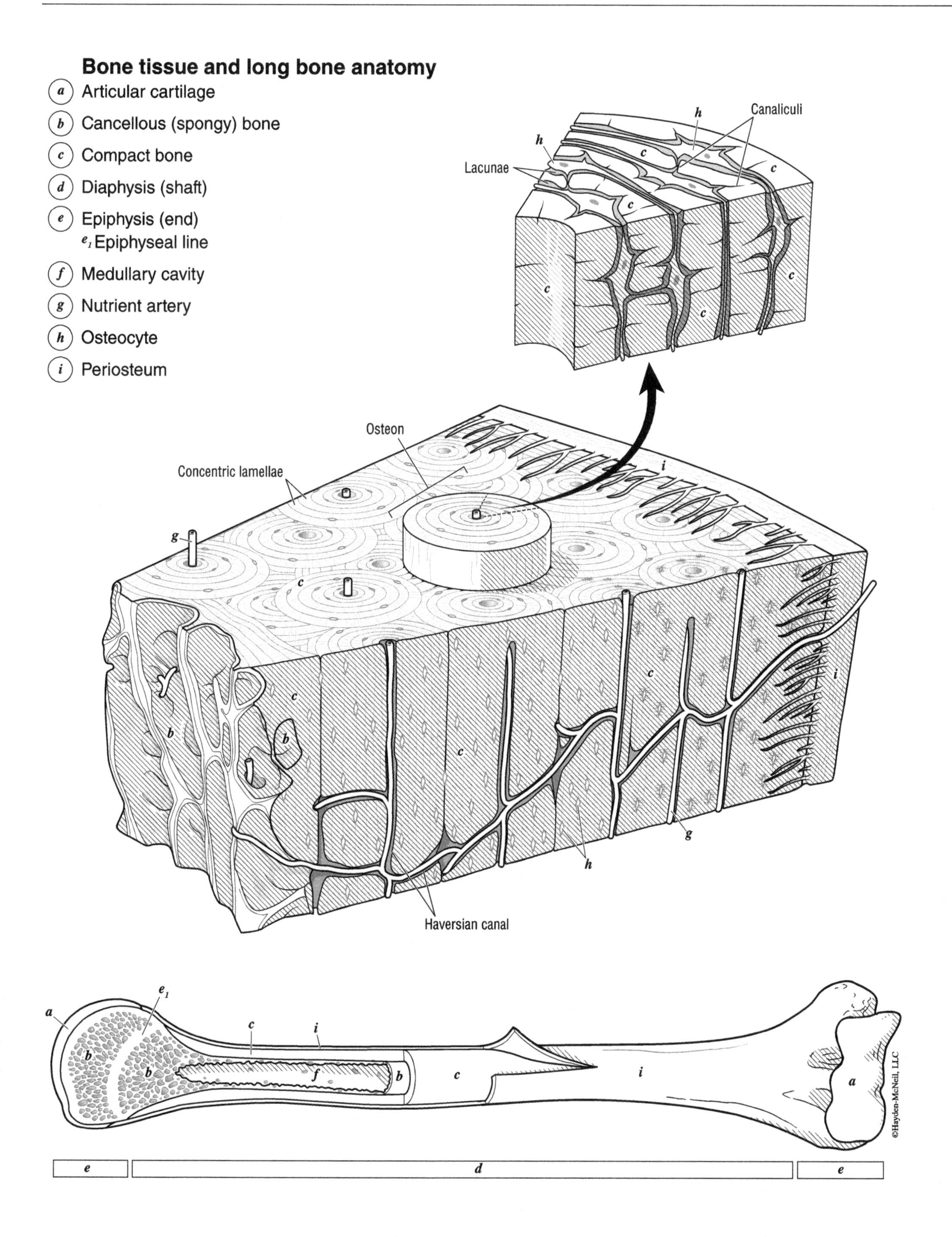
Bone tissue and long bone anatomy
a Articular cartilage
b Cancellous (spongy) bone
c Compact bone
d Diaphysis (shaft)
e Epiphysis (end)
e1 Epiphyseal line
f Medullary cavity
g Nutrient artery
h Osteocyte
i Periosteum
Canaliculi
Lacunae
Osteon
Concentric lamellae
Haversian canal
©Hayden-McNeil, LLC

Bone shapes

- (a) Flat bone
- (b) Irregular bone
- (c) Long bone
- (d) Sesamoid bone
- (e) Short bone

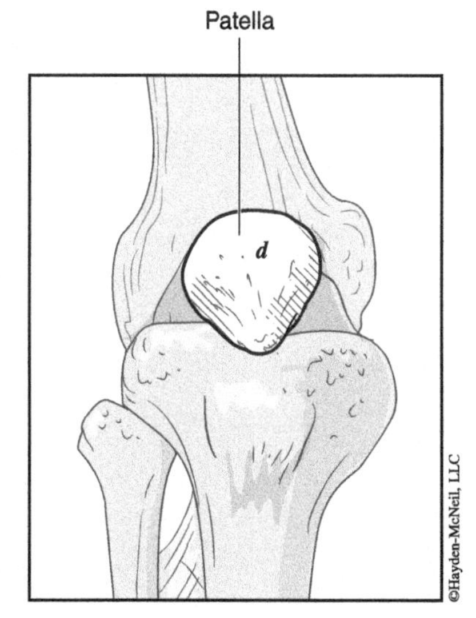

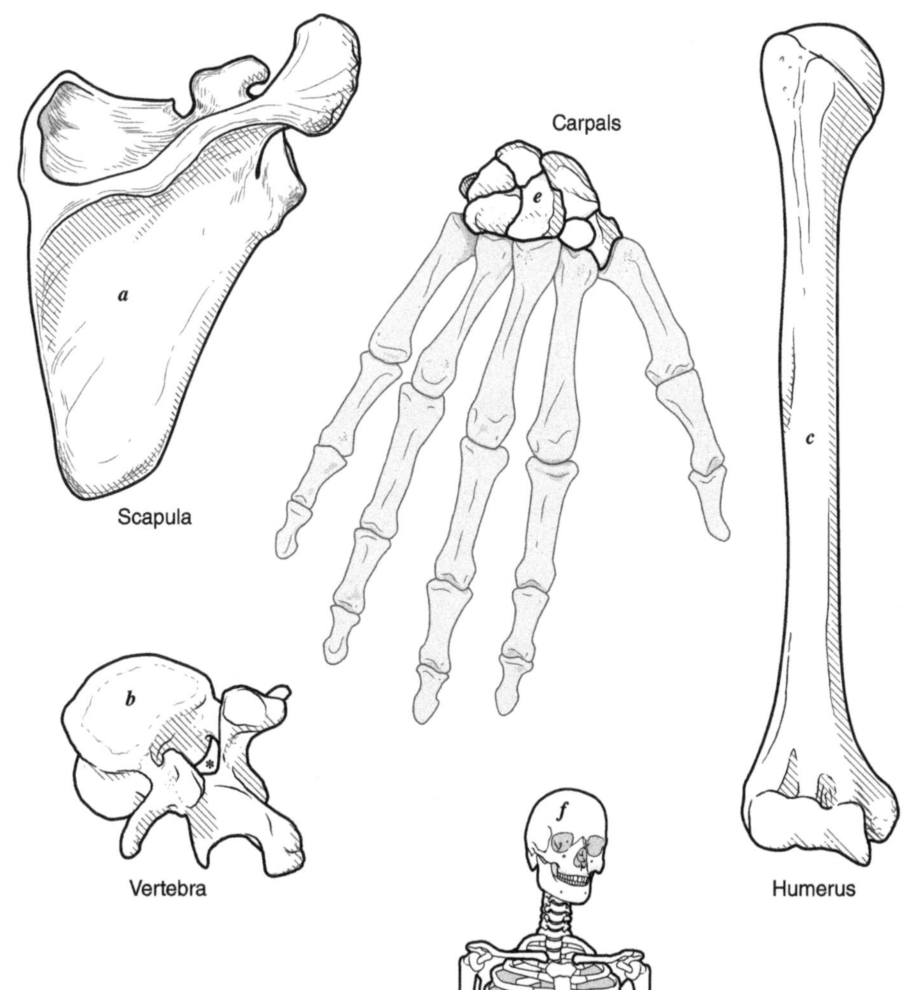

Skeletal divisions

- (f) Axial skeleton
- (g) Appendicular skeleton

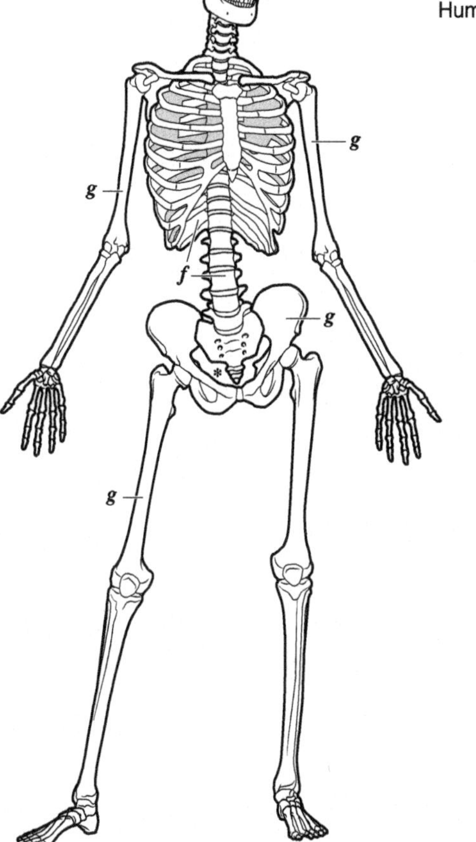

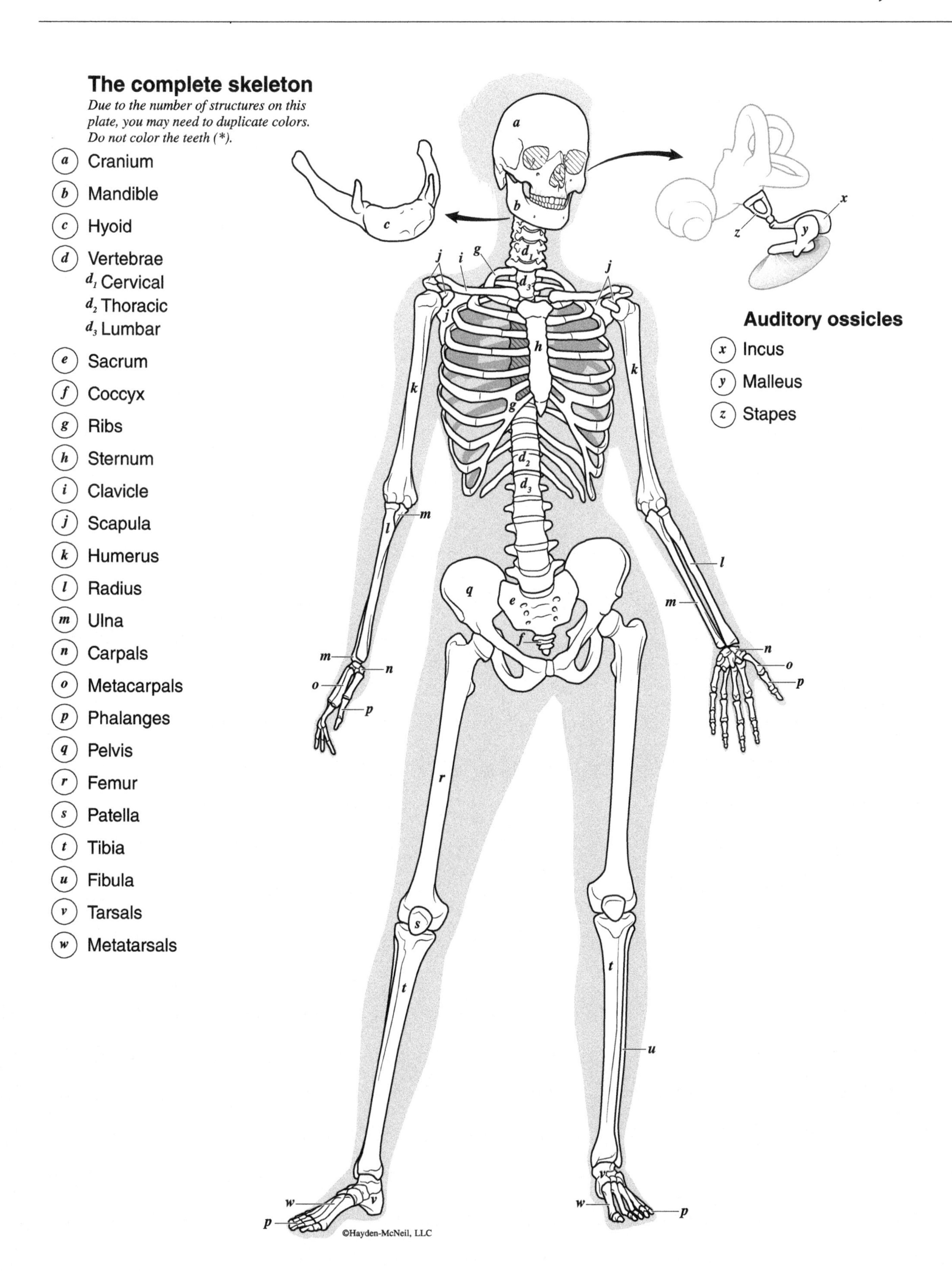
The complete skeleton
Due to the number of structures on this plate, you may need to duplicate colors. Do not color the teeth (*).
a Cranium
b Mandible
c Hyoid
d Vertebrae
d1 Cervical
d2 Thoracic
d3 Lumbar
e Sacrum
f Coccyx
g Ribs
h Sternum
i Clavicle
j Scapula
k Humerus
l Radius
m Ulna
n Carpals
o Metacarpals
p Phalanges
q Pelvis
r Femur
s Patella
t Tibia
u Fibula
v Tarsals
w Metatarsals
Auditory ossicles
x Incus
y Malleus
z Stapes
©Hayden-McNeil, LLC

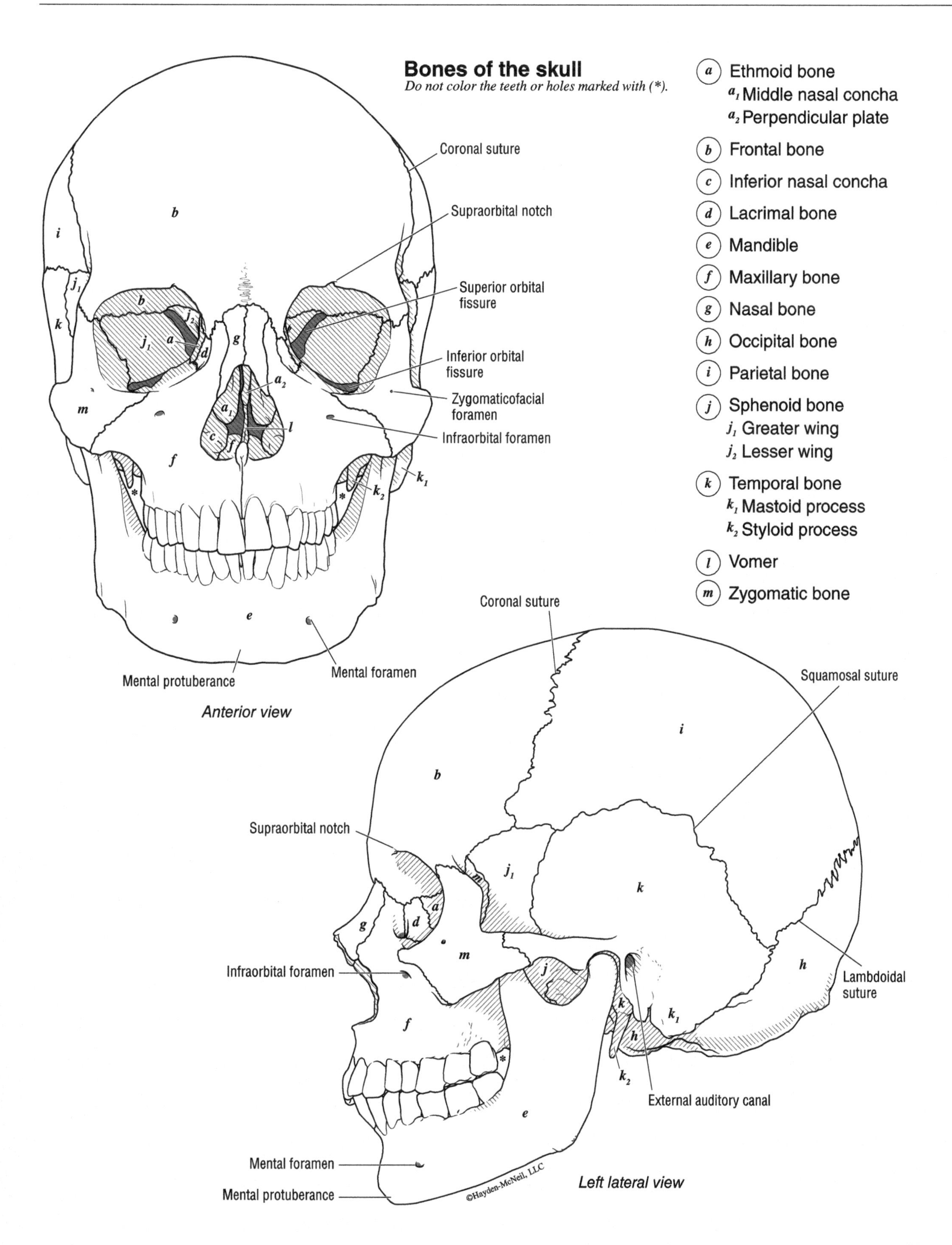

Bones of the skull
Do not color the teeth or holes marked with (*).
a Ethmoid bone
a1 Middle nasal concha
a2 Perpendicular plate
b Frontal bone
c Inferior nasal concha
d Lacrimal bone
e Mandible
f Maxillary bone
g Nasal bone
h Occipital bone
i Parietal bone
j Sphenoid bone
j1 Greater wing
j2 Lesser wing
k Temporal bone
k1 Mastoid process
k2 Styloid process
l Vomer
m Zygomatic bone
Coronal suture
Supraorbital notch
Superior orbital fissure
Inferior orbital fissure
Zygomaticofacial foramen
Infraorbital foramen
Mental protuberance
Mental foramen
Anterior view
Coronal suture
Squamosal suture
Supraorbital notch
Infraorbital foramen
Lambdoidal suture
External auditory canal
Mental foramen
Mental protuberance
Left lateral view
©Hayden-McNeil, LLC

Additional views of the skull

Do not color the teeth or holes marked with (*). Color the sutural bone (•) gray.

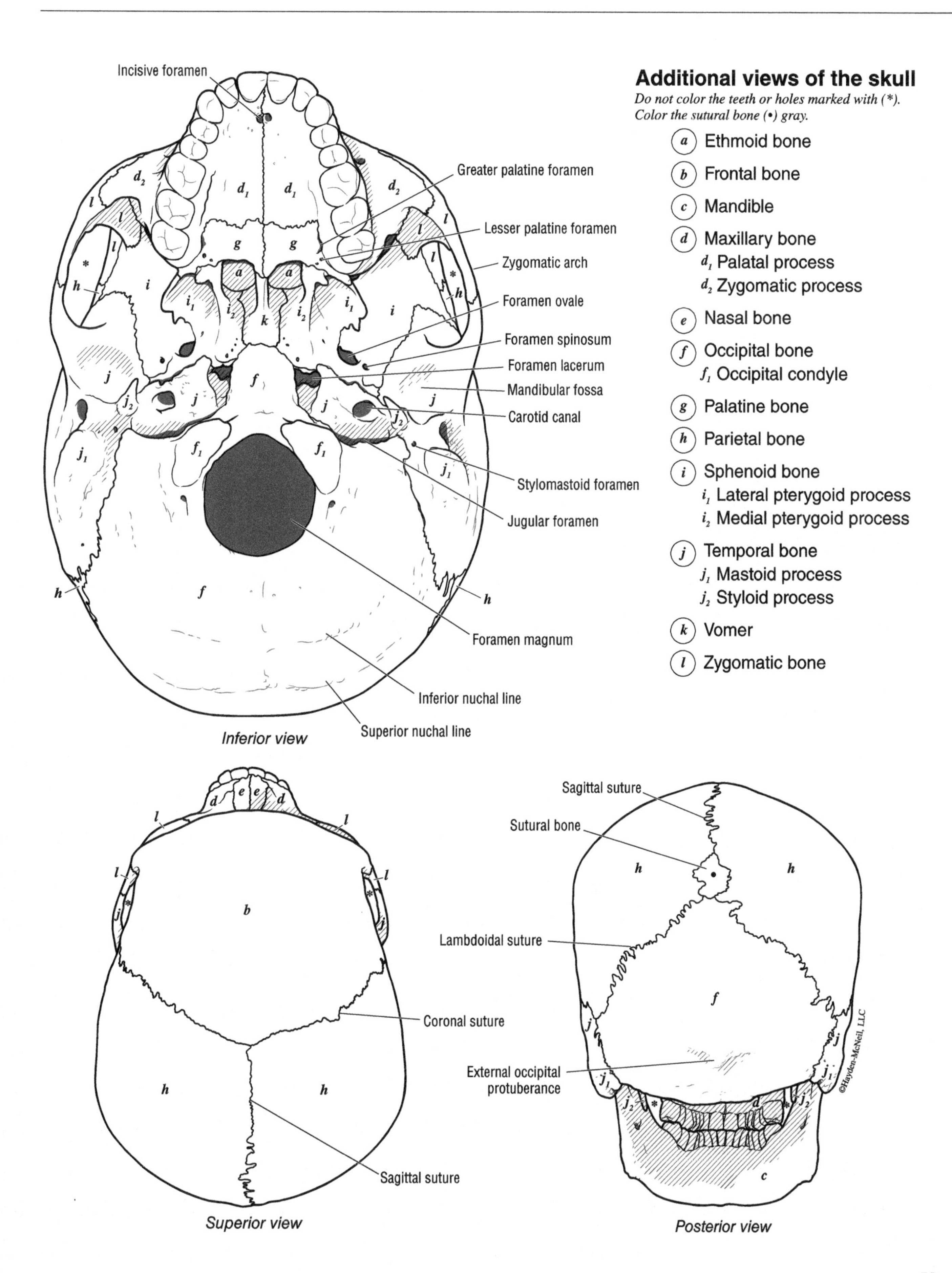

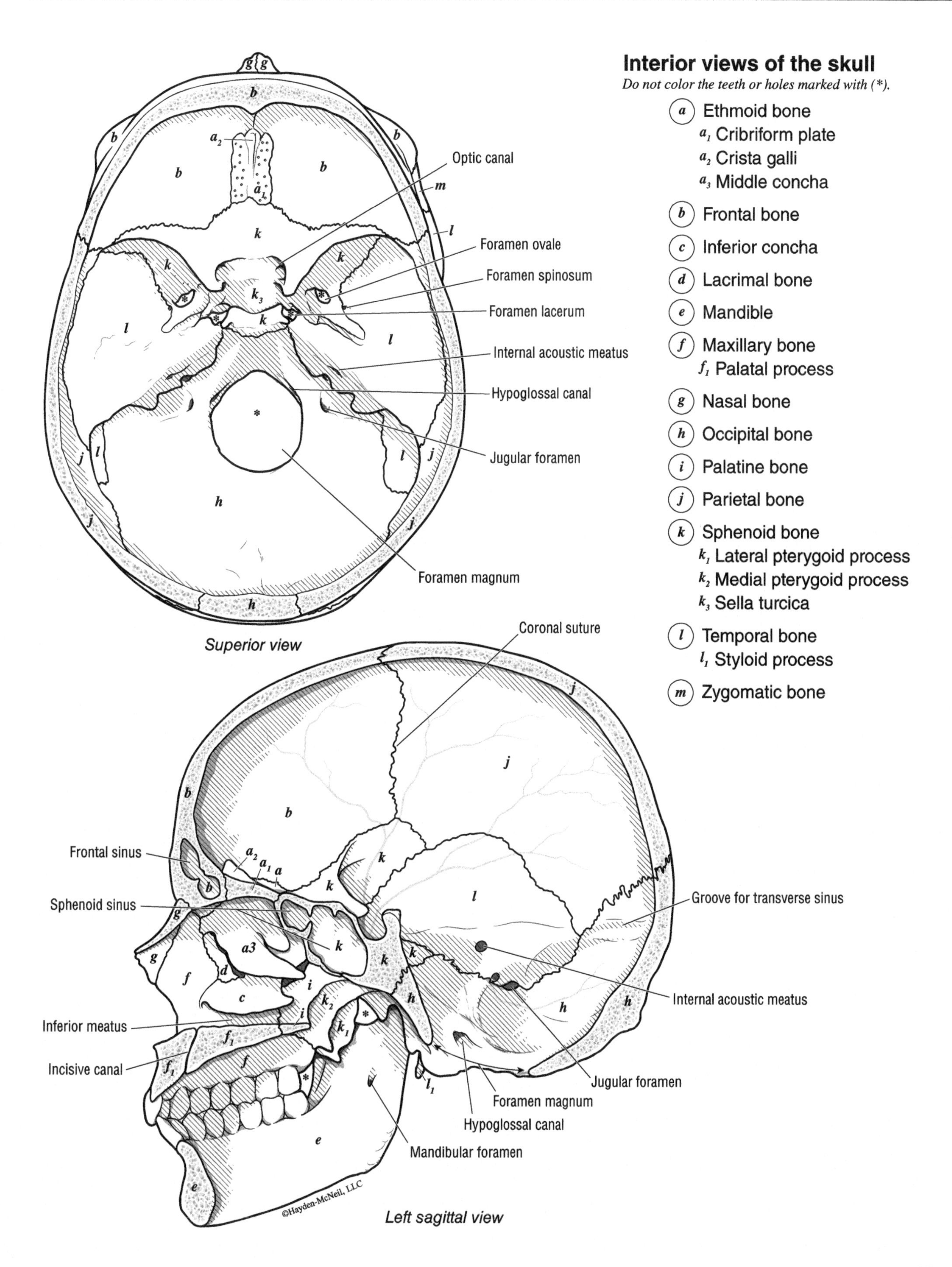

Superior view

Left sagittal view

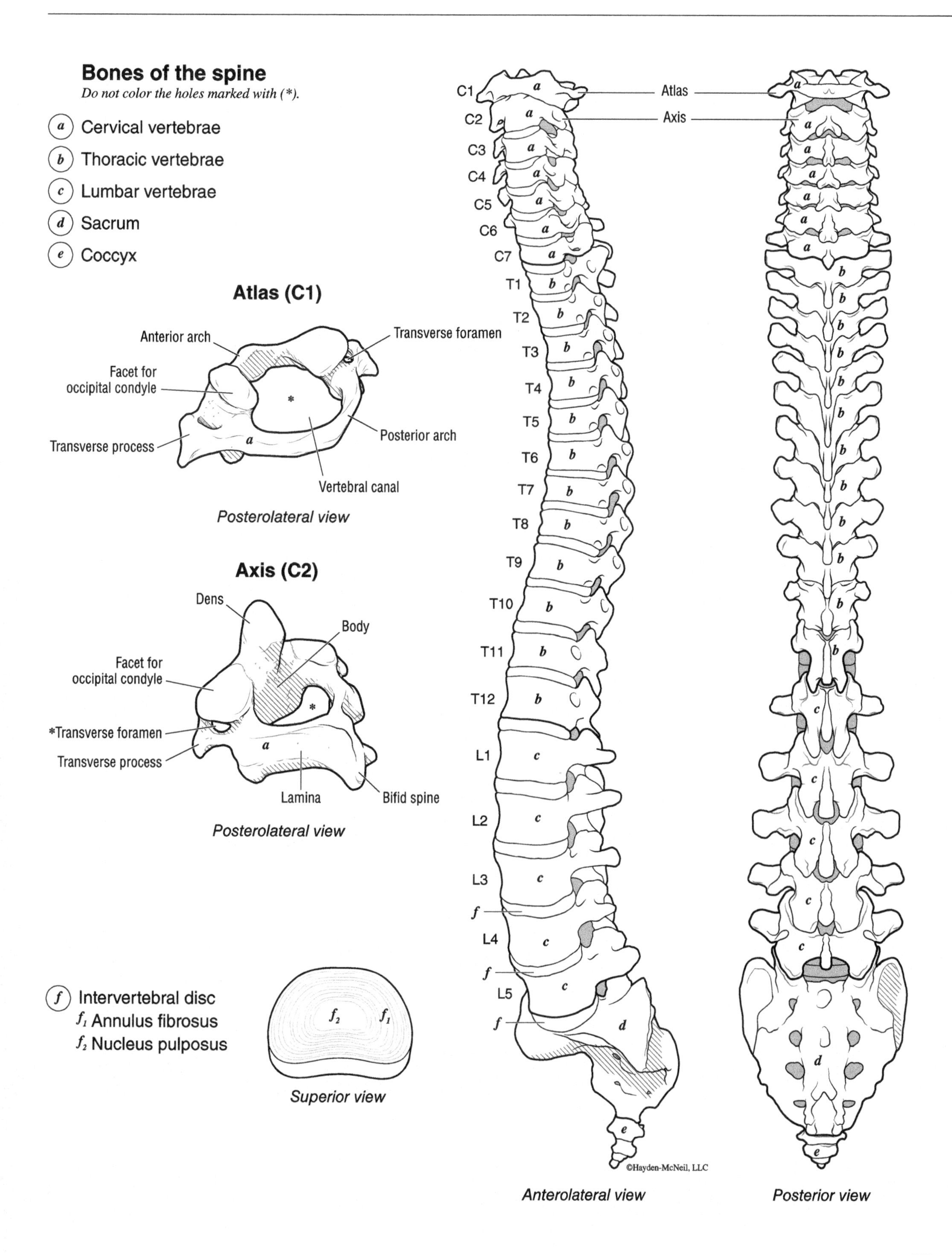
Bones of the spine
Do not color the holes marked with (*).
a Cervical vertebrae
b Thoracic vertebrae
c Lumbar vertebrae
d Sacrum
e Coccyx
Atlas (C1)
Anterior arch
Transverse foramen
Facet for occipital condyle
Posterior arch
Transverse process
Vertebral canal
Posterolateral view
Axis (C2)
Dens
Body
Facet for occipital condyle
*Transverse foramen
Transverse process
Lamina
Bifid spine
Posterolateral view
f Intervertebral disc
f1 Annulus fibrosus
f2 Nucleus pulposus
Superior view
C1
C2
C3
C4
C5
C6
C7
T1
T2
T3
T4
T5
T6
T7
T8
T9
T10
T11
T12
L1
L2
L3
L4
L5
Atlas
Axis
©Hayden-McNeil, LLC
Anterolateral view
Posterior view

Typical vertebrae

Do not color the holes marked with (*).

(a) Cervical vertebrae

(b) Thoracic vertebrae

(c) Lumbar vertebrae

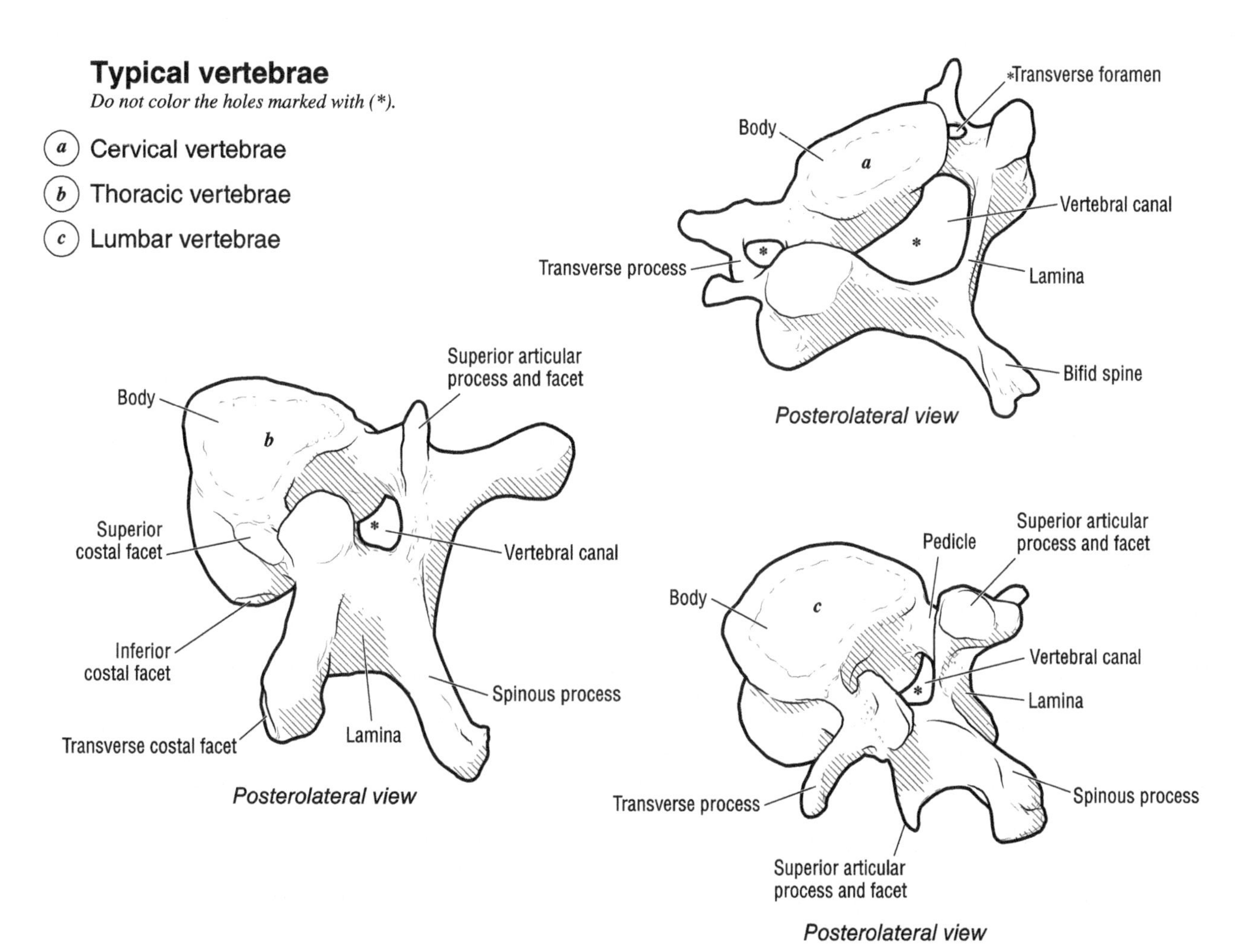

Sacrum and coccyx

(d) Sacrum

(e) Coccyx

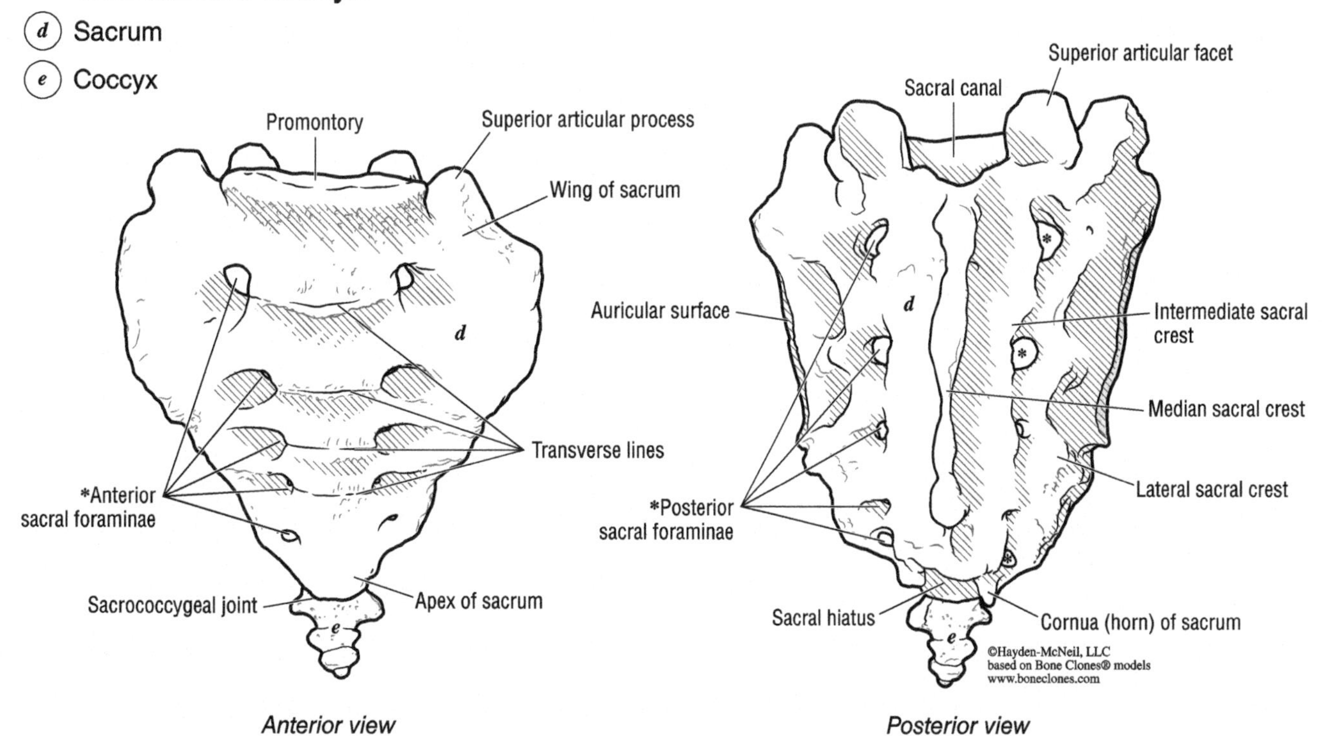

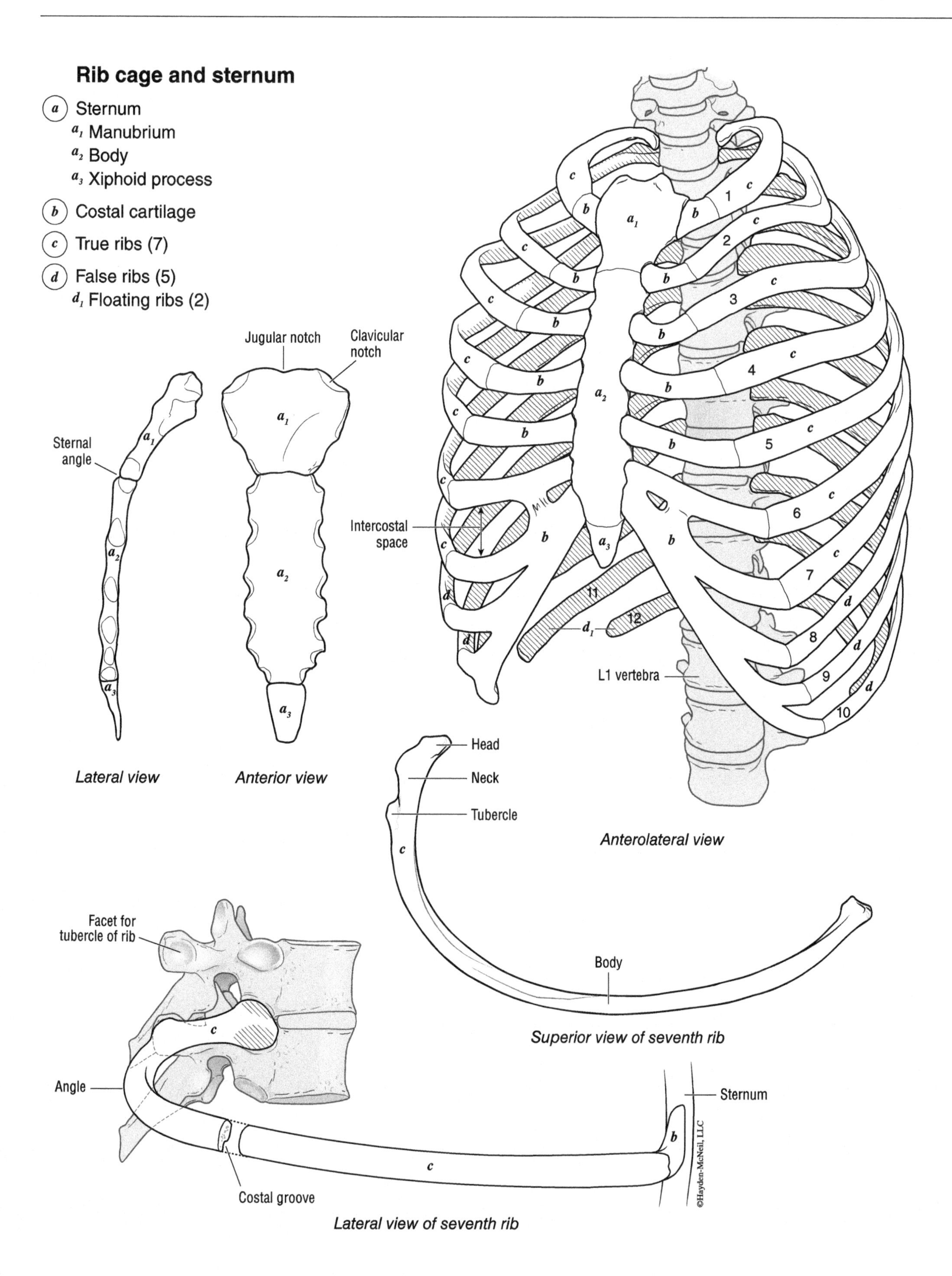
Rib cage and sternum
a Sternum
a1 Manubrium
a2 Body
a3 Xiphoid process
b Costal cartilage
c True ribs (7)
d False ribs (5)
d1 Floating ribs (2)
Jugular notch
Clavicular notch
Sternal angle
Intercostal space
L1 vertebra
Head
Neck
Tubercle
Body
Facet for tubercle of rib
Angle
Costal groove
Sternum
Lateral view
Anterior view
Anterolateral view
Superior view of seventh rib
Lateral view of seventh rib
©Hayden-McNeil, LLC

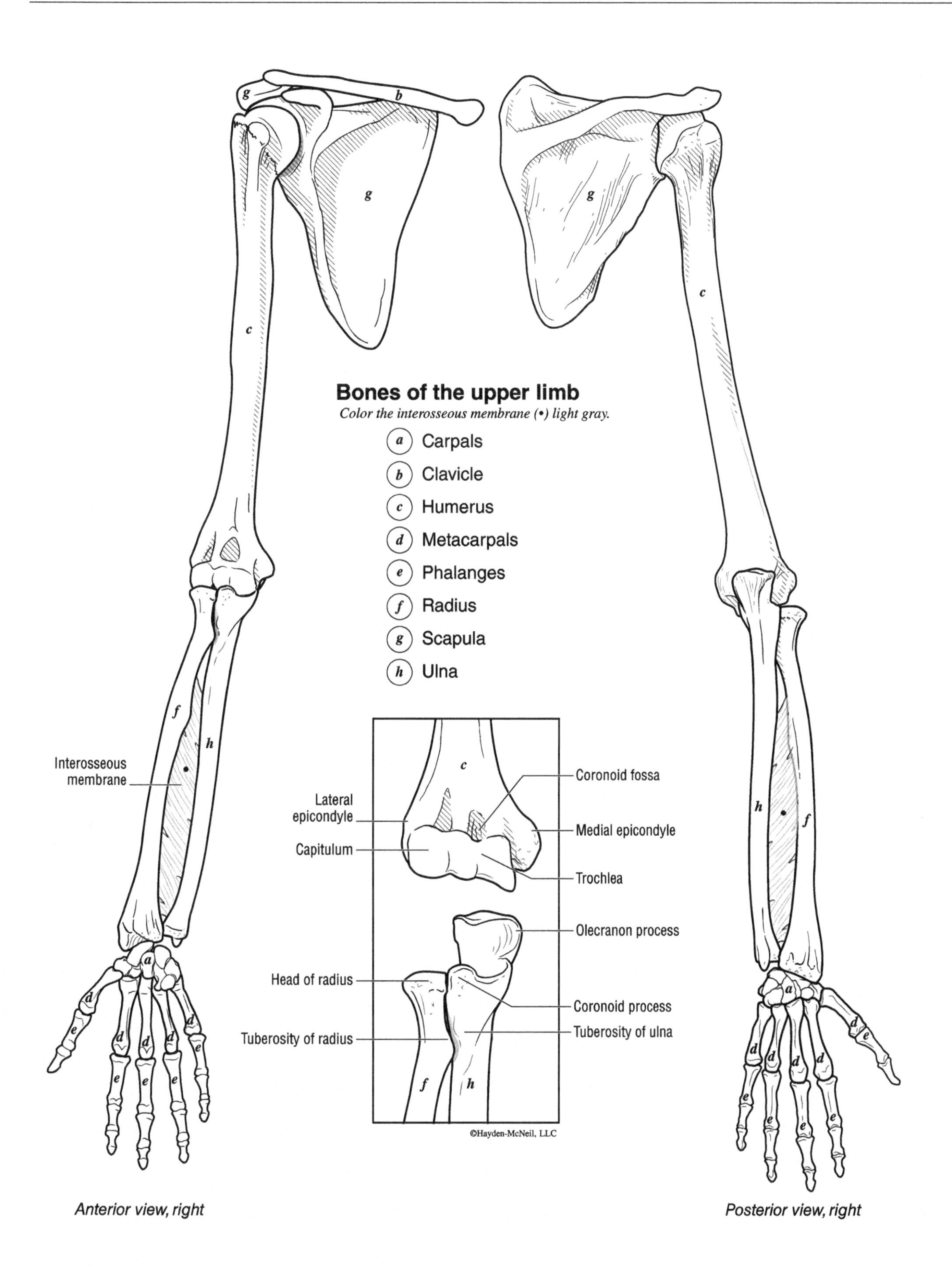

Bones of the upper limb

Color the interosseous membrane (•) light gray.

(a) Carpals
(b) Clavicle
(c) Humerus
(d) Metacarpals
(e) Phalanges
(f) Radius
(g) Scapula
(h) Ulna

Anterior view, right

Posterior view, right

Scapula and clavicle

(a) Scapula

(b) Clavicle

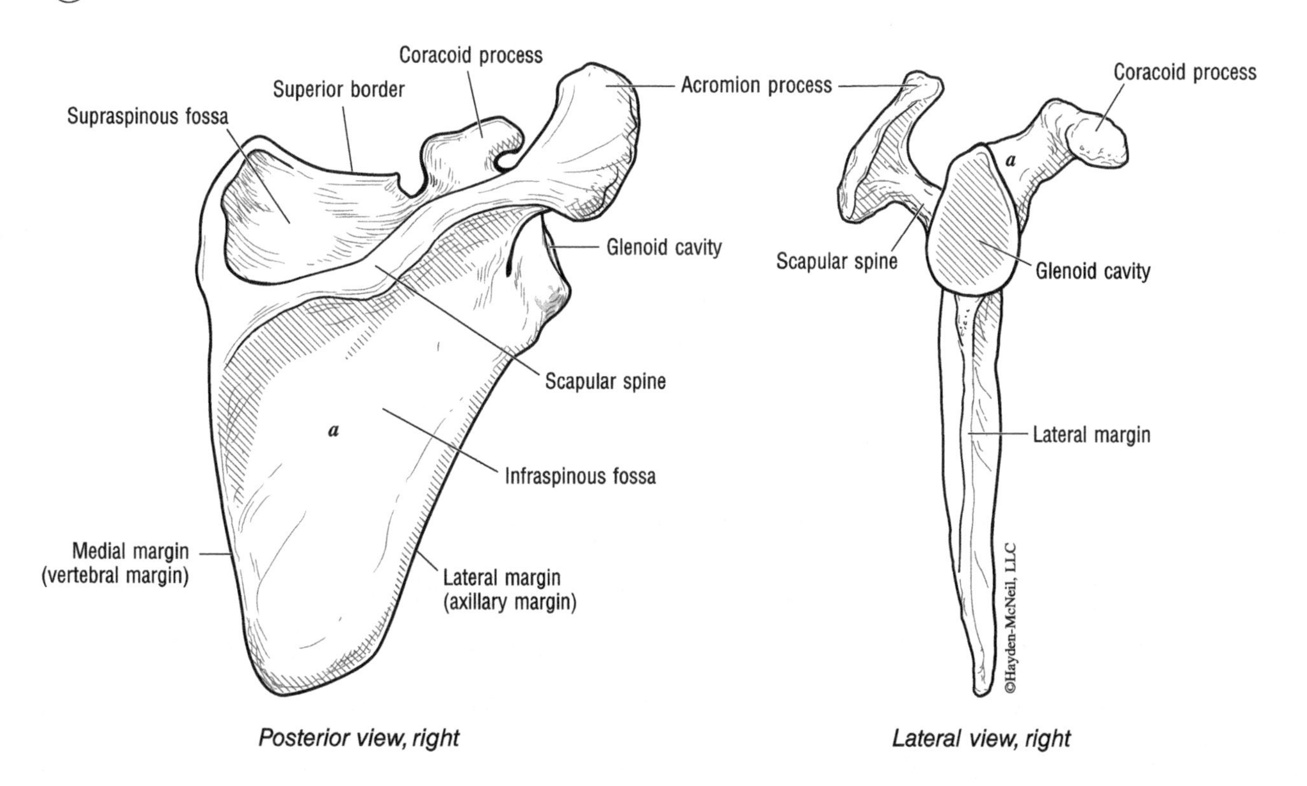

Posterior view, right

Lateral view, right

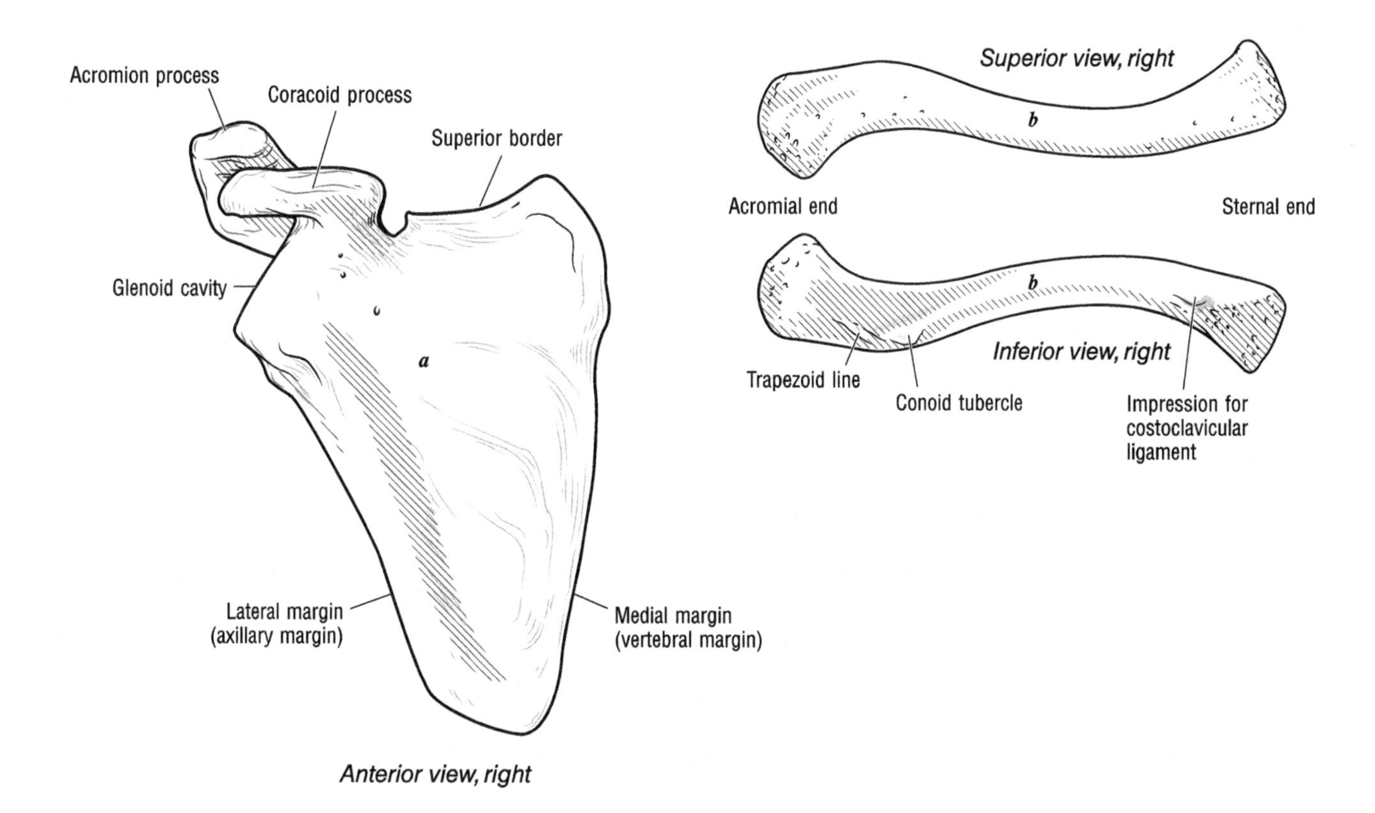

Anterior view, right

Long bones of the upper limb

Color the interosseous membrane (•) light gray.

(a) Humerus
(b) Radius
(c) Ulna

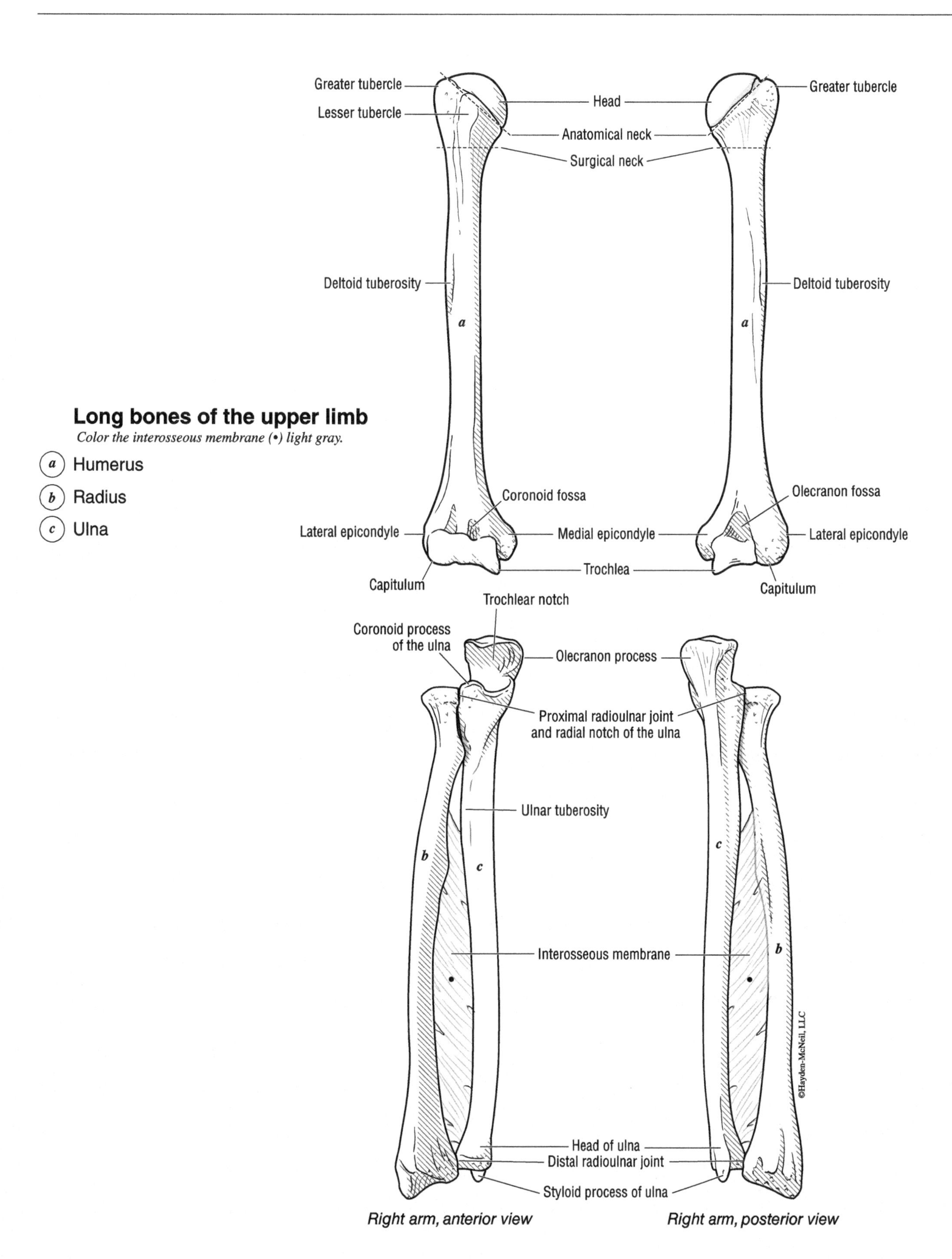

Right arm, anterior view *Right arm, posterior view*

Bones of the wrist and hand

Color the sesamoid bones (•) gray.

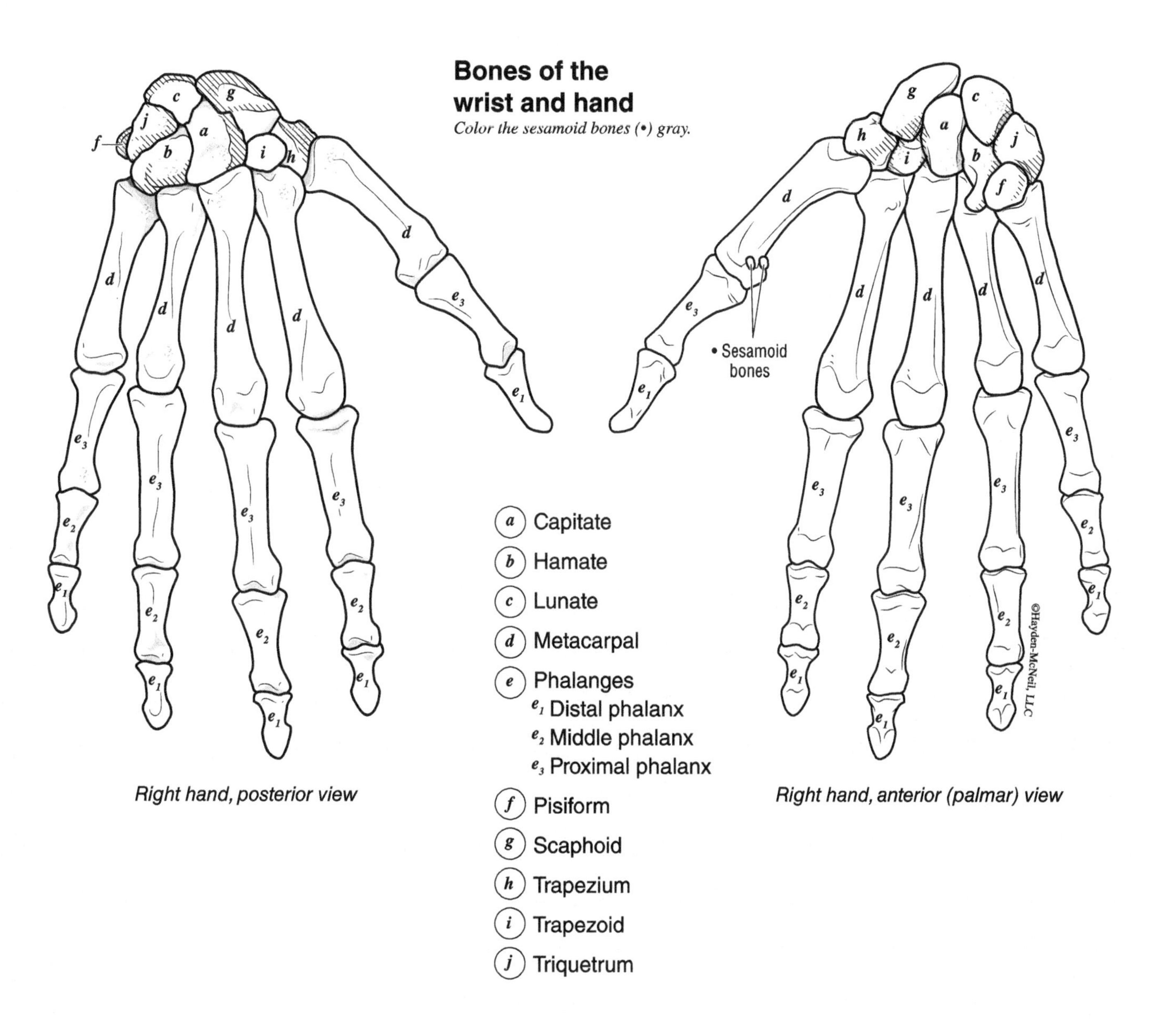

(a) Capitate
(b) Hamate
(c) Lunate
(d) Metacarpal
(e) Phalanges
e_1 Distal phalanx
e_2 Middle phalanx
e_3 Proximal phalanx
(f) Pisiform
(g) Scaphoid
(h) Trapezium
(i) Trapezoid
(j) Triquetrum

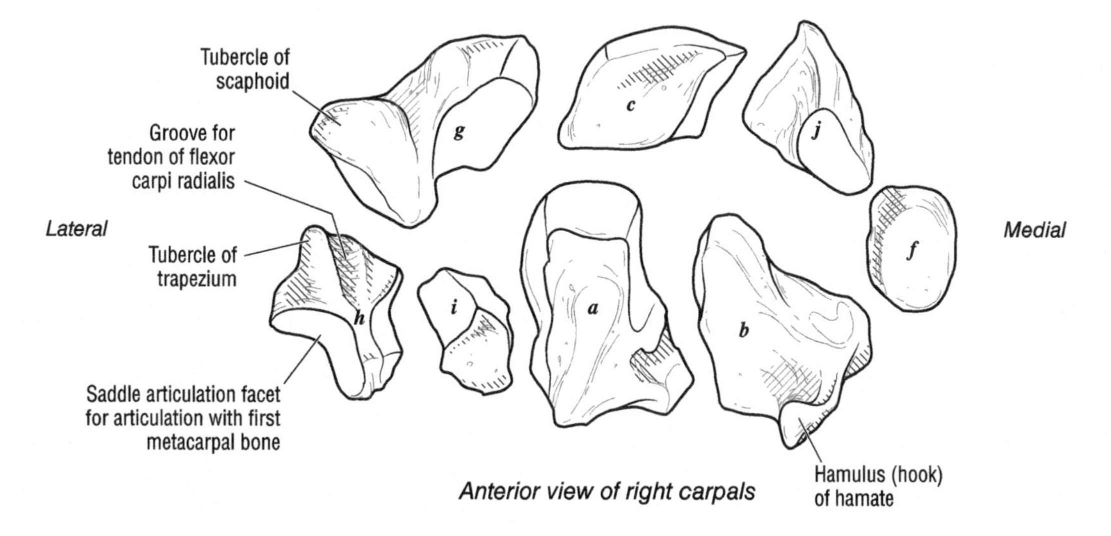

Anterior view of right carpals

Bones of the pelvis

Color the pubic symphysis (•) light gray. Note the gray dashed lines denoting the interior borders of the bones of the pelvis.

(a) Ilium

(b) Ischium

(c) Pubis

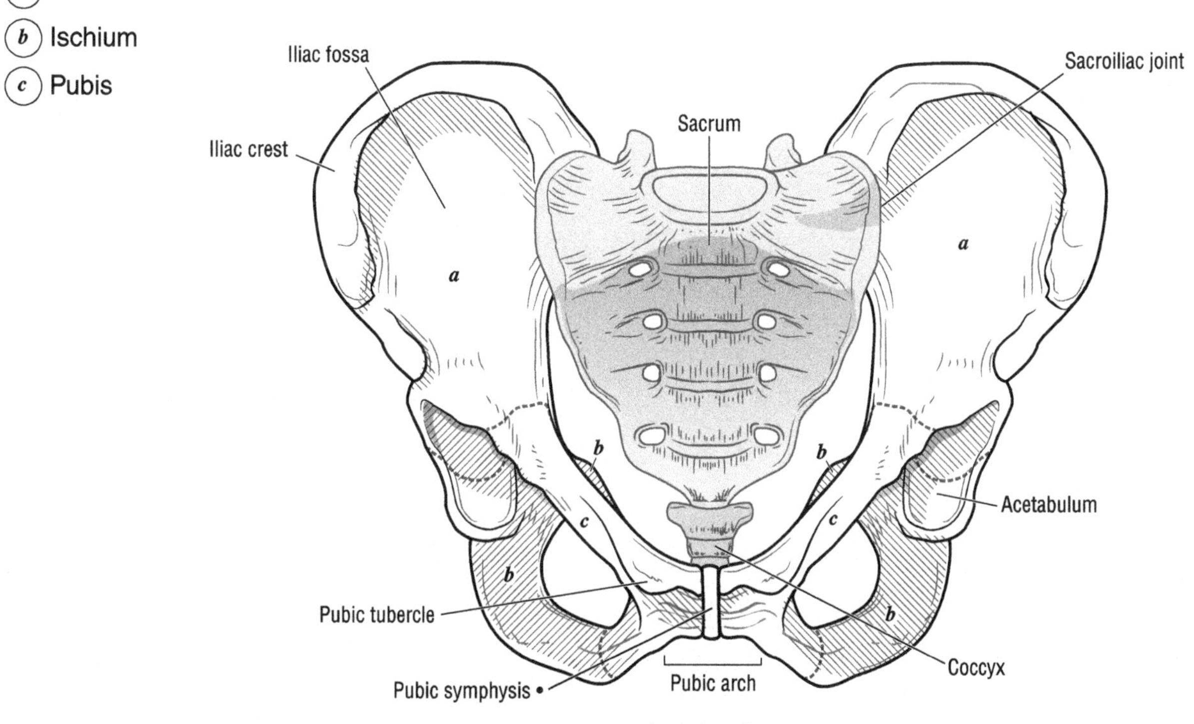

Anterior view

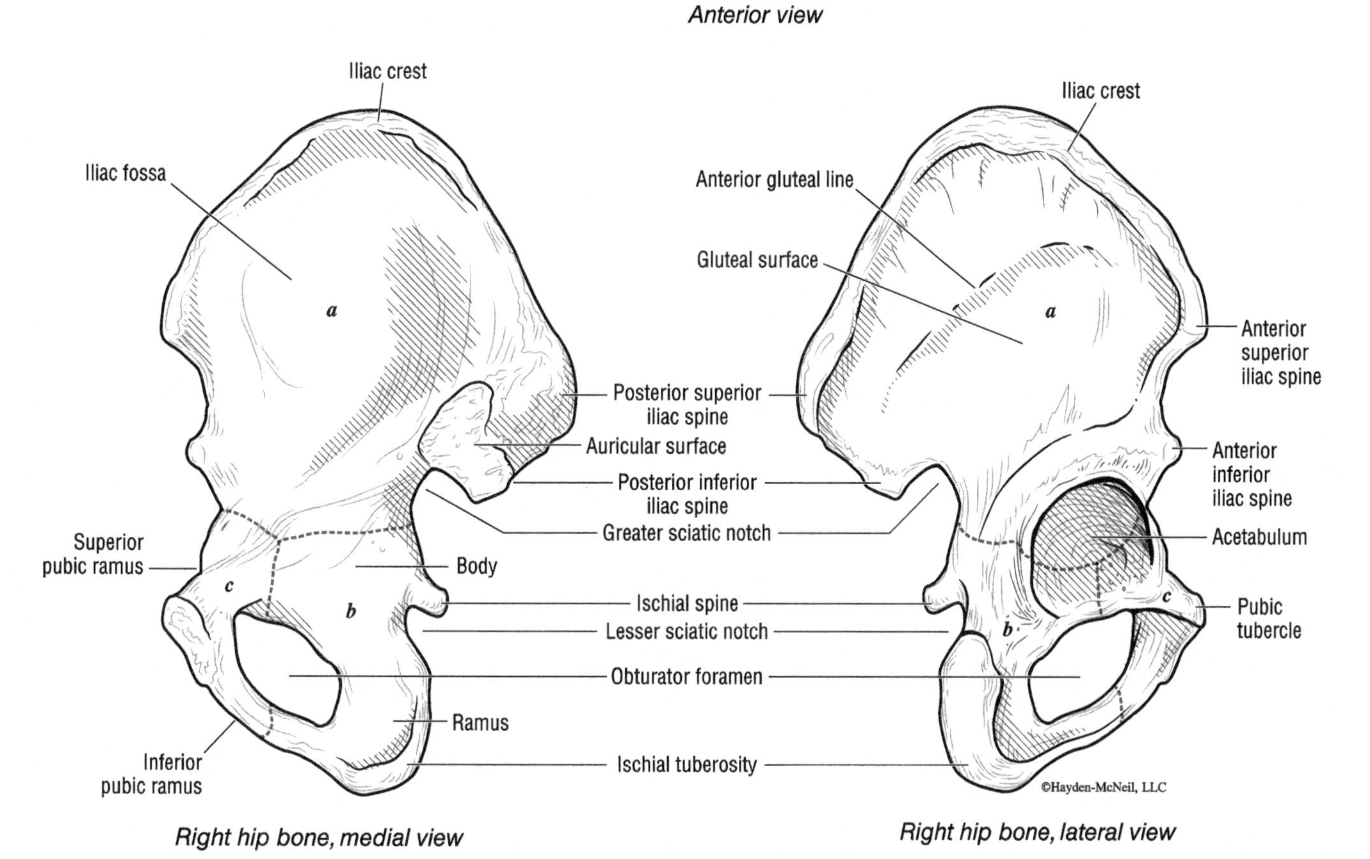

Right hip bone, medial view

Right hip bone, lateral view

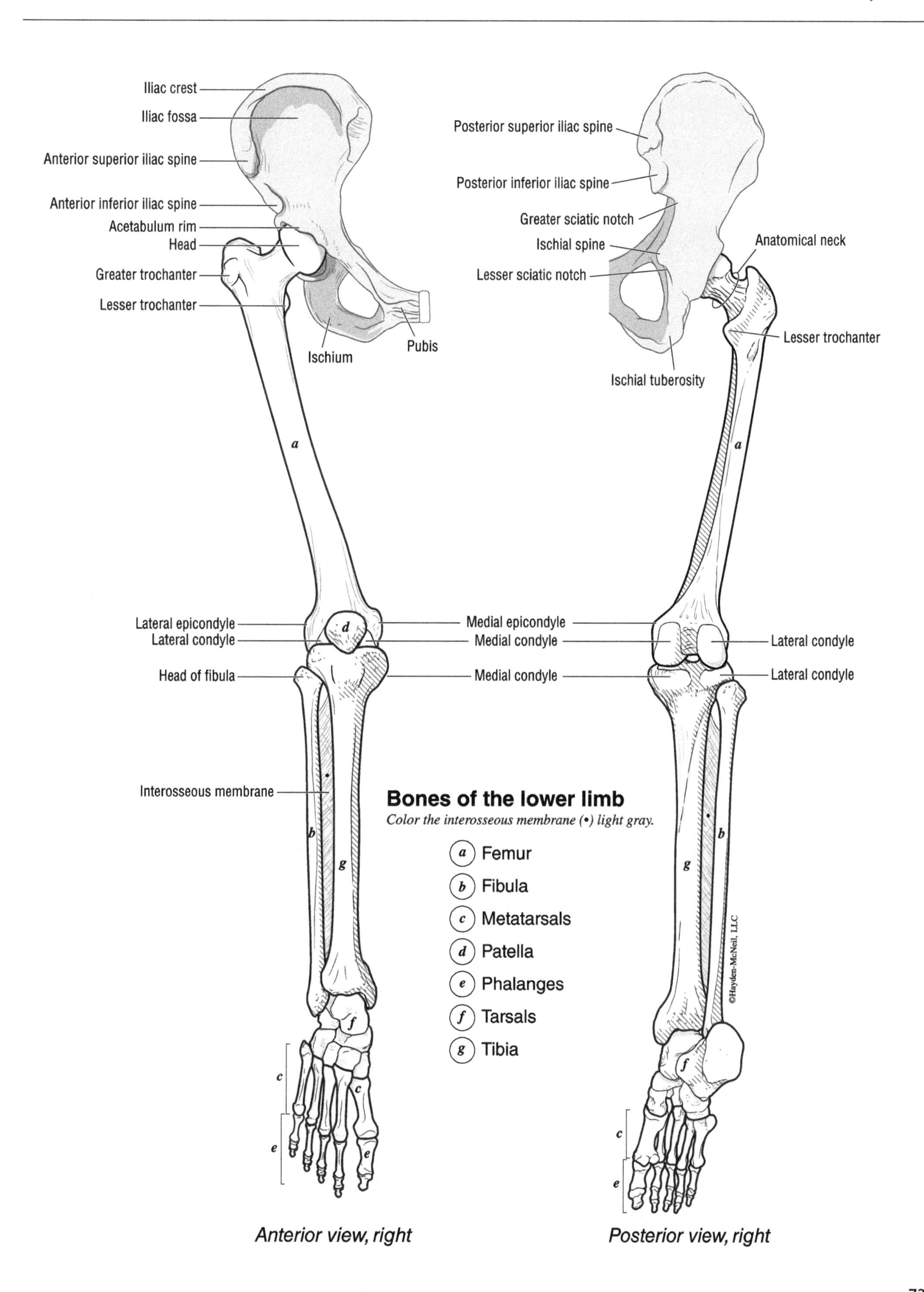

Bones of the lower limb

Color the interosseous membrane (•) light gray.

- (a) Femur
- (b) Fibula
- (c) Metatarsals
- (d) Patella
- (e) Phalanges
- (f) Tarsals
- (g) Tibia

Anterior view, right

Posterior view, right

Bones of the thigh and leg

Color the interosseous membrane (•) light gray.

- (a) Femur
- (b) Fibula
- (c) Patella
- (d) Tibia

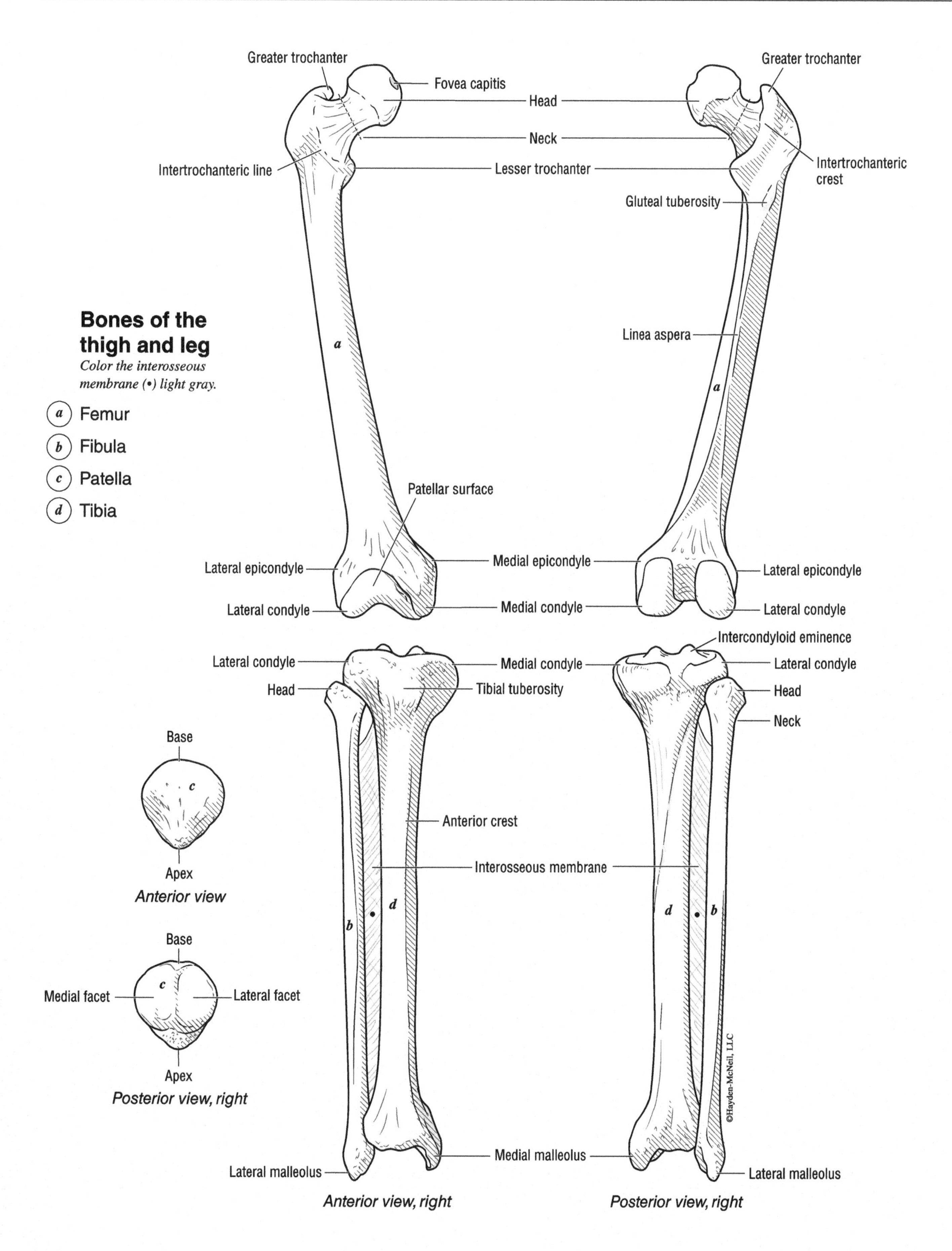

Bones of the foot

- (a) Calcaneus
- (b) Cuboid
- (c) Intermediate cuneiform
- (d) Lateral cuneiform
- (e) Medial cuneiform
- (f) Metatarsal
- (g) Navicular
- (h) Phalanges
  - h_1 Distal phalanx
  - h_2 Middle phalanx
  - h_3 Proximal phalanx
- (i) Talus

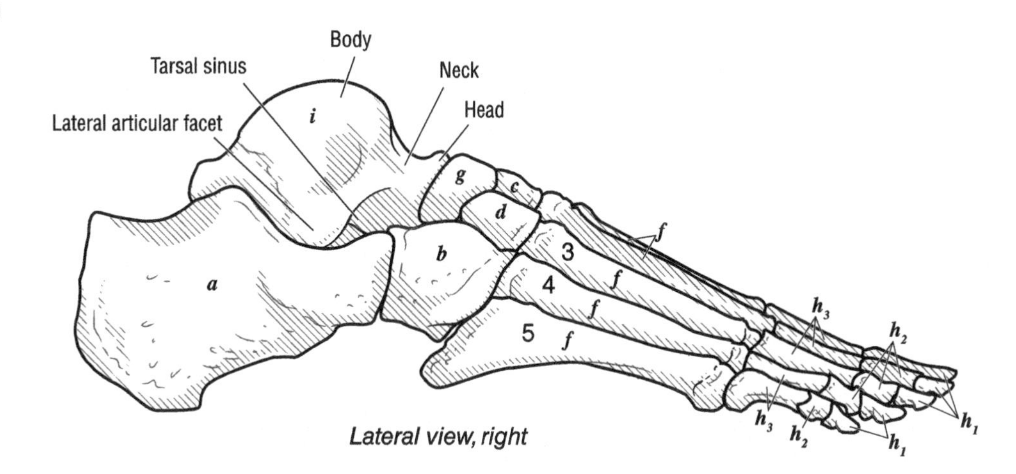

Lateral view, right

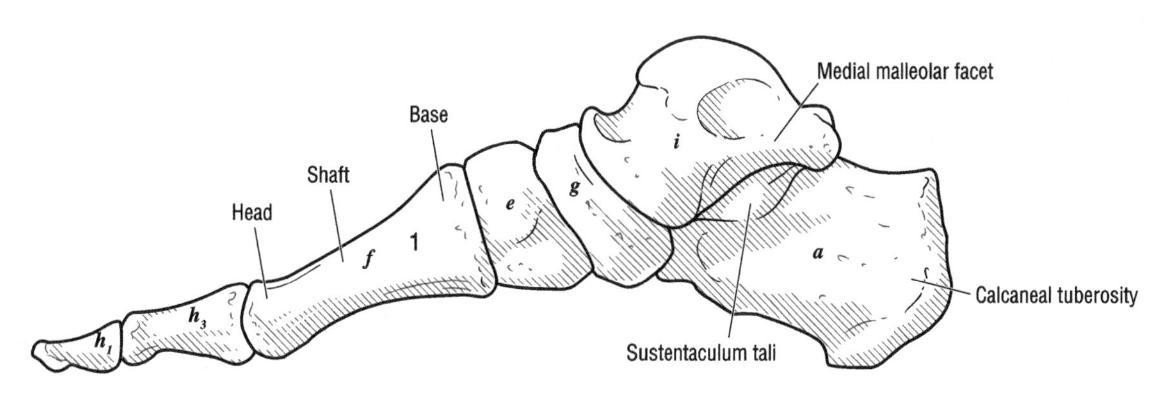

Medial view, right

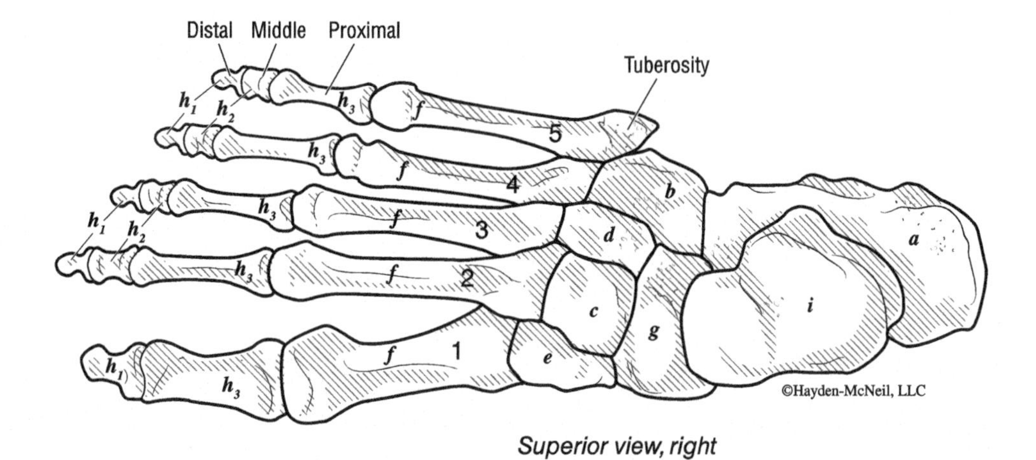

Superior view, right

6

Joints

Joint classification

Joint movement

Spinal ligaments

Shoulder joint

Elbow joint

Wrist and finger joints

Hip joint

Knee joint

Ankle and foot joints

In the upcoming chapter, you will be opening the door to our body's most impressive middlemen: **joints**! Our joints act as hinges for our body's frame, enabling us to embrace the mechanics of movement. Throughout the following sections, you will be captivated by the joints that keep us in miraculous motion, from the protective fibrous joints that make up the shield of our skull to the shock-absorbing cartilaginous joints in our spine and the versatile synovial joints. So, let's start coloring the cushioning of our cartilage and painting our body's most pivotal players!

Joint classification

(a) Fibrous joint

(b) Cartilaginous joint

Synovial joint

- (c) Articulating bones
- (d) Articular cartilage
- (e) Fibrous capsule
- (f) Synovial membrane
- (g) Synovial cavity

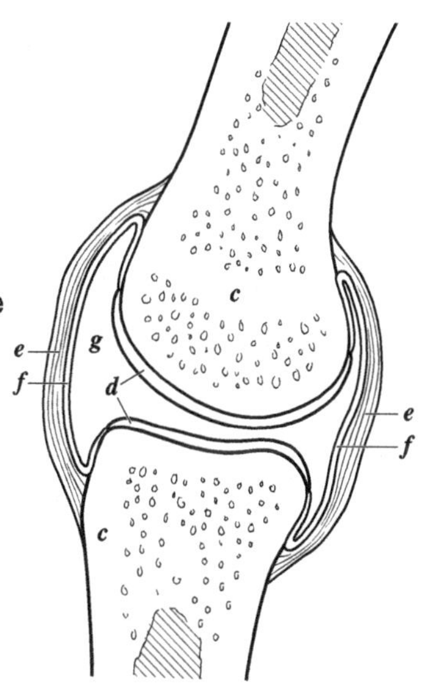

Joint movement

Color the movement arrows for each joint type.

(h) Plane joint

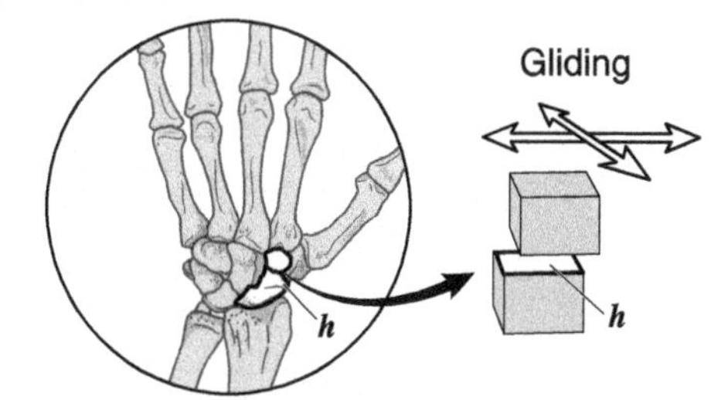

Examples: Intercarpal joints, intertarsal joints, joints between vertebral articular surfaces

(k) Condylar joint

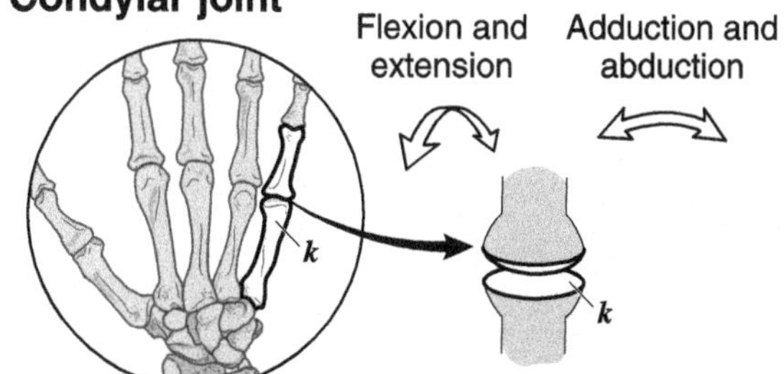

Examples: Metacarpophalangeal (knuckle) joints, wrist joints

(i) Hinge joint

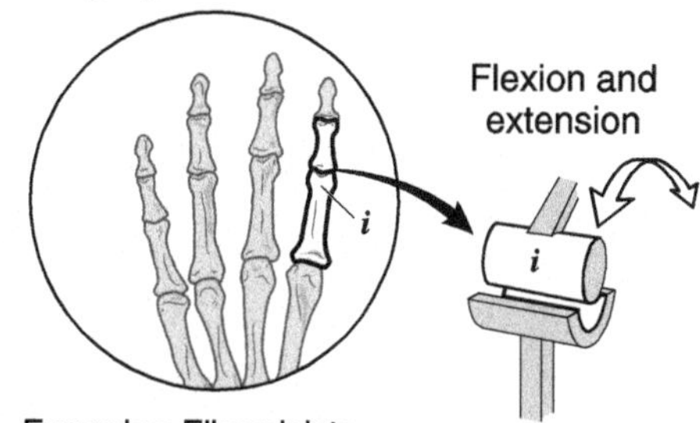

Examples: Elbow joints, interphalangeal joints

(l) Saddle joint

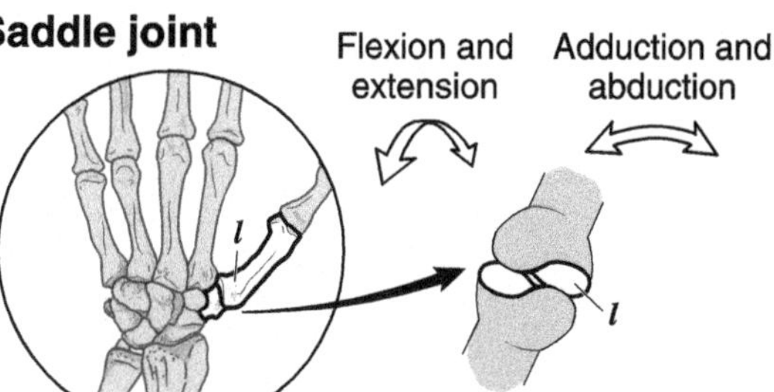

Examples: Carpometacarpal joints of the thumb

(j) Pivot joint

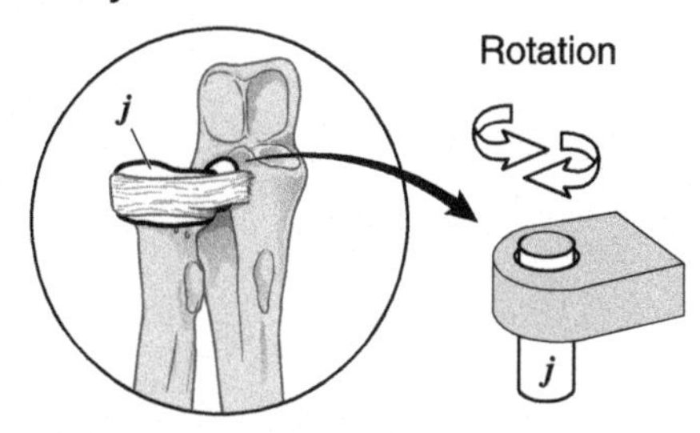

Examples: Proximal radioulnar joints, atlantoaxial joint

(m) Ball-and-socket joint

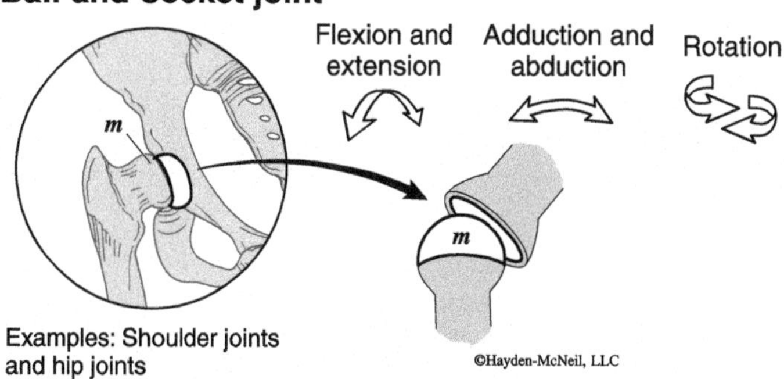

Examples: Shoulder joints and hip joints

Spinal ligaments

- (a) Anterior longitudinal ligament
- (b) Posterior longitudinal ligament
- (c) Interspinous ligament
- (d) Supraspinous ligament
- (e) Ligamenta flava
- (f) Intervertebral disk
- (g) Radiate ligaments of the head of the rib
- (h) Intertransverse ligaments
- (i) Lateral costotransverse ligaments

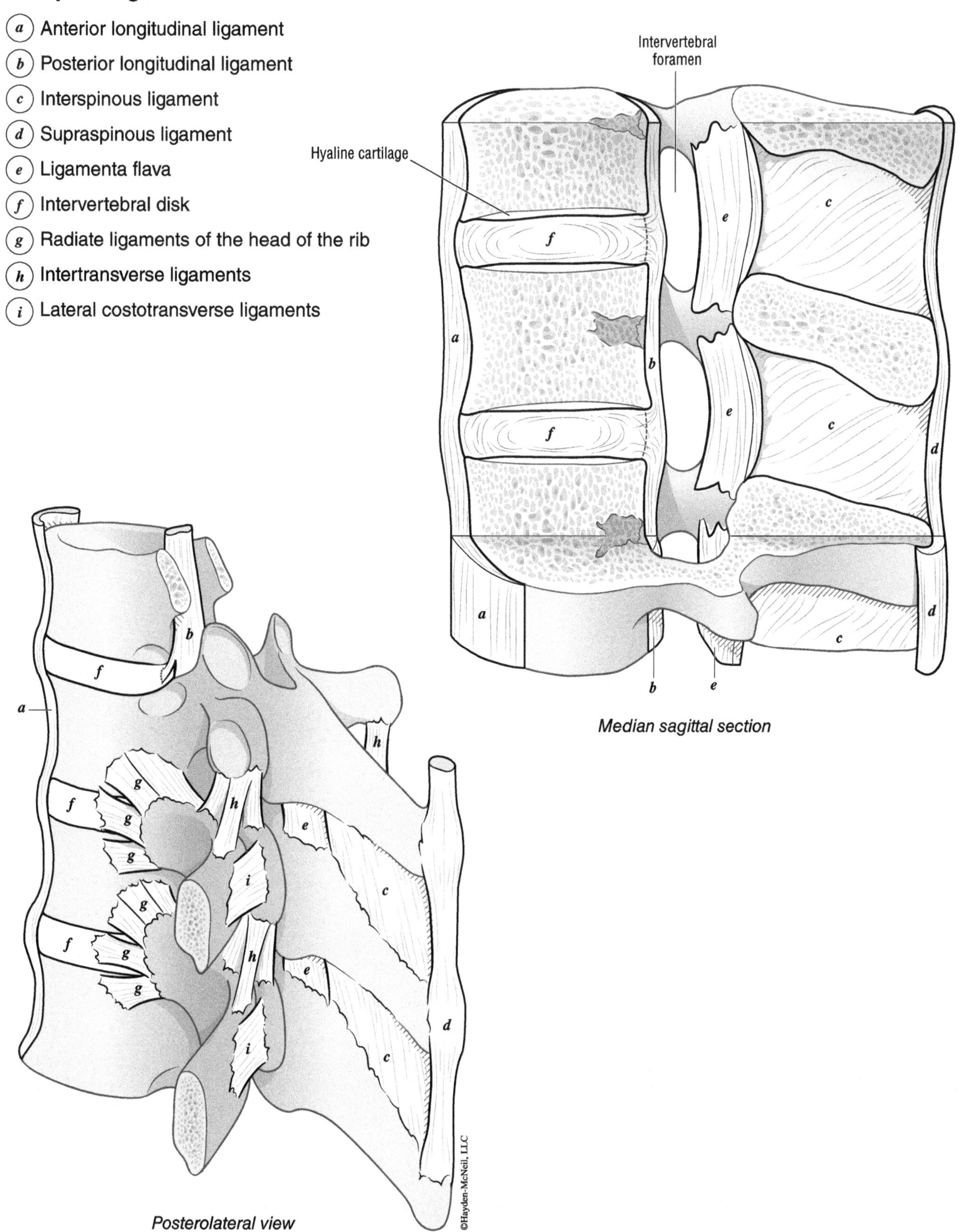

Median sagittal section

Posterolateral view

Shoulder joint

- (a) Acromioclavicular joint capsule
- (b) Coracoacromial ligament
- (c) Coracohumeral ligament
- (d) Transverse humeral ligament
- (e) Coracoclavicular ligament
- (f) Glenohumeral capsular ligament
- (g) Bursa
- (h) Glenoid cavity
- (i) Synovial membrane

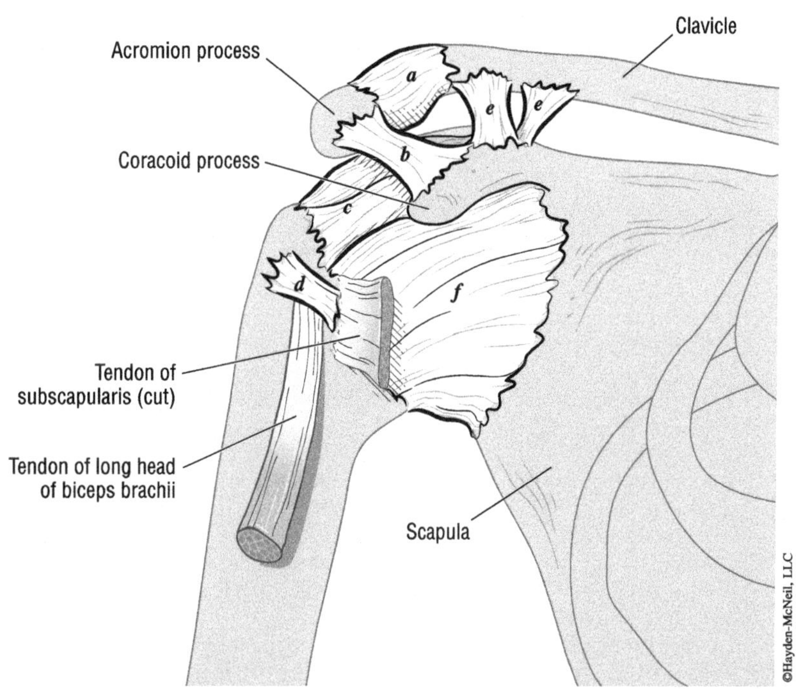

Anterior view

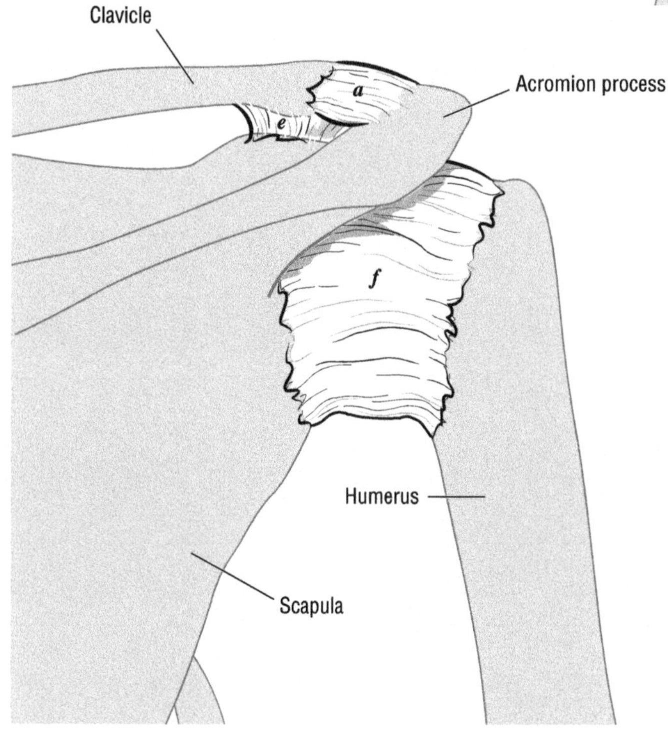

Posterior view

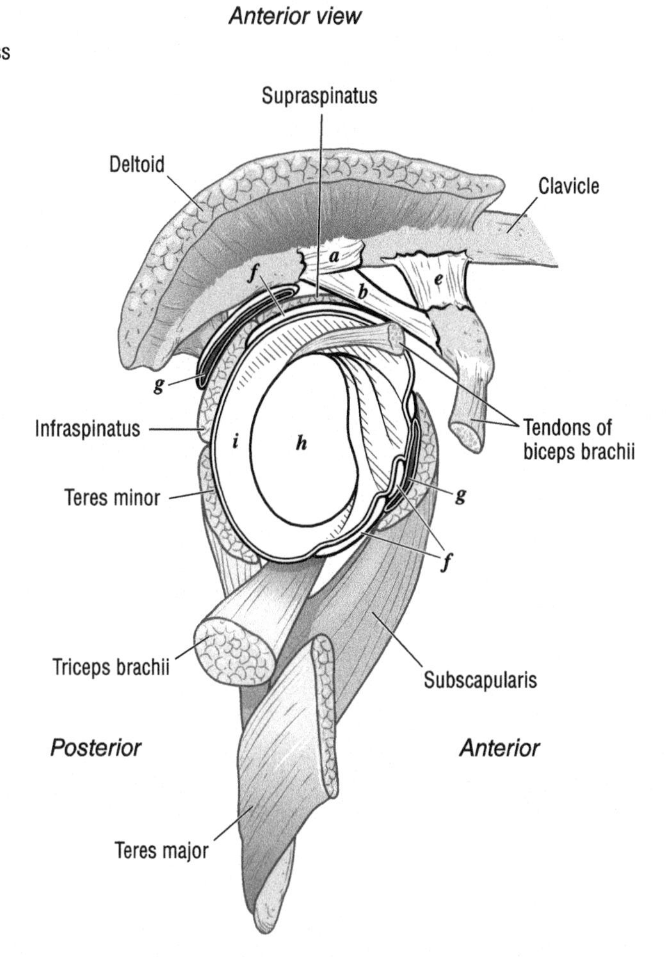

Lateral view, joint opened

Elbow joint

(a) Radial collateral ligament
(b) Anular ligament
(c) Ulnar collateral ligament
(d) Capsular ligament
(e) Olecranon bursa

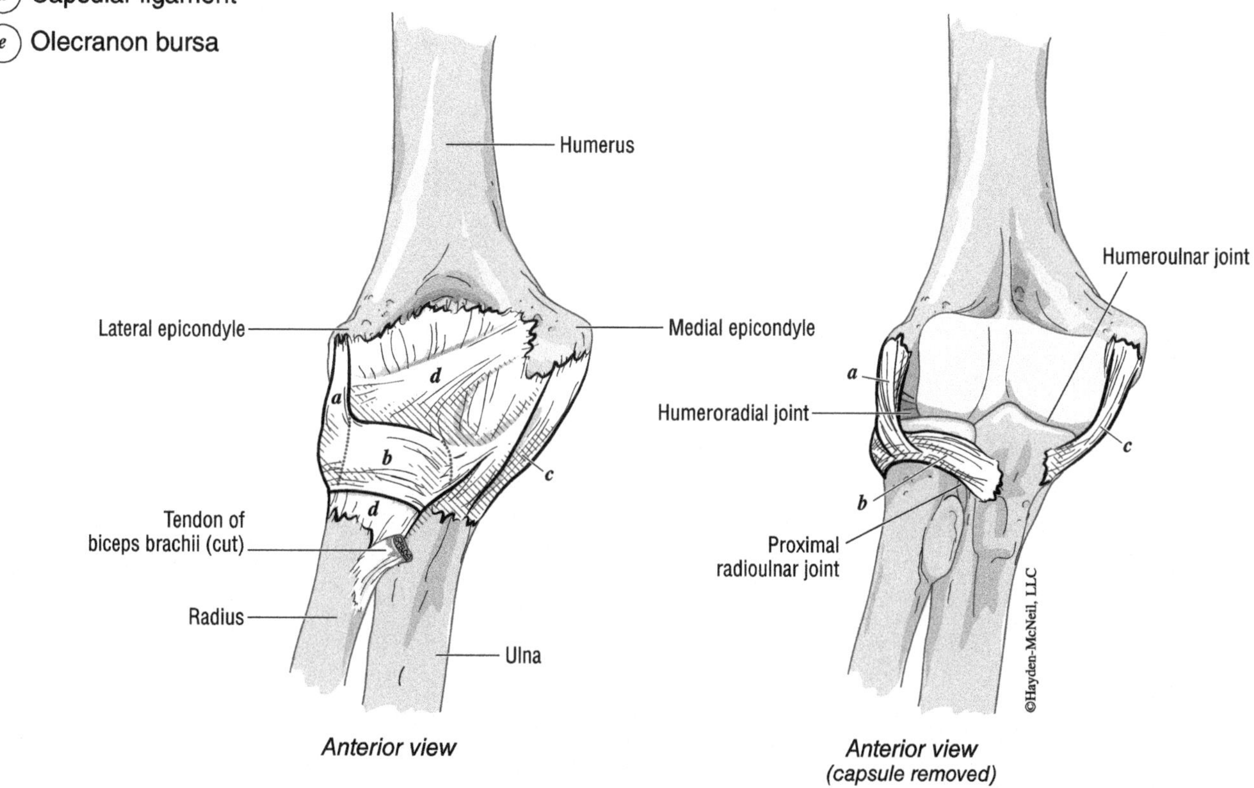

Anterior view

Anterior view (capsule removed)

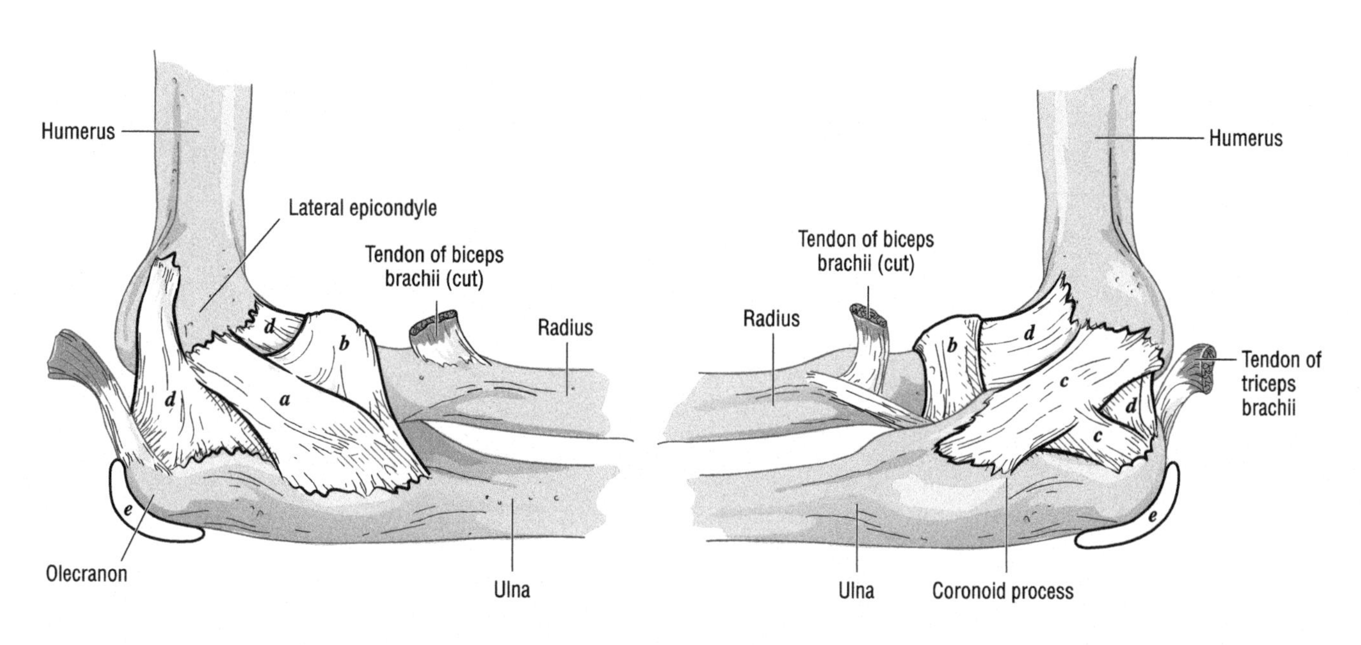

Lateral view

Medial view

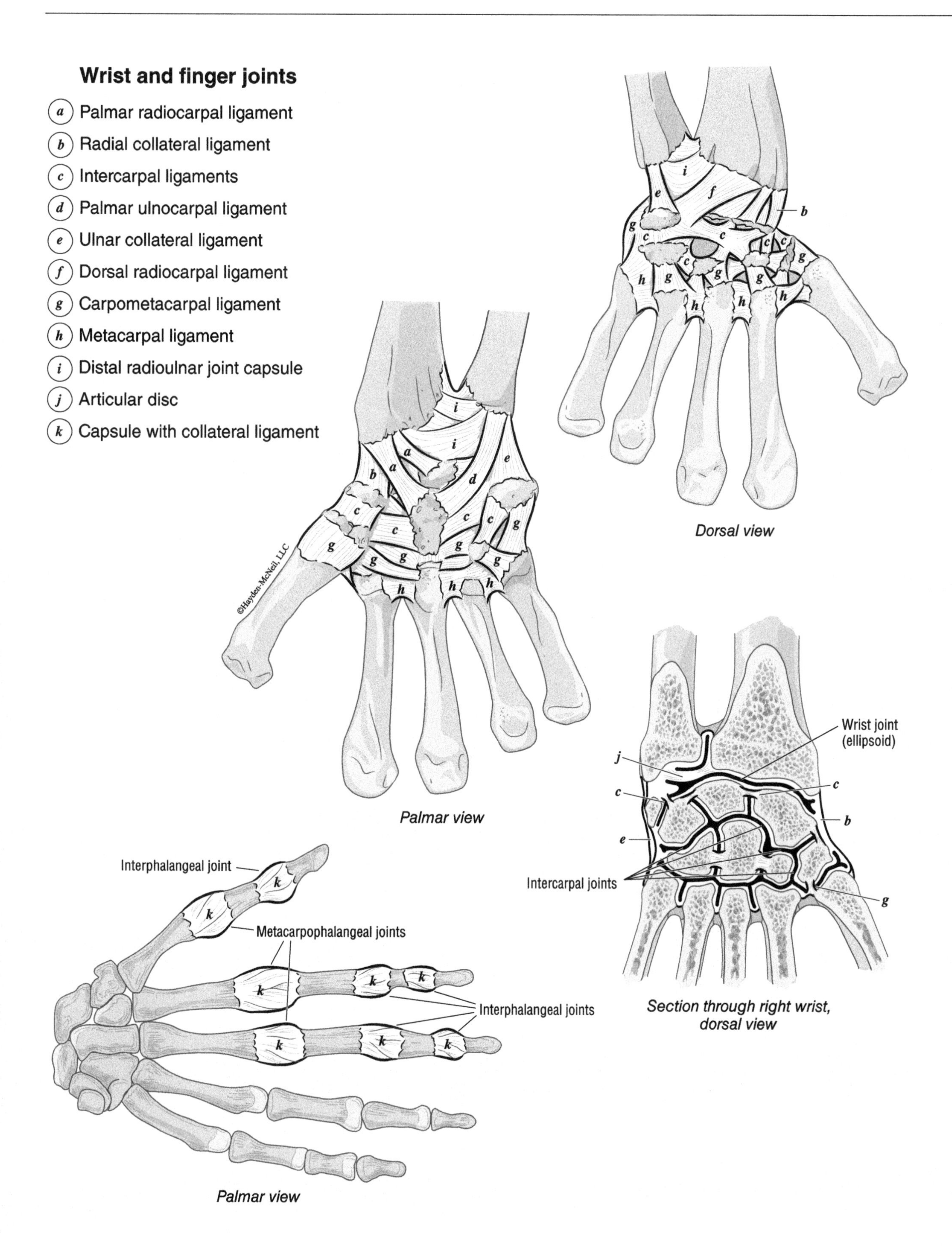

Dorsal view

Palmar view

Section through right wrist, dorsal view

Palmar view

Hip joint

(a) Iliofemoral ligament
(b) Ligamentum teres
(c) Pubofemoral ligament
(d) Acetabular labrum
(e) Ischiofemoral ligament
(f) Zona orbicularis

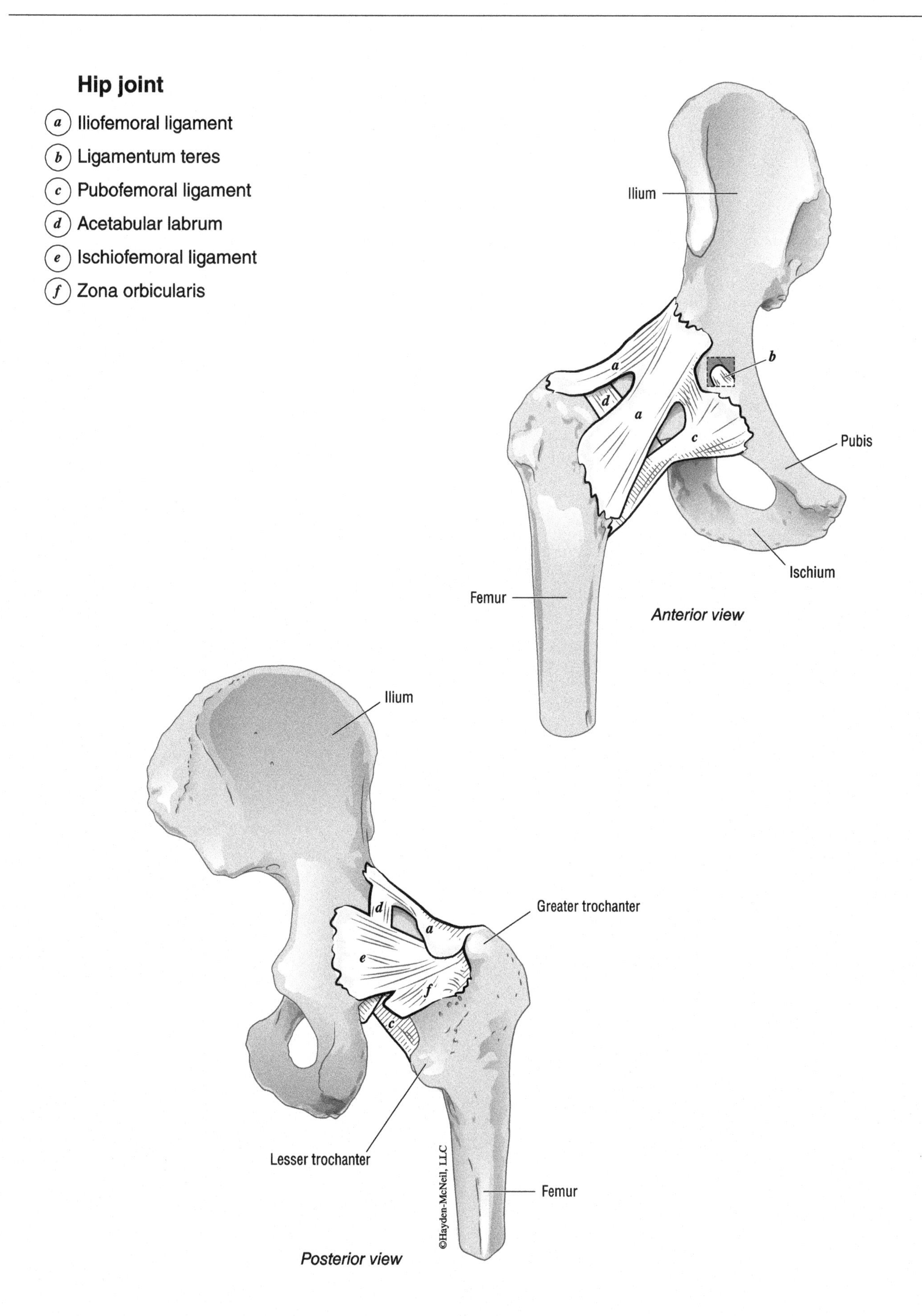

Knee joint

Color the tendon of the quadriceps femoris (•) gray.

(a) Lateral (fibular) collateral ligament
(b) Anterior cruciate ligament (ACL)
(c) Posterior cruciate ligament (PCL)
(d) Medial (tibial) collateral ligament
(e) Medial meniscus
(f) Lateral meniscus
(g) Transverse ligament of the knee
(h) Patellar ligament
(i) Capsule
(j) Lateral patellar retinaculum
(k) Medial patellar retinaculum
(l) Oblique popliteal ligament
(m) Arcuate popliteal ligament

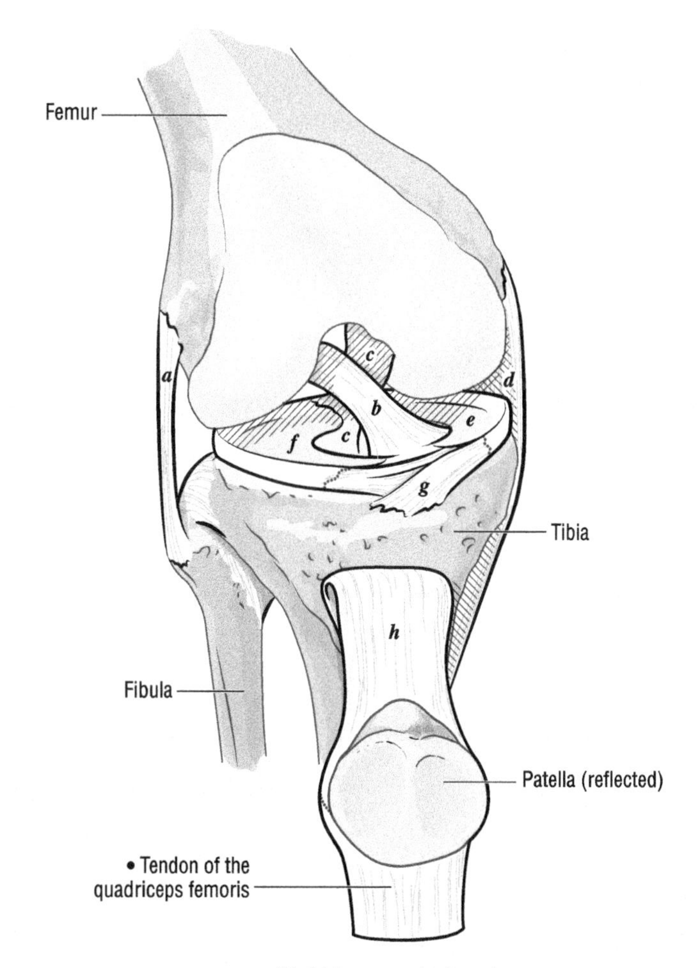

Right knee, anterior view

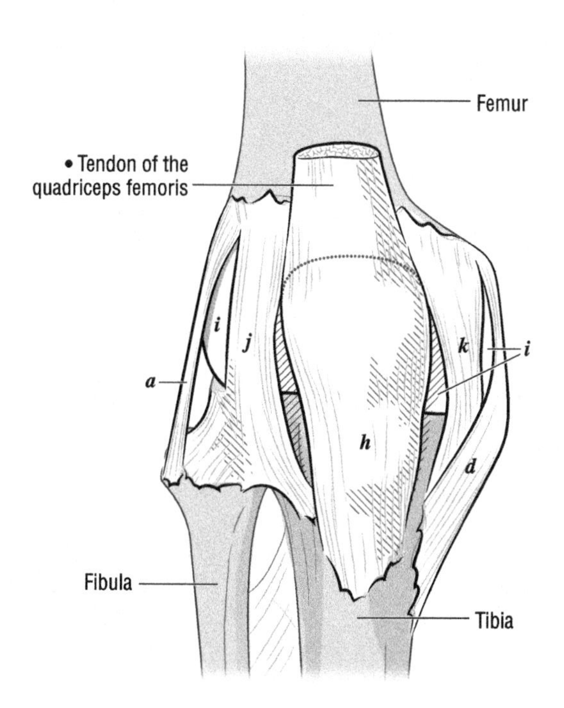

Right knee, anterior view

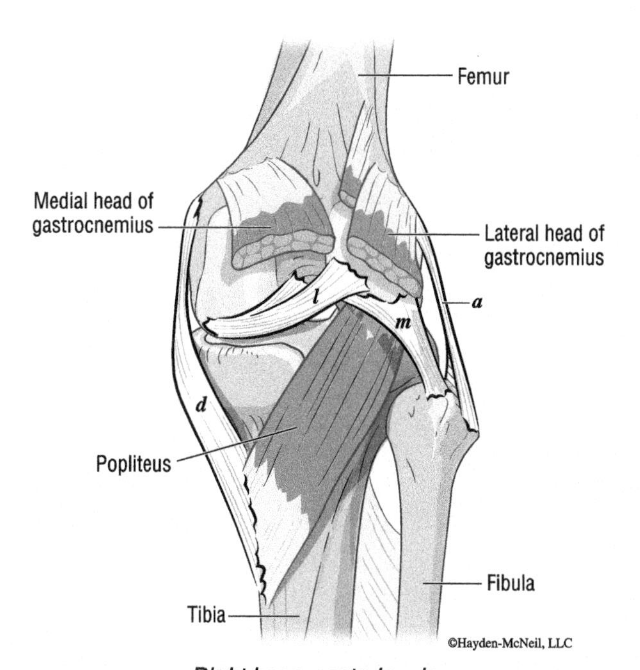

Right knee, posterior view

Ankle and foot joints

- (a) Posterior tibiofibular lig.
- (b) Posterior talofibular lig.
- (c) Calcaneofibular lig.
- (d) Medial (deltoid) lig.
- (e) Ligaments of talus and tarsals
  - e_1 Talonavicular lig.
  - e_2 Interosseous talocalcaneal lig.
  - e_3 Lateral talocalcaneal lig.
- (f) Other dorsal tarsal ligaments
  - f_1 Bifurcate lig.
- (g) Plantar tarsal ligaments
  - g_1 Long plantar lig.
  - g_2 Plantar calcaneonavicular (spring) lig.
- (h) Dorsal tarsometatarsal ligs.
- (i) Dorsal metatarsal ligs.
- (j) Plantar metatarsal ligs.
- (k) Capsule and collateral ligs.

Tibia
Fibula
Talus
Calcaneus
a
b
c
d

Posterior view

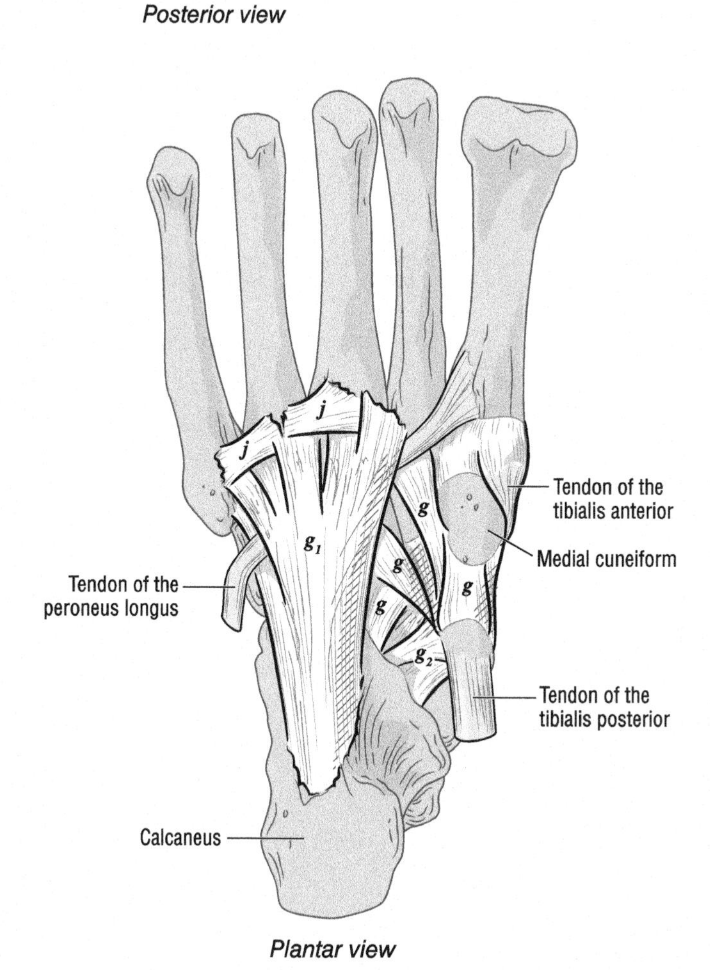

Plantar view

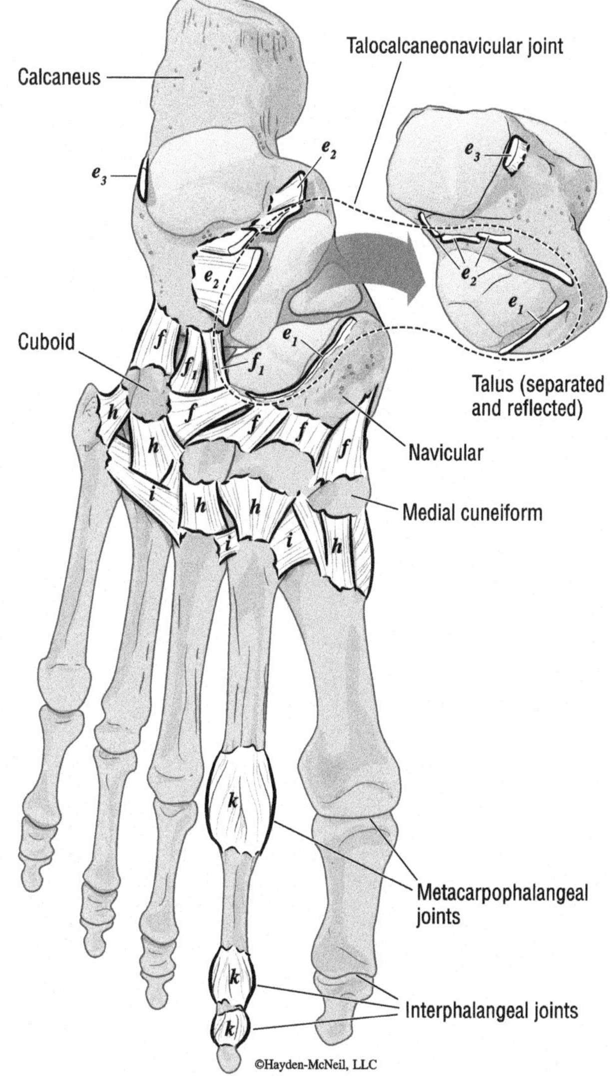

Dorsal view

7

The Muscular System

- *Neuromuscular junction*
- *Skeletal muscle shapes*
- *Sliding filament theory of muscle contraction*
- *Skeletal muscle structure*
- *Muscles of the face and neck*
- *Masticatory muscles*
- *Muscles of the mouth floor*
- *Deep neck muscles*
- *Deep back muscles*
- *Muscles of the rib cage*
- *Muscles of the abdominal wall*
- *Muscles of the pelvic diaphragm and perineum*
- *Shoulder and arm muscles*
- *Posterior muscles of the upper limb*
- *Anterior muscles of the forearm*
- *Posterior muscles of the forearm*
- *Anterior (palmar) muscles of the hand*
- *Muscular regions of the lower limb*
- *Muscles of the hip*
- *Thigh muscles*
- *Anterior muscles of the leg and foot*
- *Lateral and posterior muscles of the leg*
- *Muscles of the foot*

Get ready to flex your coloring hand as we introduce the **muscular system**! Throughout this chapter, you will thoughtfully tone your understanding and gain a new appreciation for our muscles' ability to masterfully sculpt everyday expressions and maintain the body's essential stability. From the gentle grip of a father's hand to the powerful leap of a professional ballerina, you will unveil the magic behind every movement, one muscle at a time.

Neuromuscular junction

- (a) Motor neuron
- (b) Axon terminal of motor neuron
- (c) Sarcolemma
- (d) Myofibrils
- (e) Sarcoplasmic reticulum
- (f) Transverse tubules
- (g) Acetylcholine
- (h) Acetylcholine receptors
- (i) Mitochondrion
- (j) Synaptic vesicle
- (k) Motor end plate
- (l) Junctional fold

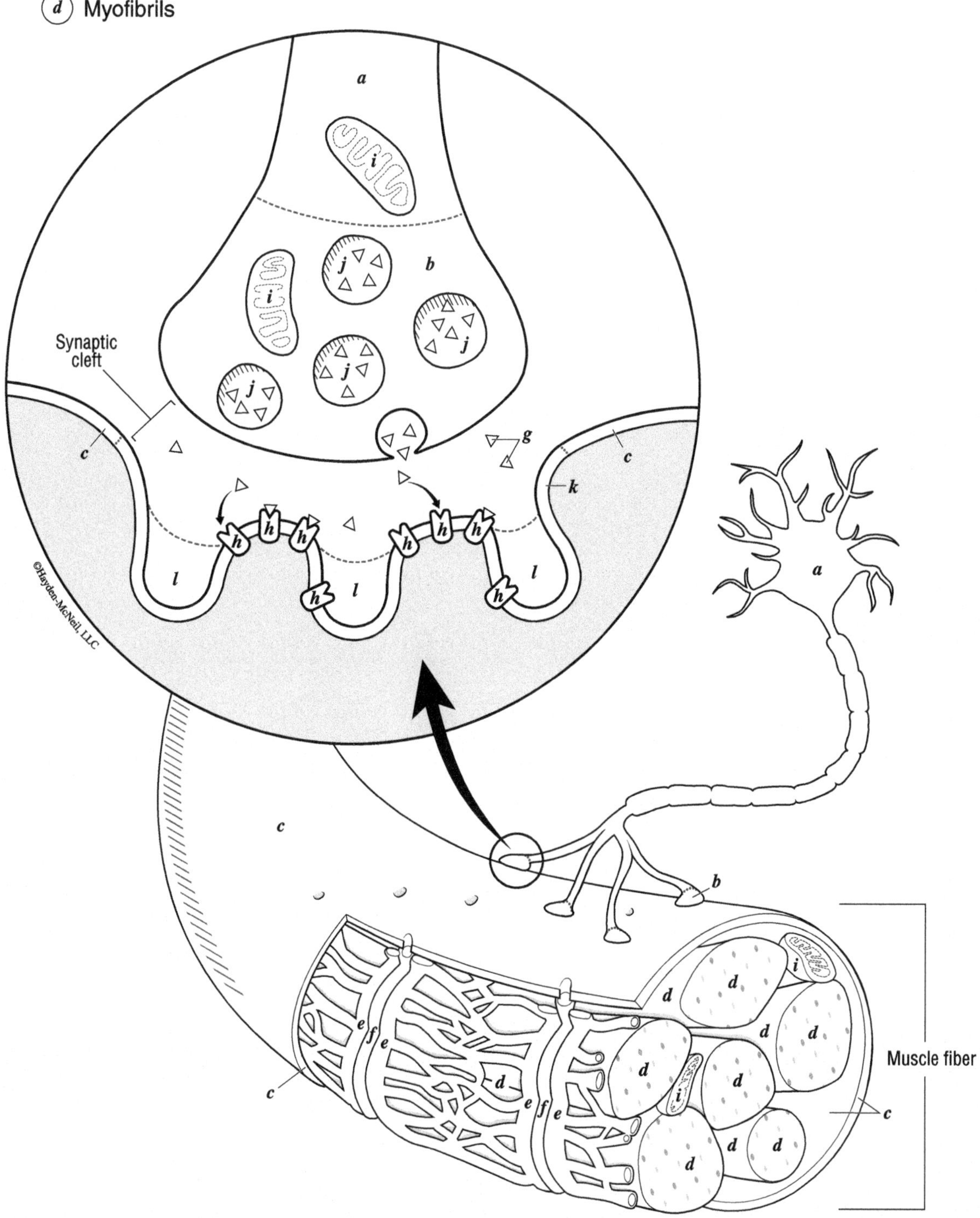

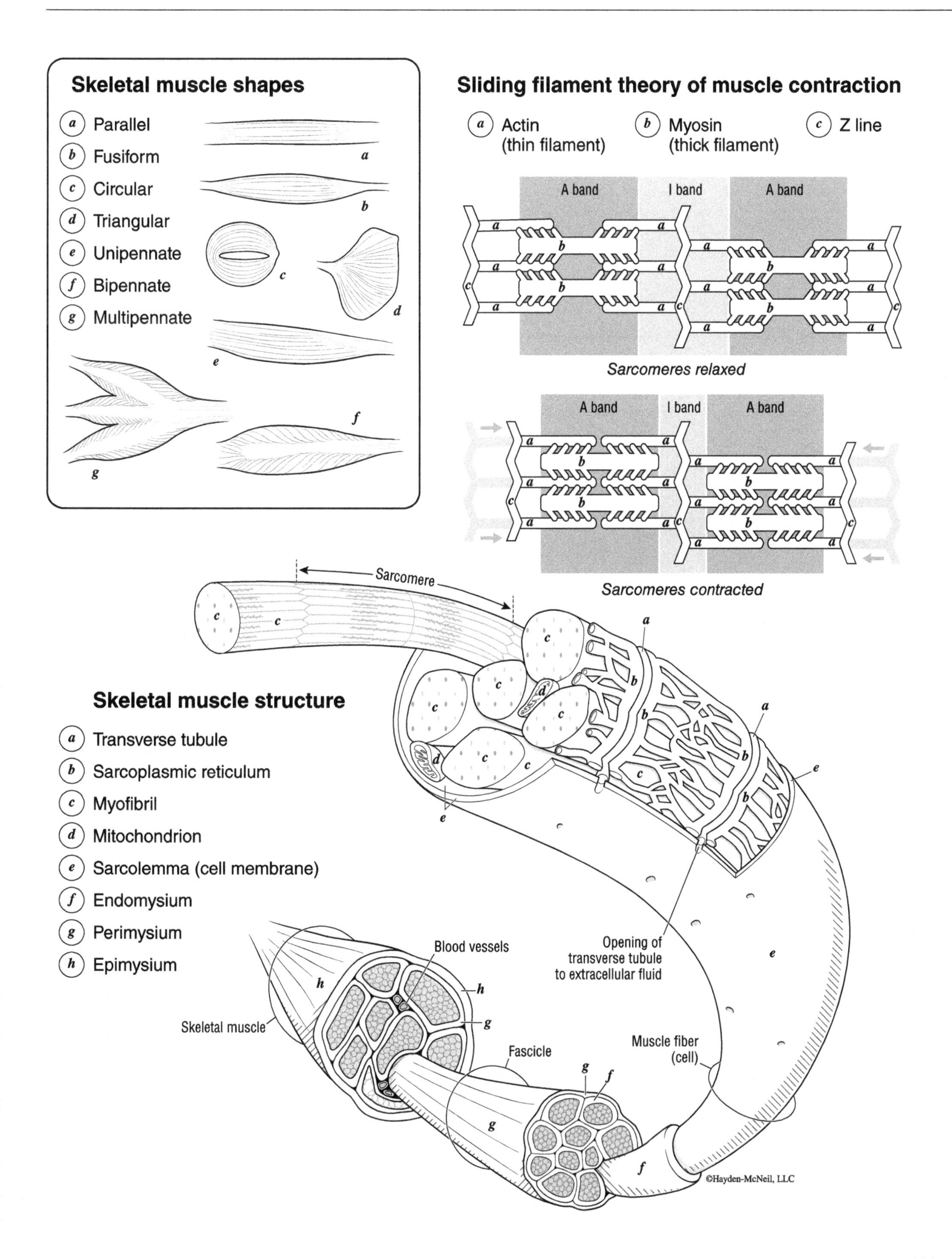

Skeletal muscle shapes
a Parallel
b Fusiform
c Circular
d Triangular
e Unipennate
f Bipennate
g Multipennate
Sliding filament theory of muscle contraction
a Actin (thin filament)
b Myosin (thick filament)
c Z line
A band
I band
A band
Sarcomeres relaxed
A band
I band
A band
Sarcomeres contracted
Sarcomere
Skeletal muscle structure
a Transverse tubule
b Sarcoplasmic reticulum
c Myofibril
d Mitochondrion
e Sarcolemma (cell membrane)
f Endomysium
g Perimysium
h Epimysium
Blood vessels
Opening of transverse tubule to extracellular fluid
Skeletal muscle
Fascicle
Muscle fiber (cell)
©Hayden-McNeil, LLC

Muscles of the face and neck

- (a) Frontalis
- (b) Corrugator supercilii
- (c) Orbicularis oculi (partially cut away in anterior view)
- (d) Levator anguli oris
- (e) Orbicularis oris
- (f) Mentalis
- (g) Sternohyoid
- (h) Epicranial aponeurosis
- (i) Temporalis
- (j) Levator labii superioris
- (k) Levator labii superioris alaeque nasi
- (l) Zygomaticus major
- (m) Zygomaticus minor
- (n) Masseter
- (o) Risorius
- (p) Depressor anguli oris
- (q) Depressor labii inferioris
- (r) Sternocleidomastoid
  - r_1 Sternal head
  - r_2 Clavicular head
- (s) Omohyoid
  - s_1 Superior belly
  - s_2 Inferior belly
- (t) Nasalis
- (u) Trapezius
- (v) Anterior scalene
- (w) Occipitalis
- (x) Buccinator
- (y) Platysma (partially cut away in anterior view)

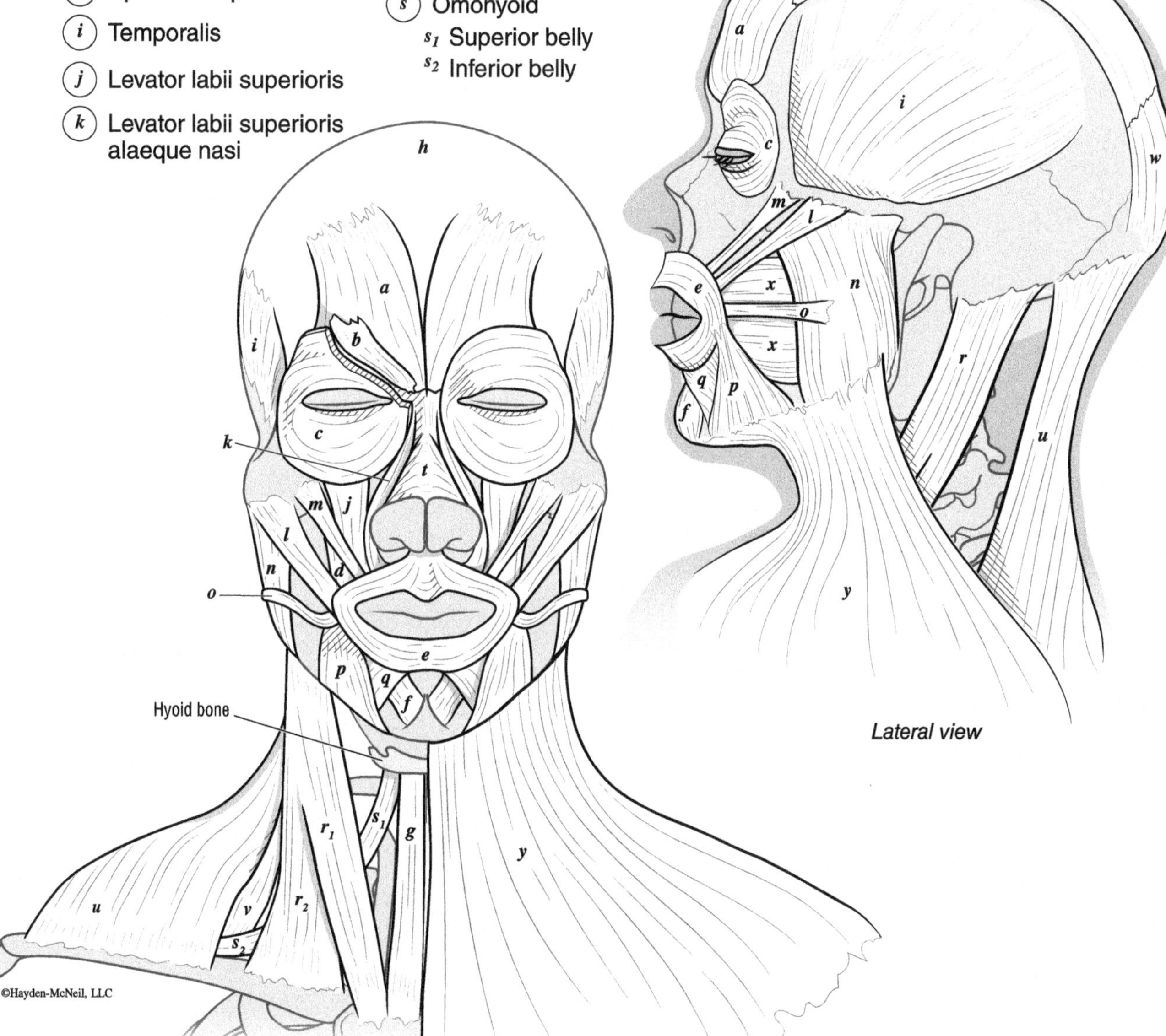

Lateral view

Anterior view

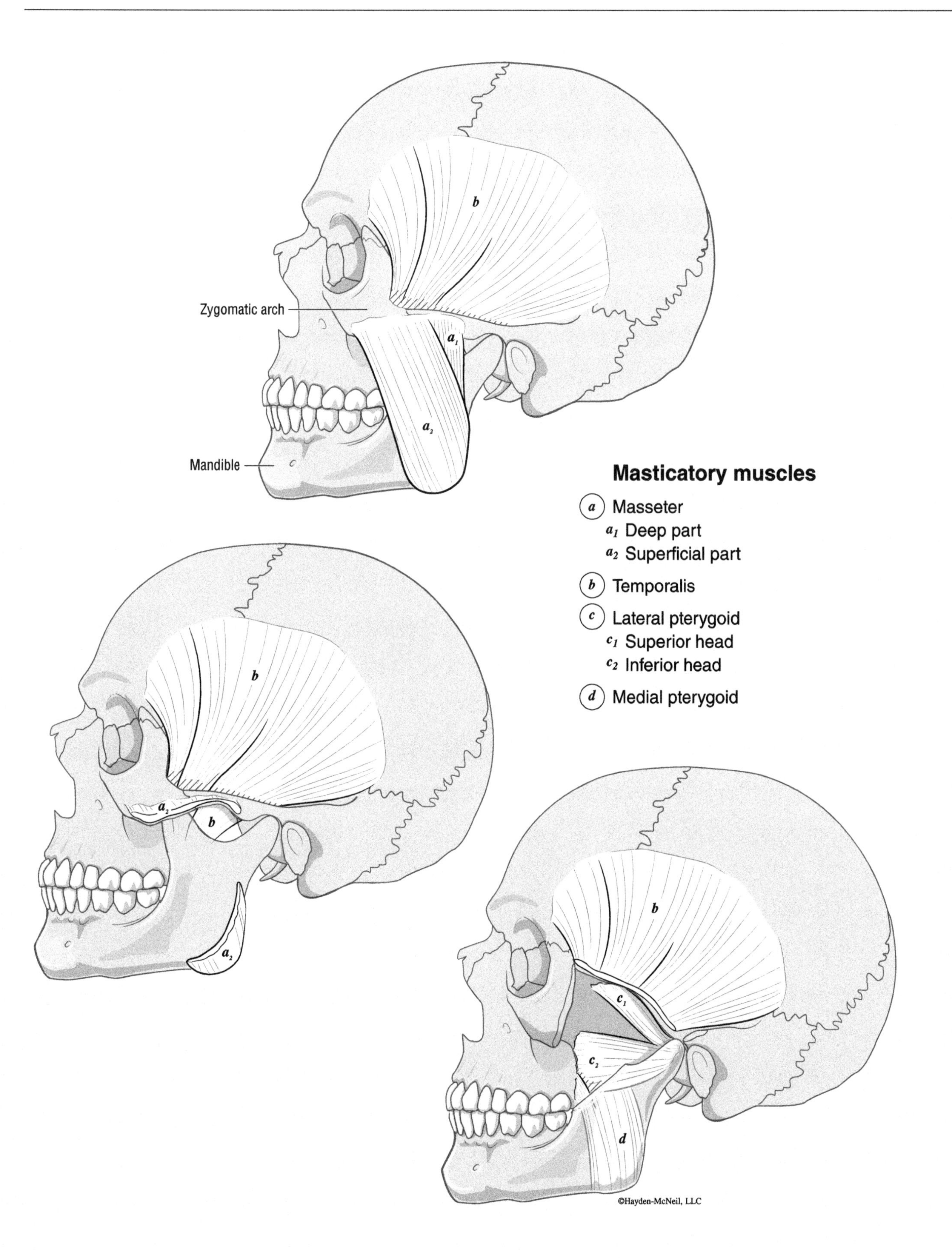

Zygomatic arch
Mandible
b
a1
a2
b
a2
b
a2
b
c1
c2
d
Masticatory muscles
a Masseter
a1 Deep part
a2 Superficial part
b Temporalis
c Lateral pterygoid
c1 Superior head
c2 Inferior head
d Medial pterygoid
©Hayden-McNeil, LLC

Muscles of the mouth floor

Do not color the tendinous sling for the digastric muscle marked with (*).

- (a) Hyoglossus
- (b) Cricothyroid
- (c) Stylohyoid
- (d) Geniohyoid
- (e) Digastric
  - e_1 Posterior belly
  - e_2 Anterior belly
- (f) Mylohyoid
- (g) Genioglossus

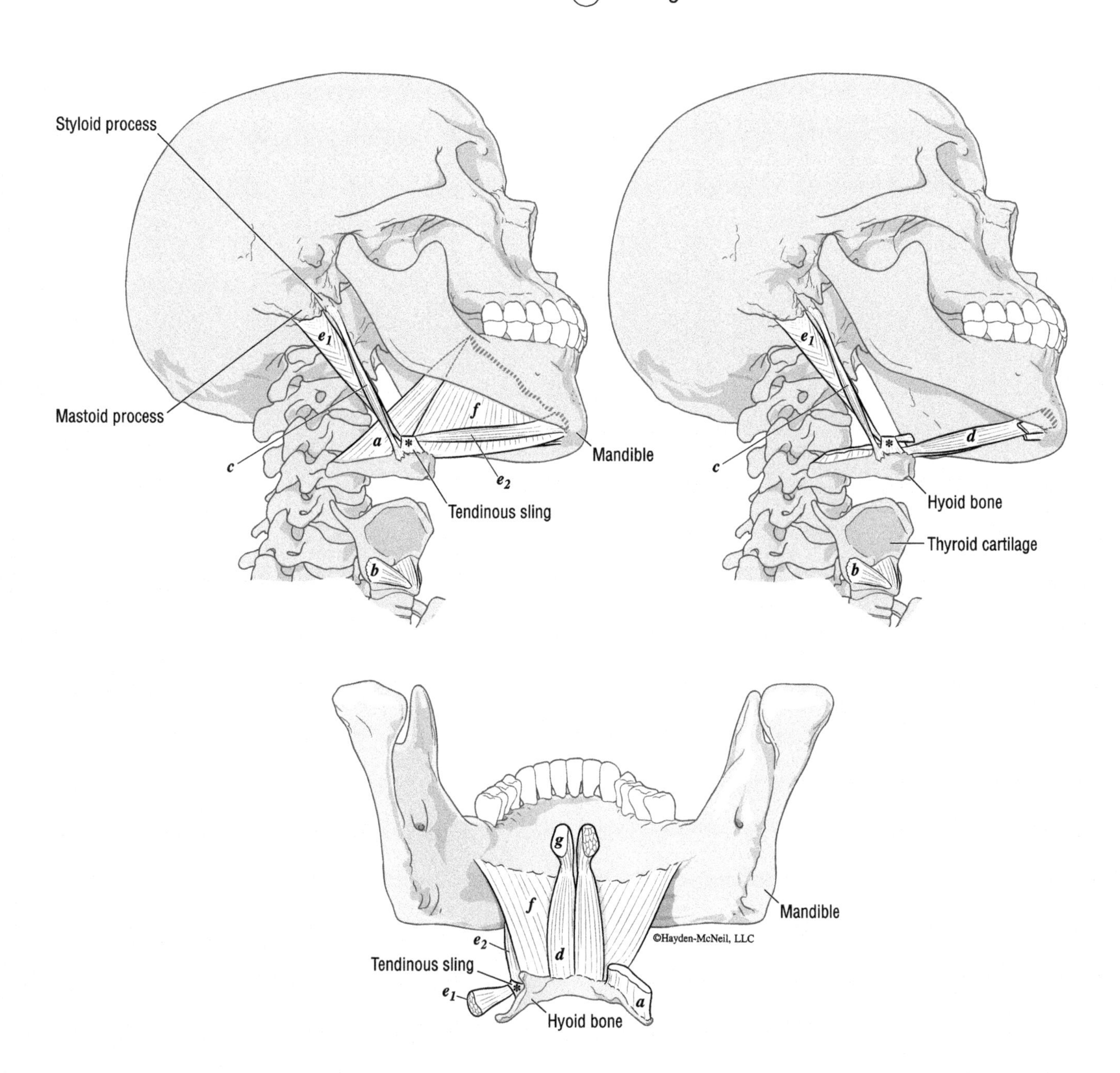

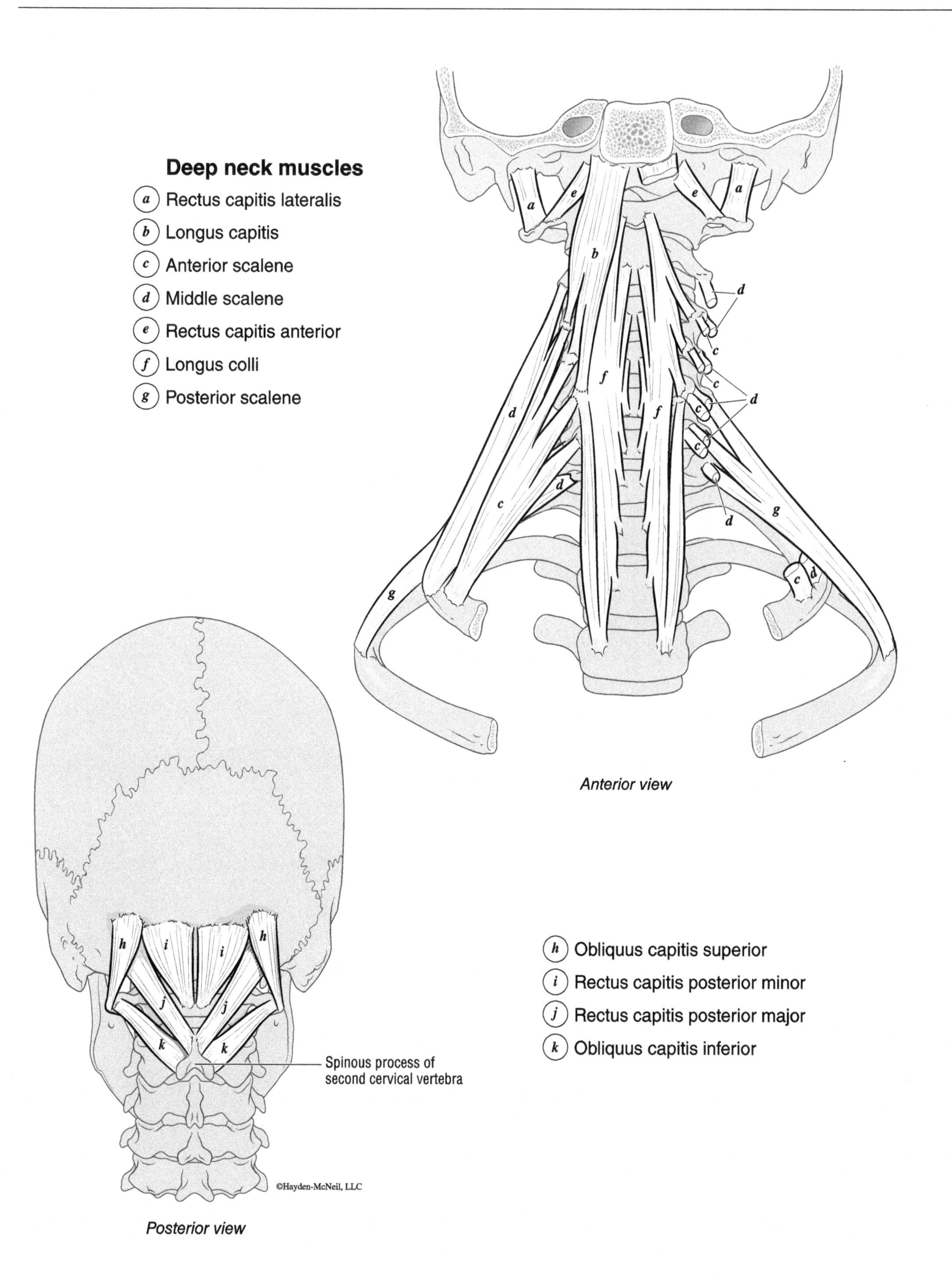

Anterior view

Posterior view

Deep back muscles

- (a) Splenius
  - a_1 Splenius capitis
  - a_2 Splenius cervicis
- (b) Iliocostalis *(part of erector spinae group)*
  - b_1 Iliocostalis cervicis
  - b_2 Iliocostalis thoracis
  - b_3 Iliocostalis lumborum
- (c) Longissimus *(part of erector spinae group)*
  - c_1 Longissimus capitis
  - c_2 Longissimus cervicis
  - c_3 Longissimus thoracis
- (d) Spinalis *(part of erector spinae group)*
  - d_1 Spinalis cervicis
  - d_2 Spinalis thoracis
- (e) Semispinalis capitis *(part of transversospinalis group)*
- (f) Multifidus *(part of transversospinalis group)*
- (g) Rotatores *(part of transversospinalis group)*
  - g_1 Rotatores brevis
  - g_2 Rotatores longus
- (h) Quadratus lumborum

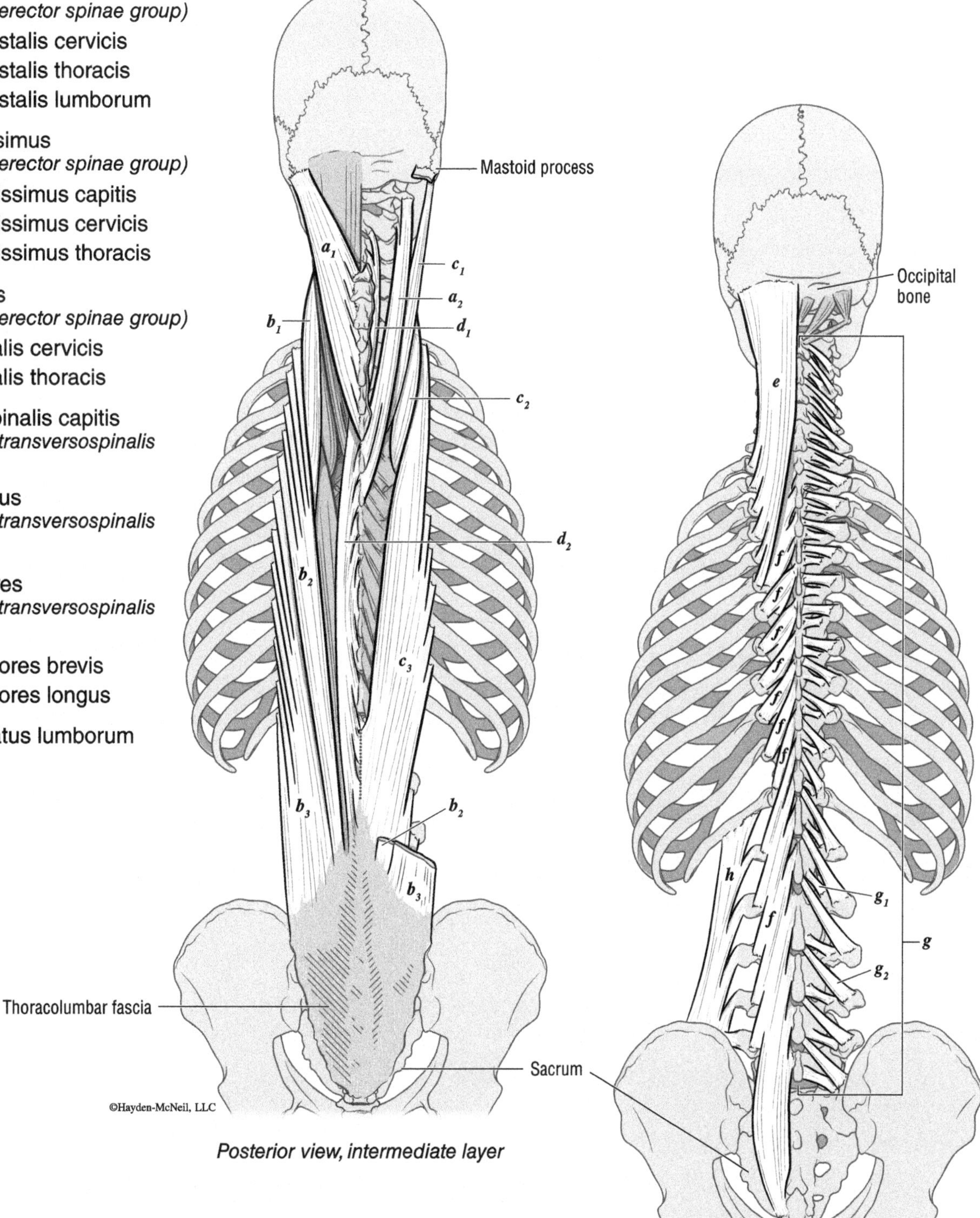

Posterior view, intermediate layer

Posterior view, deep layer

Muscles of the rib cage

- (a) External intercostals
- (b) Internal intercostals
- (c) Transversus thoracis
- (d) Diaphragm

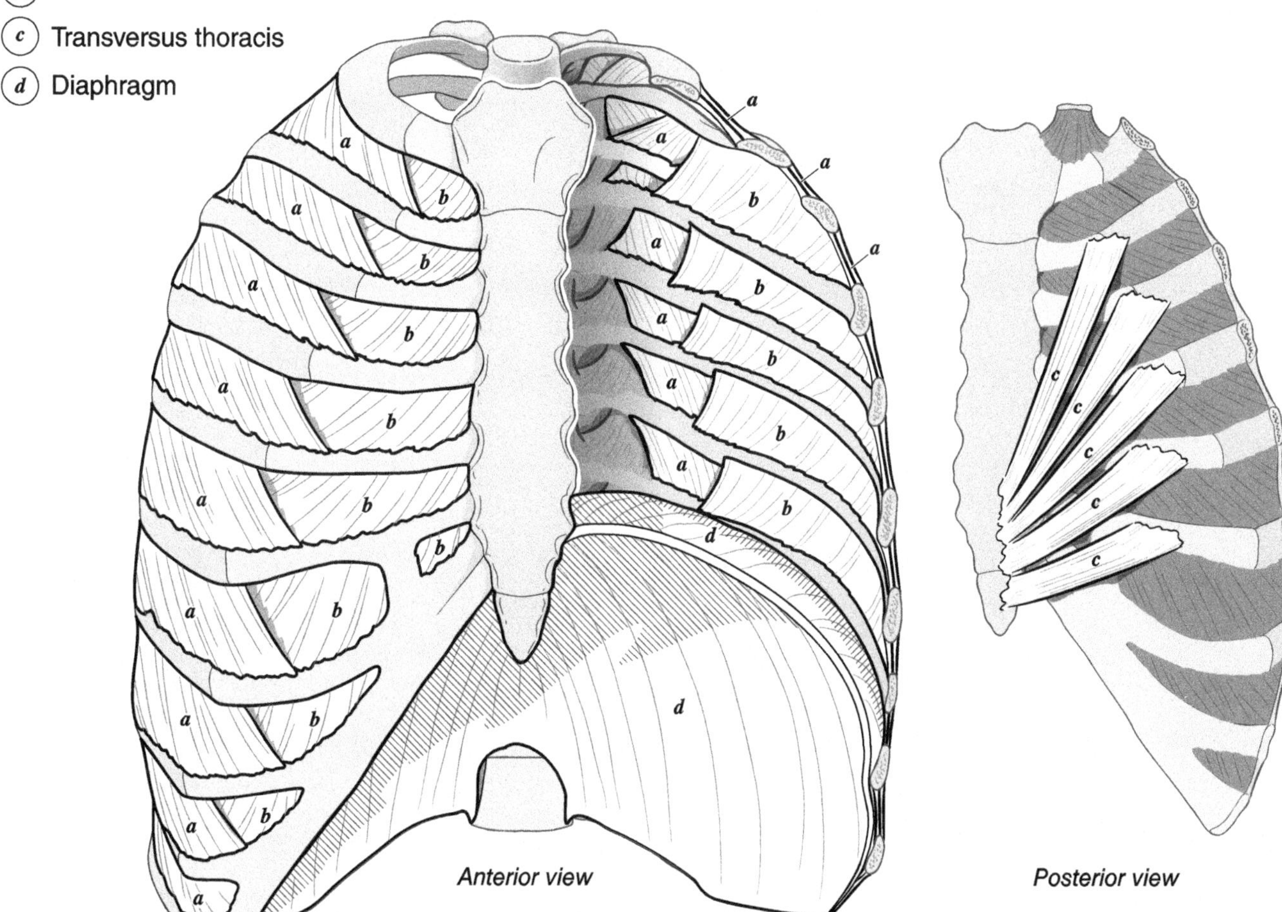

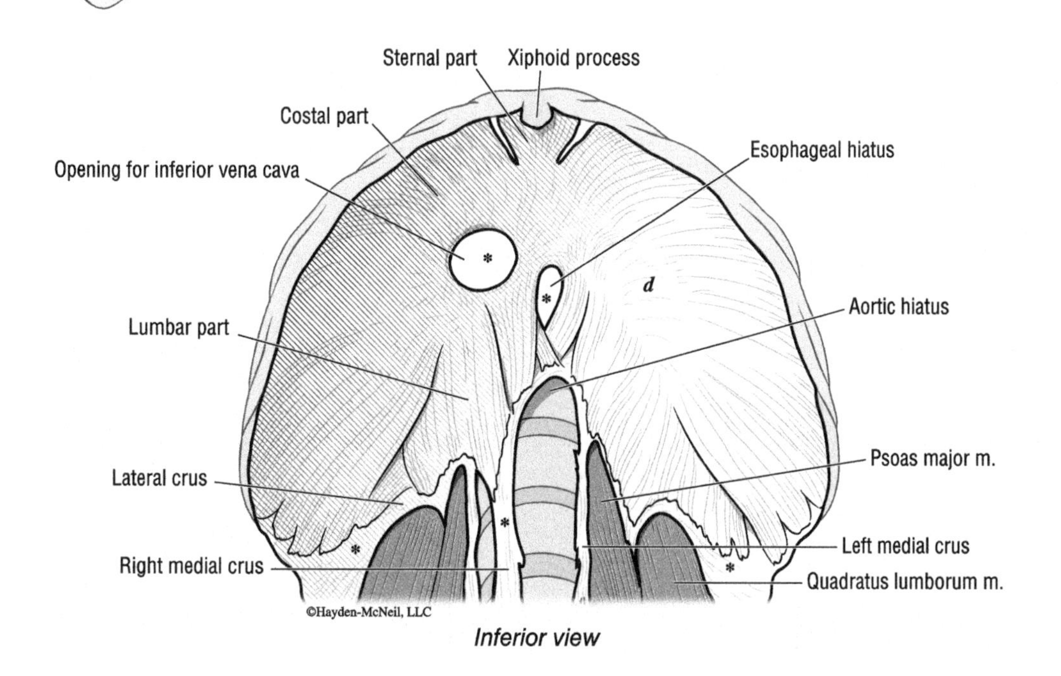

Inferior view

Muscles of the abdominal wall

Do not color the aponeuroses marked with (*). Be sure to color the cut edges of the obliques in the intermediate and deep views to better understand the layering.

- (*a*) External abdominal oblique
- (*b*) Internal abdominal oblique
- (*c*) Transversus abdominis
- (*d*) Rectus abdominis

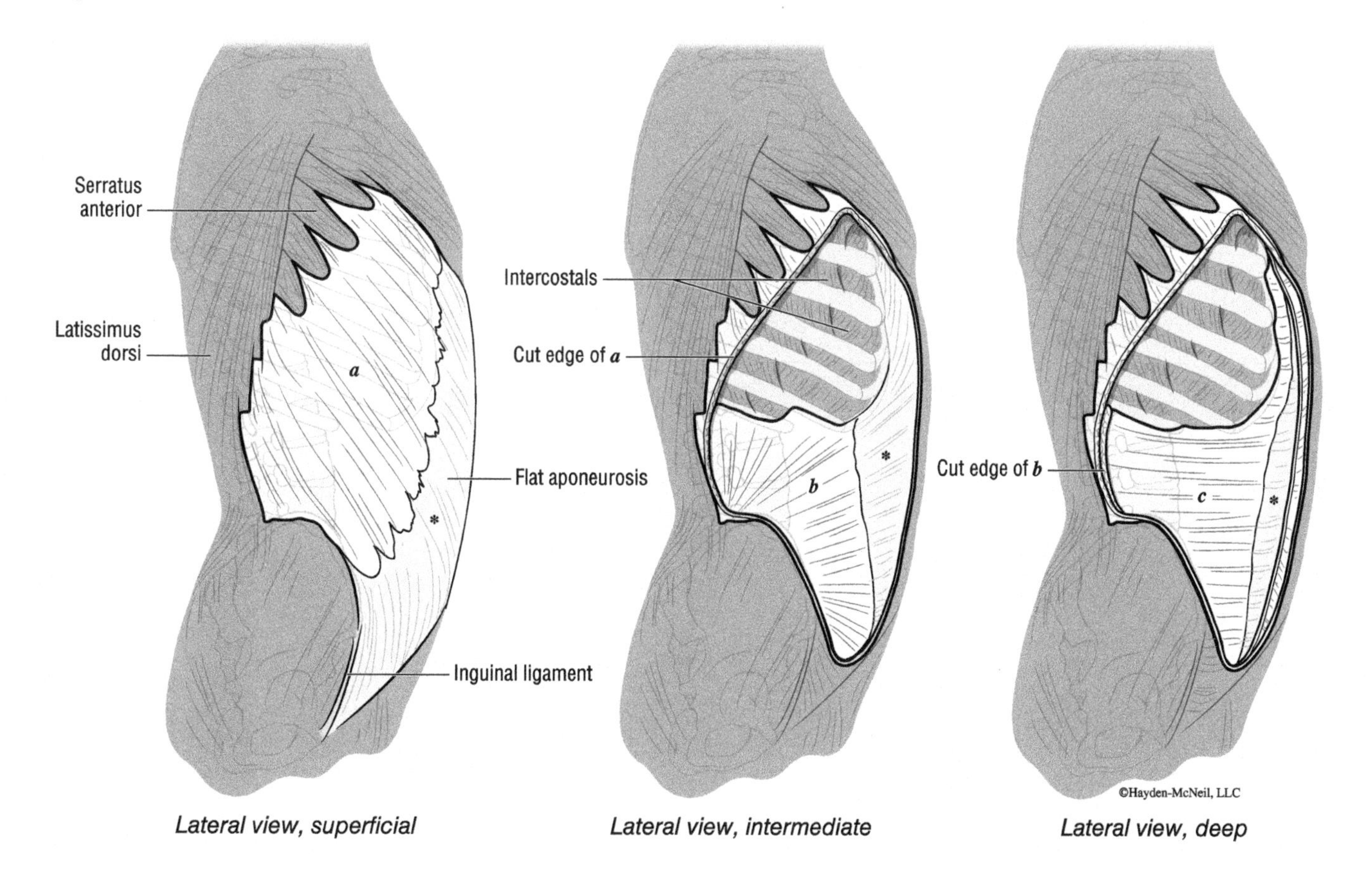

Lateral view, superficial *Lateral view, intermediate* *Lateral view, deep*

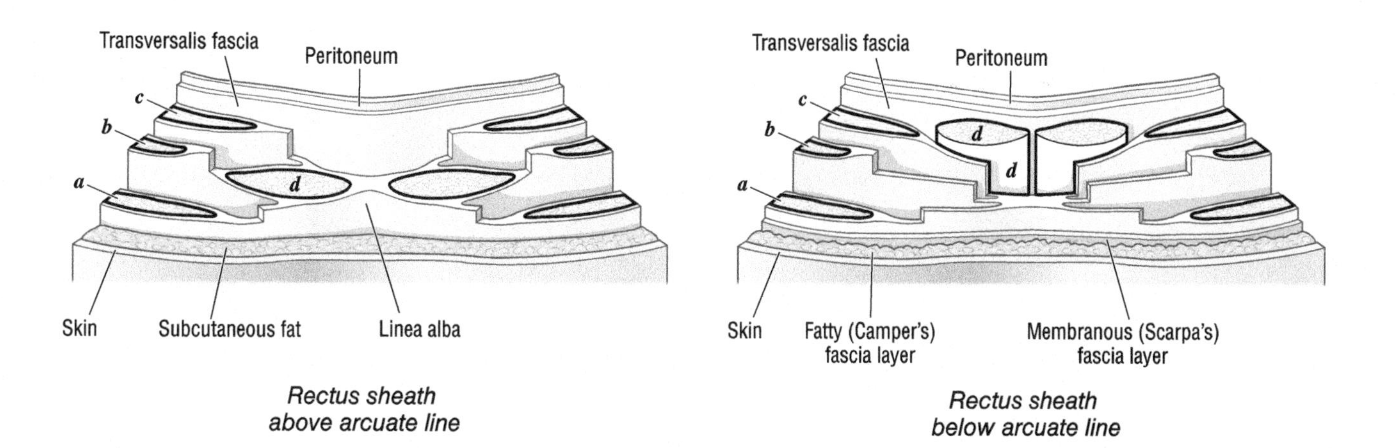

Rectus sheath above arcuate line *Rectus sheath below arcuate line*

Muscles of the pelvic diaphragm and perineum

Do not color holes and tendons marked with (*).

- (a) Deep transverse perineal
- (b) Levator ani
  - b_1 Puborectalis
  - b_2 Pubococcygeus
  - b_3 Iliococcygeus
- (c) Coccygeus
- (d) Superficial transverse perineal
- (e) External anal sphincter
- (f) Bulbospongiosus
- (g) Ischiocavernosus

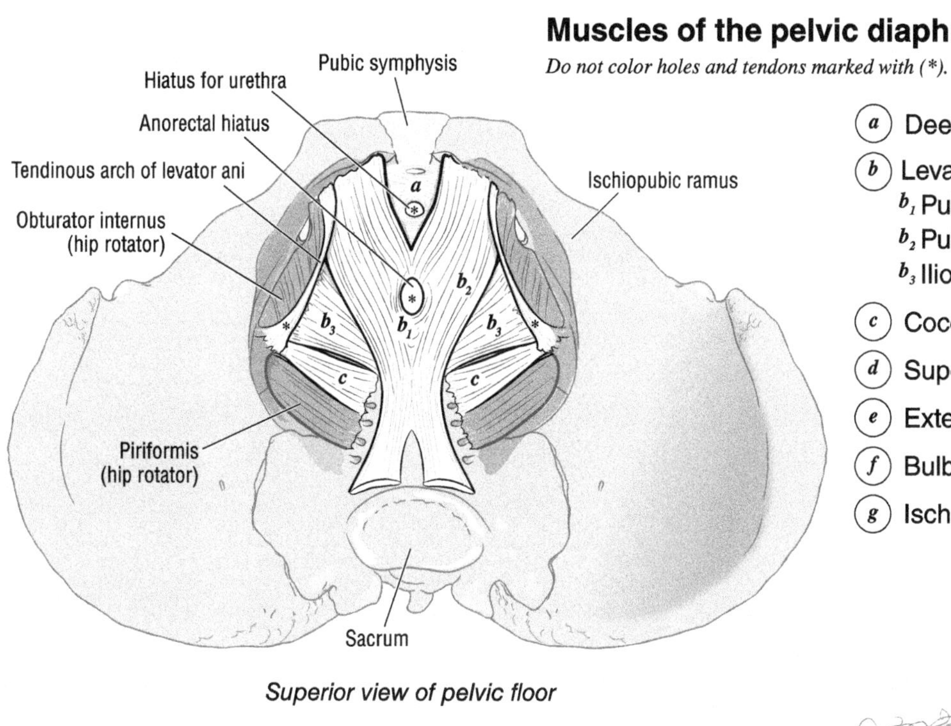

Superior view of pelvic floor

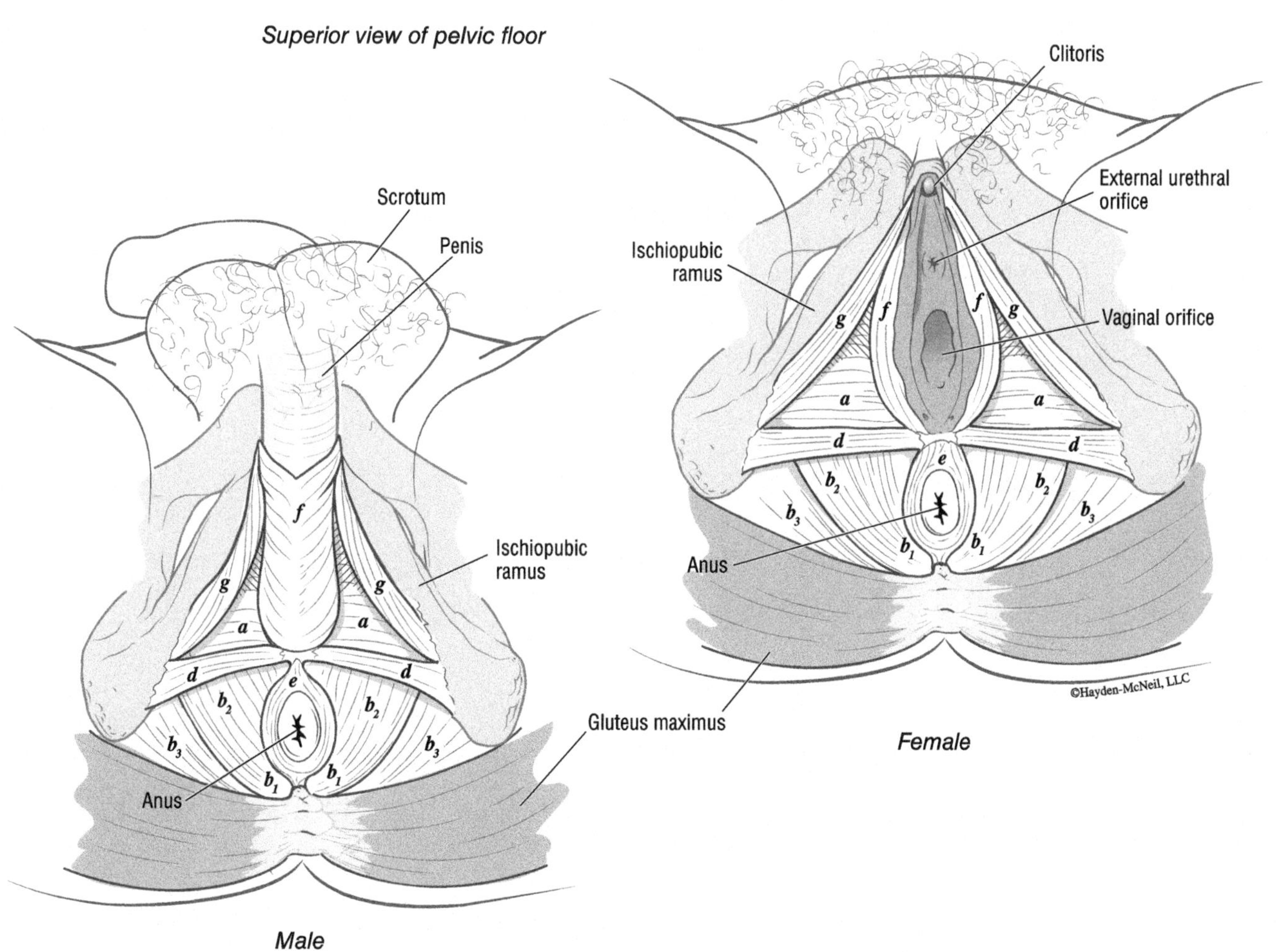

Male

Female

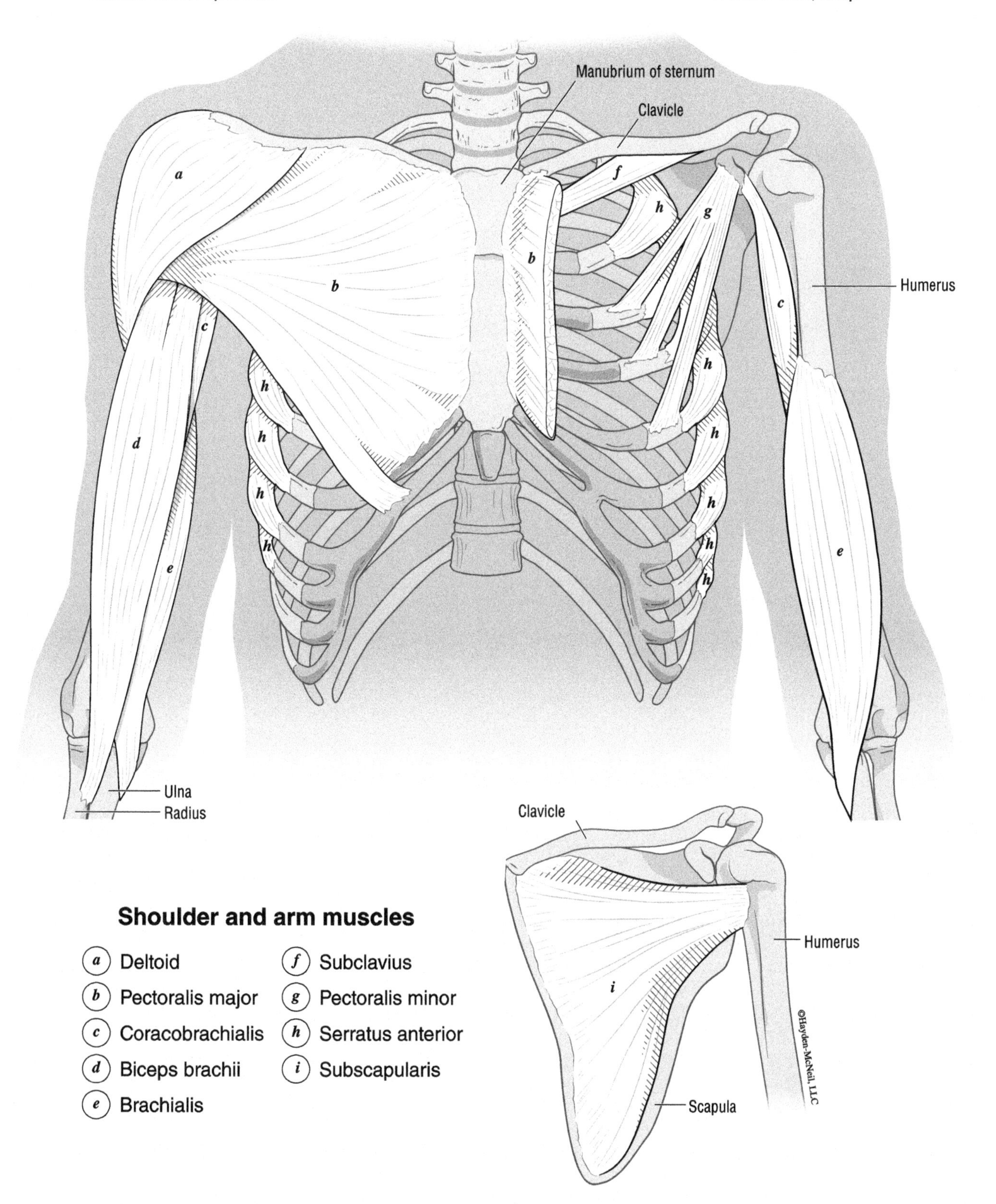
Anterior view, superficial
Anterior view, deep
Manubrium of sternum
Clavicle
Humerus
Ulna
Radius
a
b
c
d
e
f
g
h
i
Clavicle
Humerus
Scapula
Shoulder and arm muscles
a Deltoid
b Pectoralis major
c Coracobrachialis
d Biceps brachii
e Brachialis
f Subclavius
g Pectoralis minor
h Serratus anterior
i Subscapularis
©Hayden-McNeil, LLC

Posterior muscles of the upper limb

- (a) Deltoid
- (b) Infraspinatus
- (c) Latissimus dorsi
- (d) Levator scapulae
- (e) Rhomboid major
- (f) Rhomboid minor
- (g) Serratus anterior
- (h) Subscapularis
- (i) Supraspinatus
- (j) Teres major
- (k) Teres minor
- (l) Trapezius
- (m) Triceps brachii

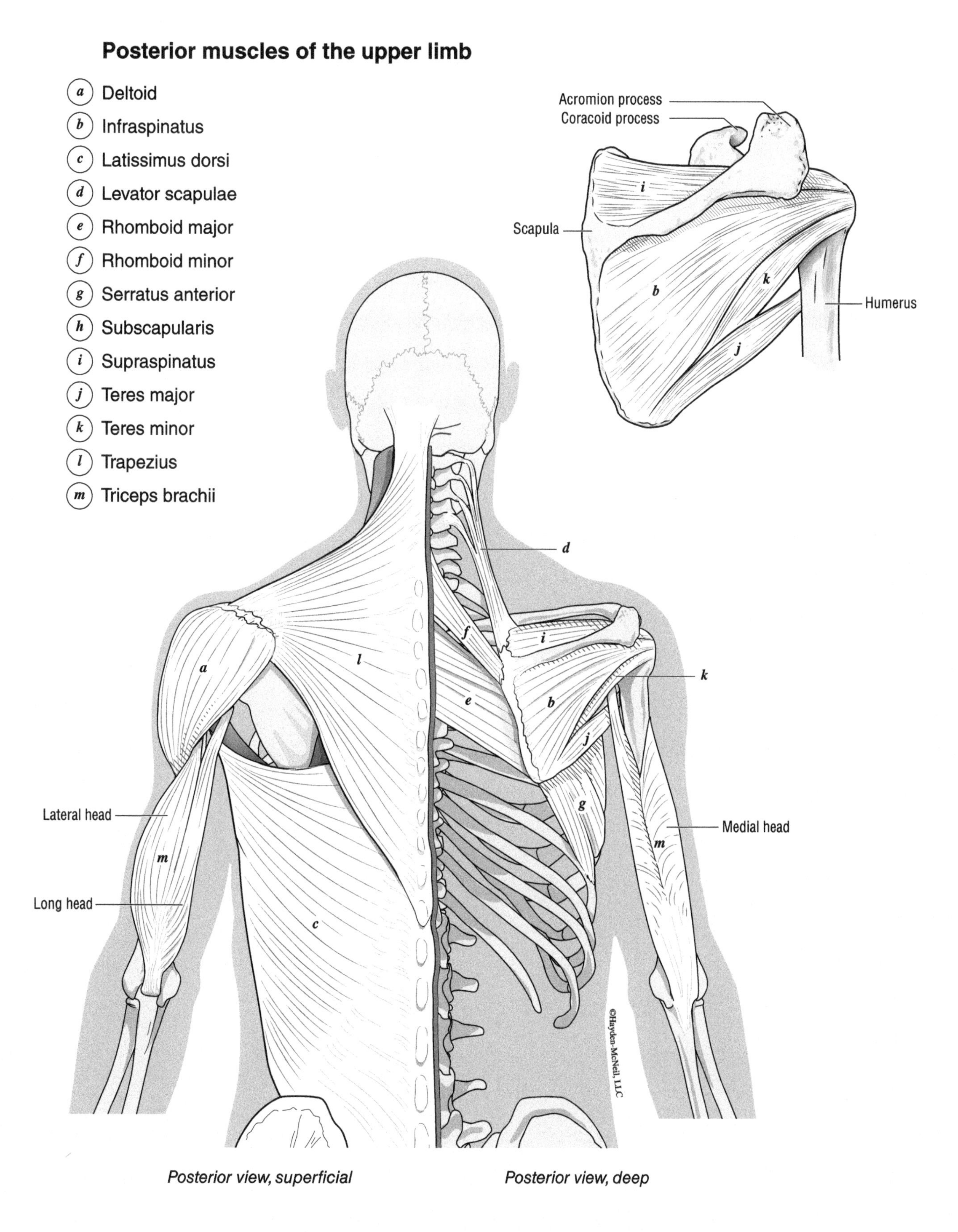

Posterior view, superficial

Posterior view, deep

Anterior muscles of the forearm

a	Pronator teres	*f*	Flexor pollicis longus
b	Brachioradialis	*g*	Flexor digitorum superficialis
c	Flexor carpi radialis	*h*	Flexor digitorum profundus
d	Palmaris longus	*i*	Pronator quadratus
e	Flexor carpi ulnaris	*j*	Supinator

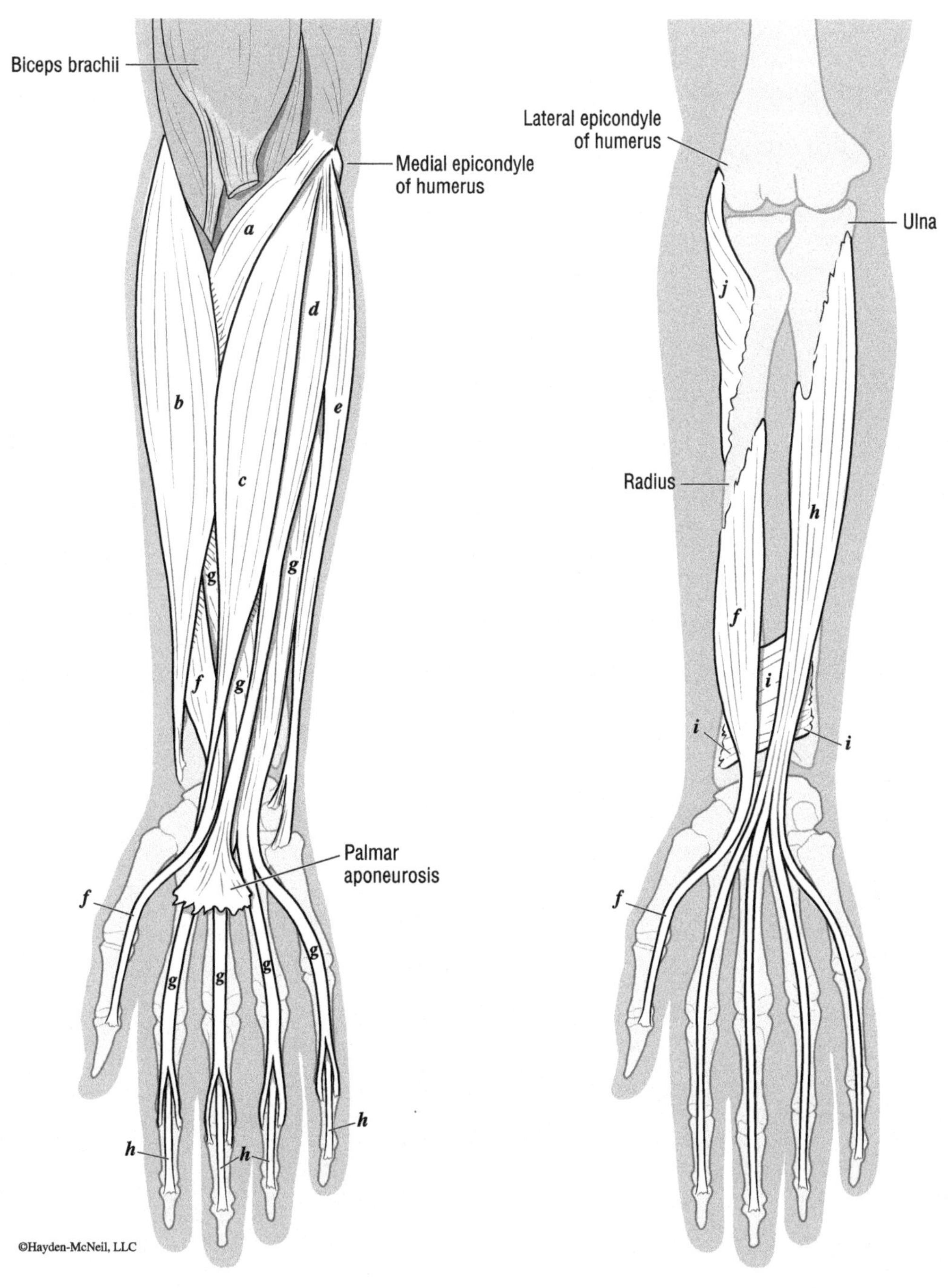

Anterior view, superficial

Anterior view, deep

Posterior muscles of the forearm

- (a) Brachioradialis
- (b) Extensor carpi radialis longus
- (c) Extensor carpi radialis brevis
- (d) Extensor digitorum
- (e) Abductor pollicis longus
- (f) Extensor pollicis brevis
- (g) Extensor pollicis longus
- (h) Extensor digiti minimi
- (i) Extensor carpi ulnaris
- (j) Flexor carpi ulnaris
- (k) Anconeus
- (l) Extensor indices
- (m) Supinator

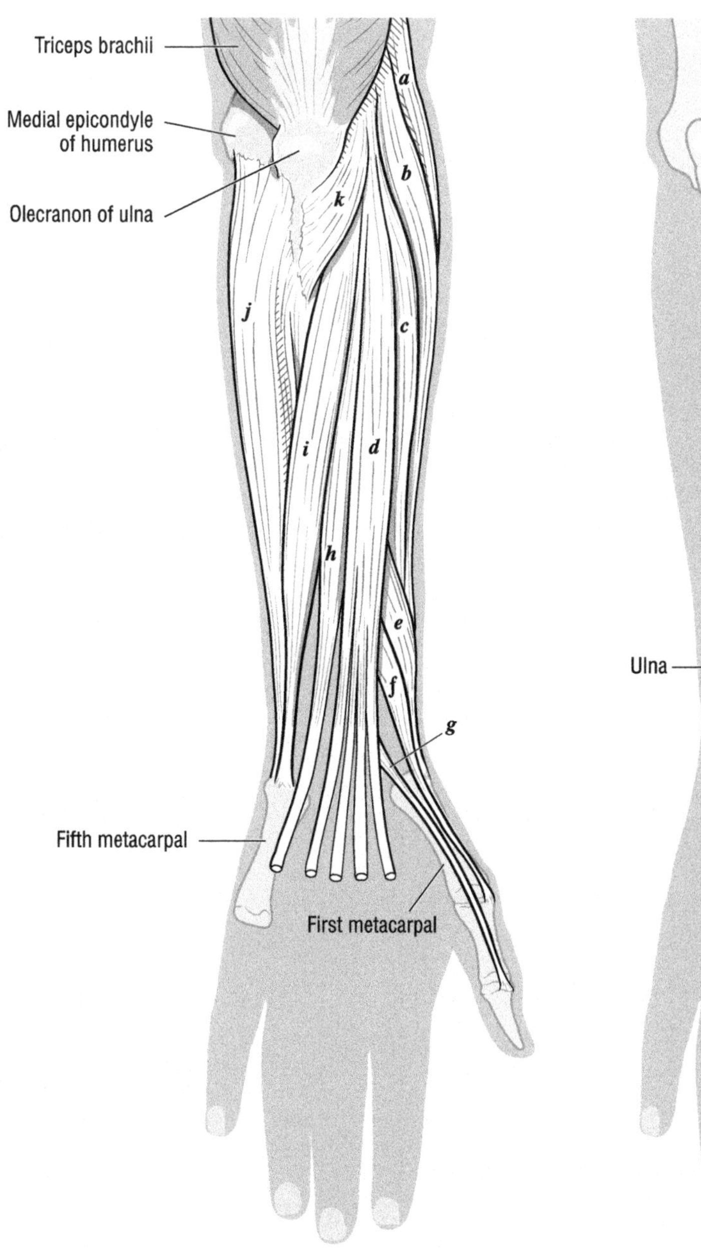

Posterior view, superficial

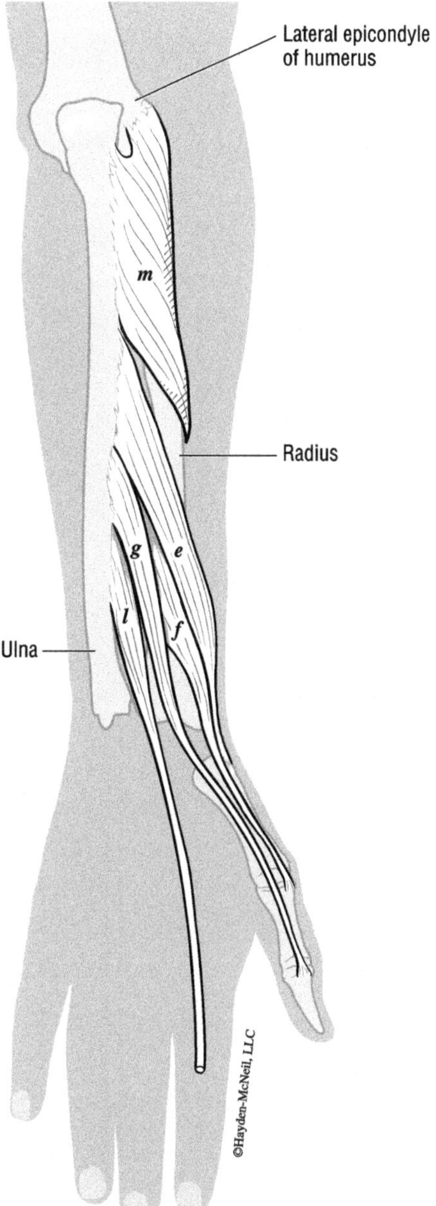

Posterior view, deep

Anterior (palmar) muscles of the hand

Do not color connective tissues and tendons marked with (*).

- (a) Abductor pollicis brevis
- (b) Flexor pollicis brevis
- (c) Flexor digiti minimi brevis
- (d) Palmaris brevis
- (e) Abductor digiti minimi brevis
- (f) Opponens pollicis
- (g) Opponens digiti minimi
- (h) Lumbrical muscles
- (i) Adductor pollicis

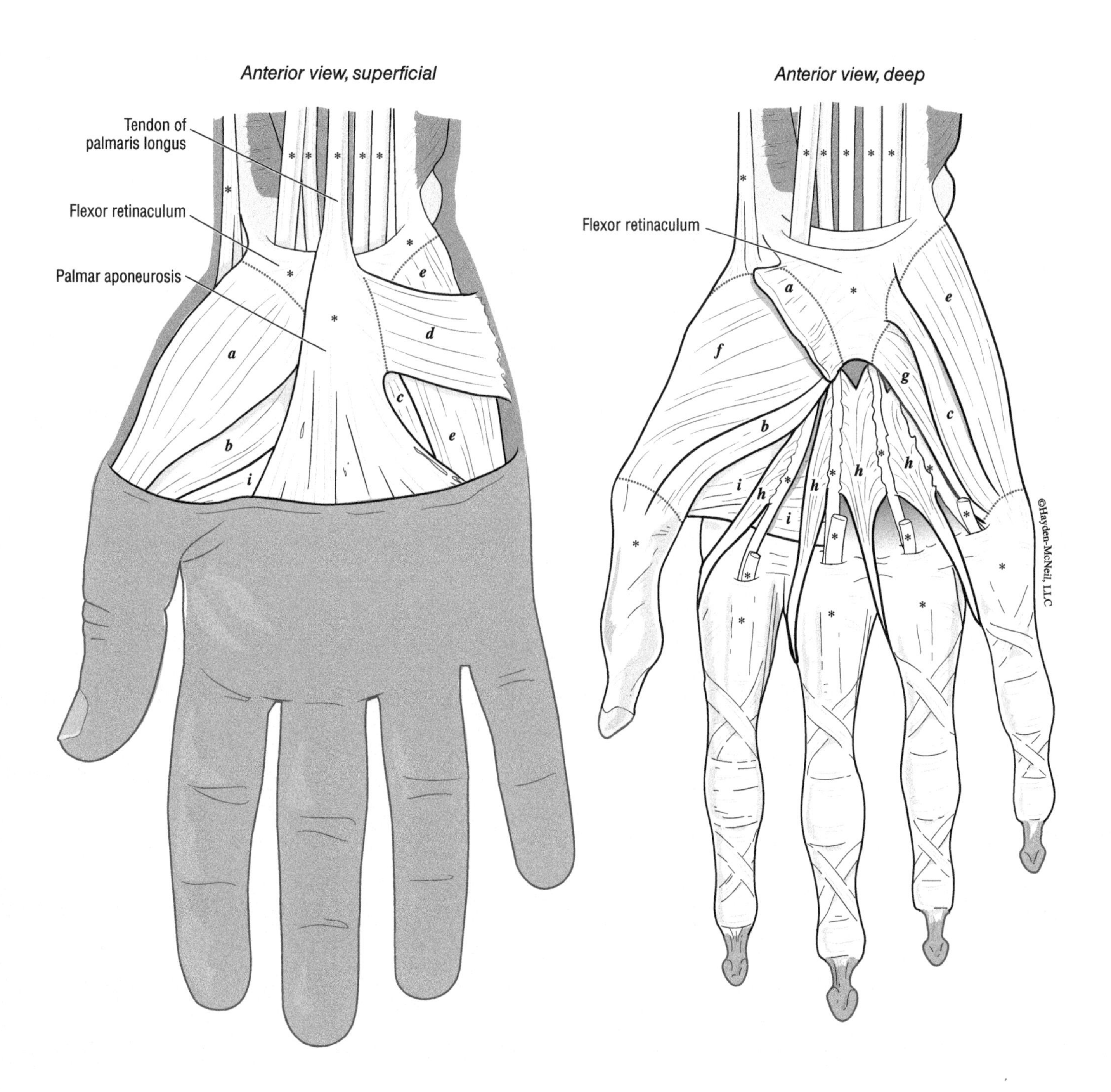

Muscular regions of the lower limb

Do not color the iliotibial tract (*).

- (a) Gluteal
- (b) Anterior thigh
- (c) Posterior thigh
- (d) Medial thigh
- (e) Anterior leg
- (f) Lateral leg
- (g) Posterior leg
- (h) Foot

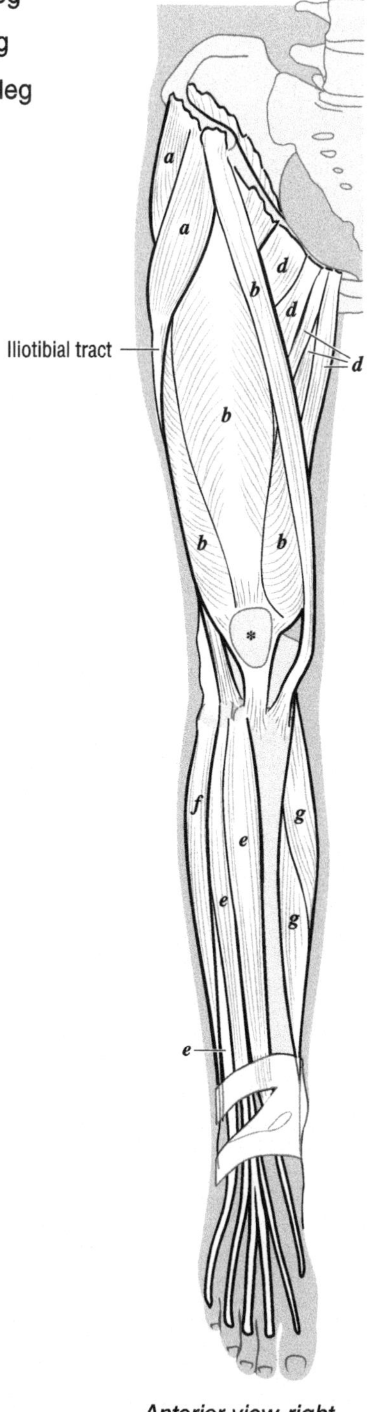

Anterior view, right

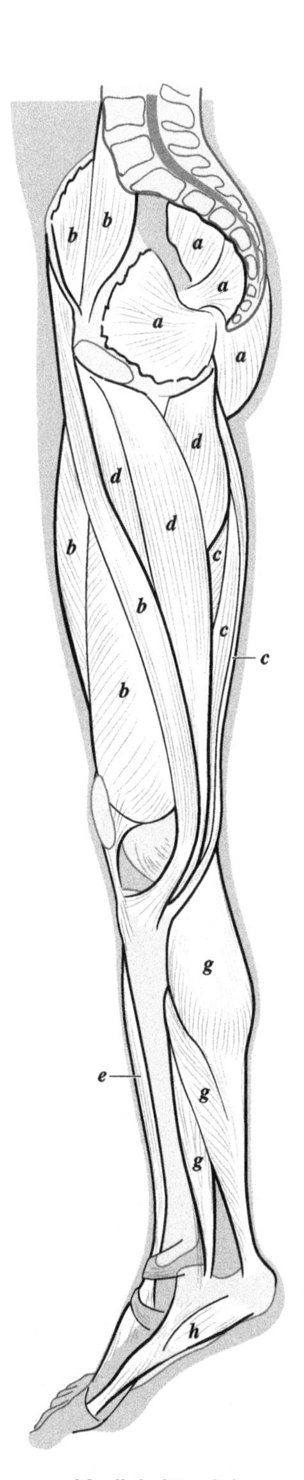

Medial view, right

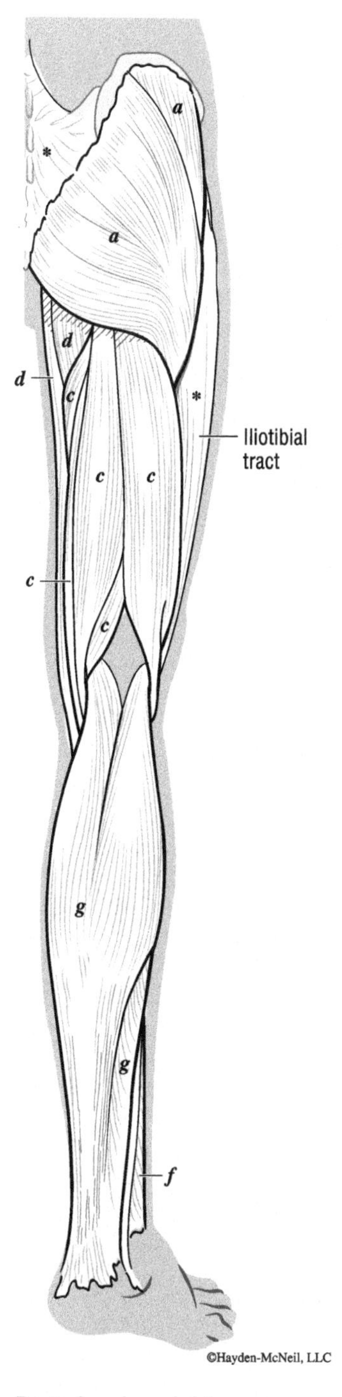

Posterior view, right

Muscles of the hip

Do not color the iliotibial tract or the sacrotuberous ligament marked with (*).

- (a) Gluteus medius
- (b) Tensor fasciae latae
- (c) Gluteus maximus
- (d) Gluteus minimus
- (e) Piriformis
- (f) Superior gemellus
- (g) Obturator internus
- (h) Inferior gemellus
- (i) Quadratus femoris
- (j) Obturator externus
- (k) Psoas major *(part of iliopsoas)*
- (l) Iliacus *(part of iliopsoas)*
- (m) Pectineus
- (n) Adductor brevis
- (o) Adductor longus

Iliac crest
a
b
c
Iliotibial tract
*
Femur

Posterior view, superficial

Iliac crest
l
k
d
m
n
o
Lesser trochanter
Femur

Anterior view

a
d
c
e
a
f
g
h
Greater trochanter
*
g
i
i
j
c
Lesser trochanter
Sacrotuberous ligament
Ishial tuberosity
Femur

Posterior view, deep

Thigh muscles

Do not color the iliotibial tract marked with (*).

- (a) Sartorius
- (b) Tensor fasciae latae
- (c) Vastus lateralis *(part of quadriceps femoris)*
- (d) Vastus intermedius *(part of quadriceps femoris)*
- (e) Vastus medialis *(part of quadriceps femoris)*
- (f) Rectus femoris (cut) *(part of quadriceps femoris)*
- (g) Quadratus femoris
- (h) Biceps femoris
- (i) Semitendinosus
- (j) Semimembranosus
- (k) Pectineus
- (l) Adductor longus
- (m) Adductor magnus
- (n) Adductor brevis
- (o) Gracilis

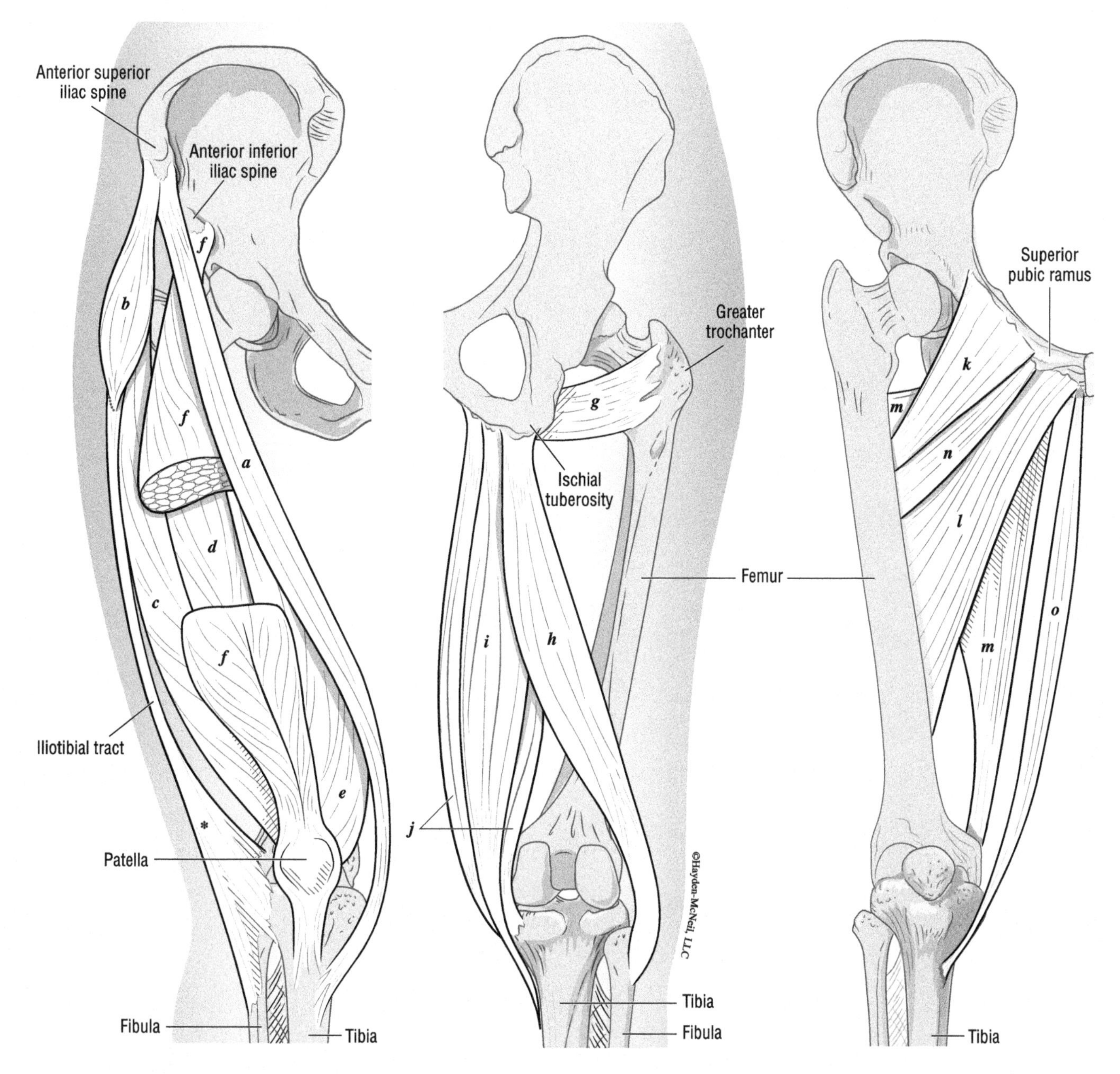

Anterior view, superficial *Posterior view* *Anterior view, deep (adductors of the thigh)*

Anterior muscles of the leg and foot

- (a) Tibialis anterior
- (b) Extensor digitorum longus
- (c) Extensor hallucis longus
- (d) Fibularis tertius
- (e) Extensor digitorum brevis
- (f) Interosseous muscles
- (g) Extensor hallucis brevis

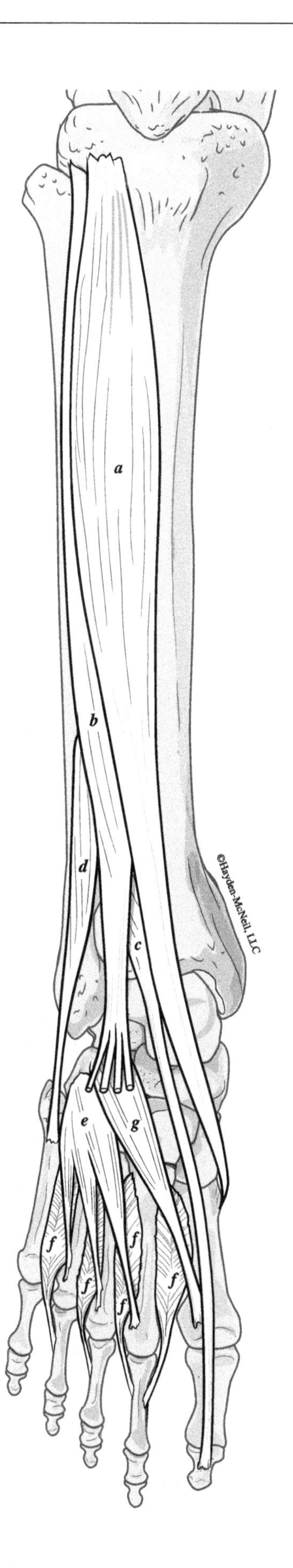

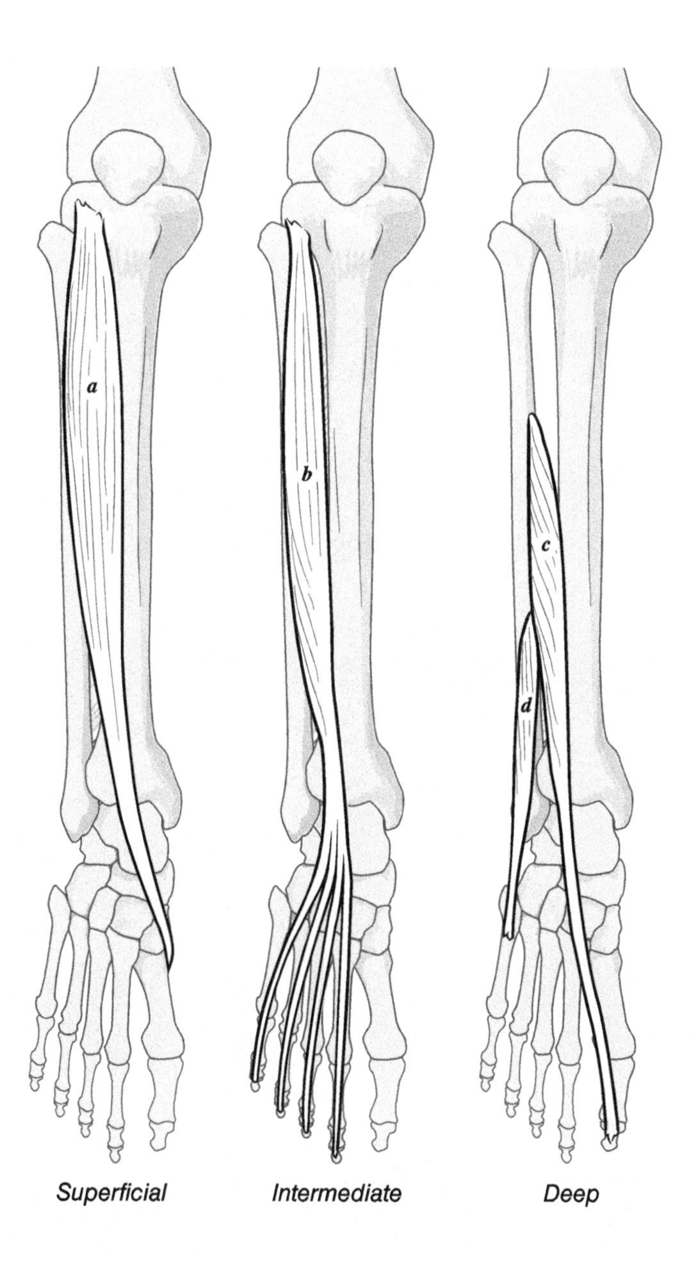

Lateral and posterior muscles of the leg

- (a) Tibialis anterior
- (b) Extensor digitorum longus
- (c) Extensor hallucis longus
- (d) Fibularis longus
- (e) Fibularis brevis
- (f) Gastrocnemius
- (g) Soleus
- (h) Calcaneal (Achilles) tendon
- (i) Plantaris
- (j) Popliteus
- (k) Tibialis posterior
- (l) Flexor hallucis longus
- (m) Flexor digitorum longus

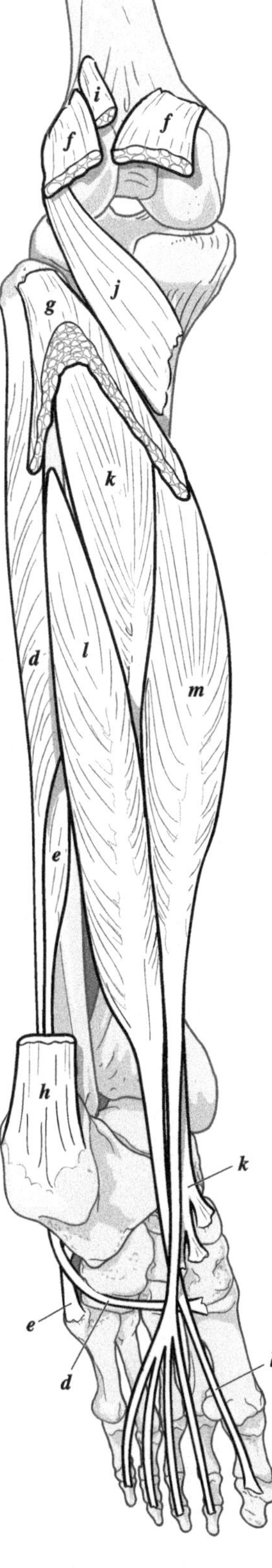

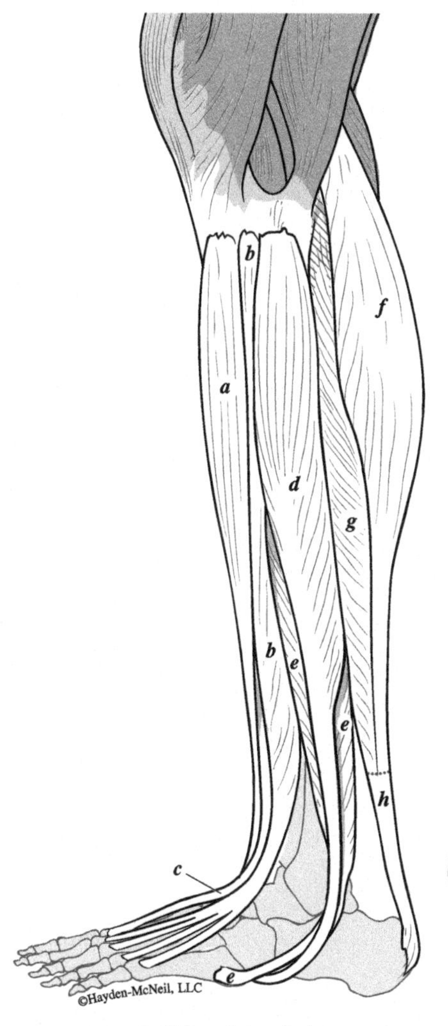

Left leg, lateral view, superficial

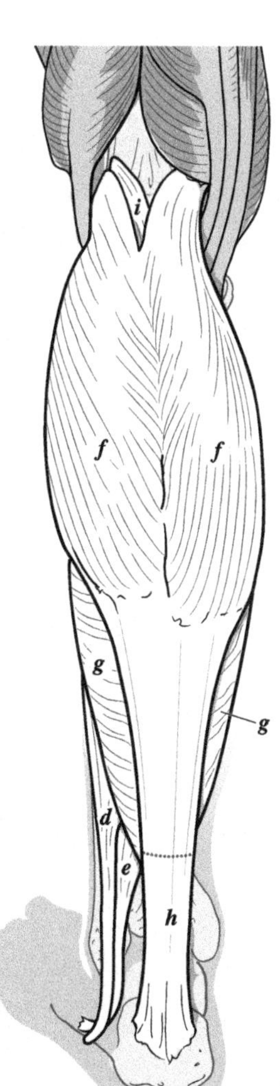

Left leg, posterolateral view, superficial

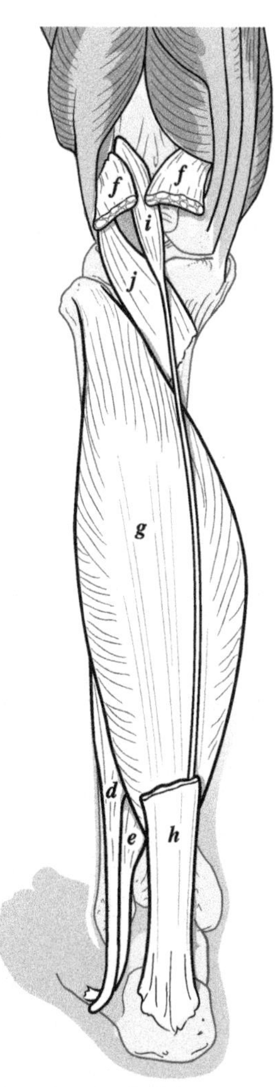

Left leg, posterolateral view, intermediate

Left leg, posteromedial view, deep

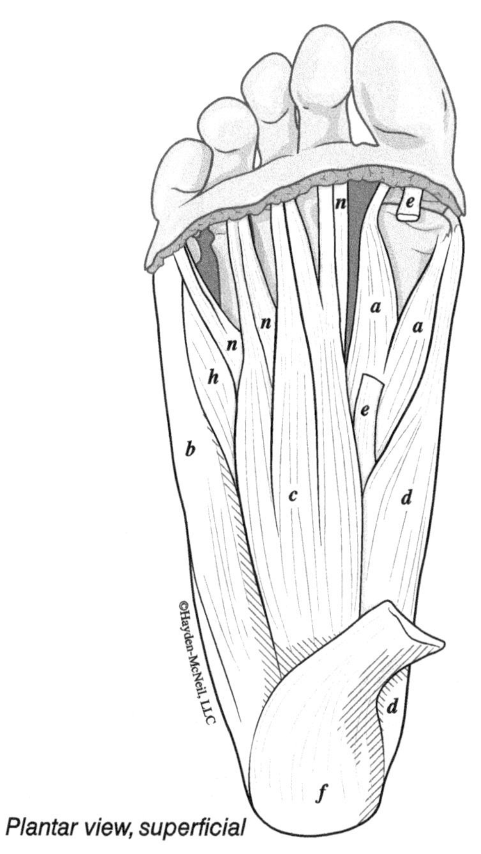

Plantar view, superficial

Muscles of the foot

Do not color the joint capsules marked with (*).

- (a) Flexor hallucis brevis
- (b) Abductor digiti minimi
- (c) Flexor digitorum brevis
- (d) Abductor hallucis
- (e) Flexor hallucis longus
- (f) Plantar aponeurosis
- (g) Opponens digiti minimi
- (h) Flexor digiti minimi brevis
- (i) Lumbrical muscles
- (j) Quadratus plantae
- (k) Interosseous muscles
- (l) Flexor hallucis
  - l_1 Medial head
  - l_2 Lateral head
- (m) Adductor hallucis
  - m_1 Transverse head
  - m_2 Oblique head
- (n) Flexor digitorum longus
- (o) Long plantar ligament

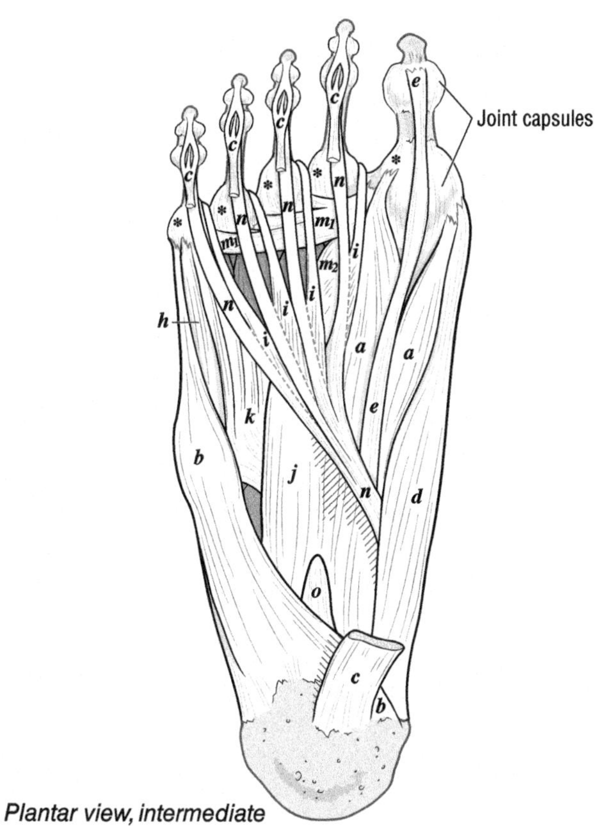

Plantar view, intermediate

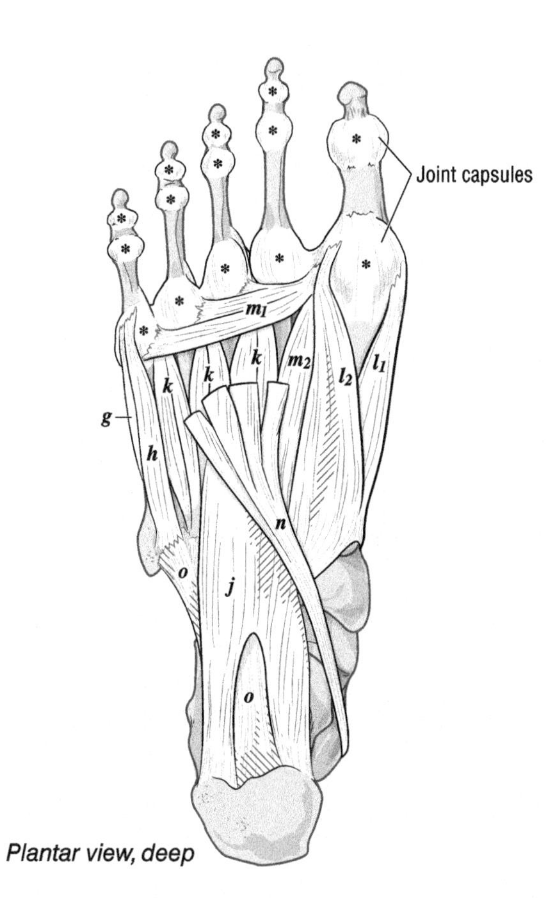

Plantar view, deep

8

The Nervous System

Neuron structure
Glial cells
Synapse
Reflex arcs
Cerebrum
Cerebellum
Brainstem and midbrain
Ventricles of the brain
Circulation of cerebrospinal fluid
Fiber tracts
Cerebral basal ganglia
Limbic system
Hypothalamic nuclei
Brain meninges
Precentral and postcentral regions
Cranial nerves
Spinal cord
Spinal cord tracts
Spinal cord and meninges
Neural pathways: Ascending tracts
Neural pathways: Descending tracts
Cervical plexus
Nerves of the upper limb
Brachial plexus
Lumbosacral plexus
Nerves of the lower limb
Dermatomes
Autonomic nervous system

Grab your ticket! In this chapter, you will be navigating through the bustling train stations of our body's most active and complex communication network: the **nervous system**. You will be witnessing the rapid-fire reactions of our neurons as they zip throughout our body like perfectly timed trains on well-oiled tracks, from the central hub of the brain to the far-reaching terminals of the peripheral nervous system. As you color their electrifying parallel pathways, you will be learning how our nerves serve as the most reliable of messagers, relaying sensations and movements at lightning speed. All aboard to Nervous System Station!

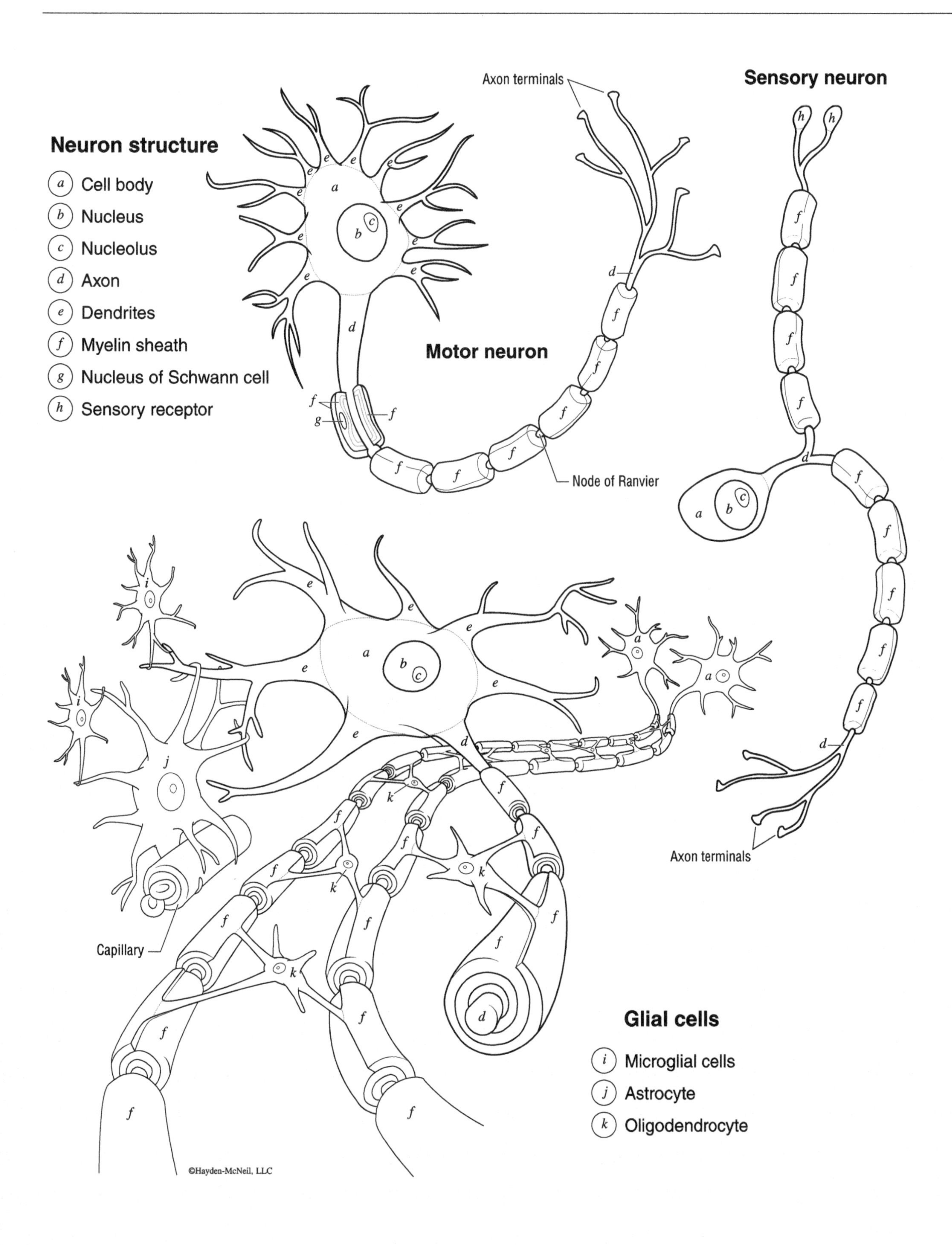
Neuron structure
a Cell body
b Nucleus
c Nucleolus
d Axon
e Dendrites
f Myelin sheath
g Nucleus of Schwann cell
h Sensory receptor
Axon terminals
Sensory neuron
Motor neuron
Node of Ranvier
Axon terminals
Capillary
Glial cells
i Microglial cells
j Astrocyte
k Oligodendrocyte
©Hayden-McNeil, LLC

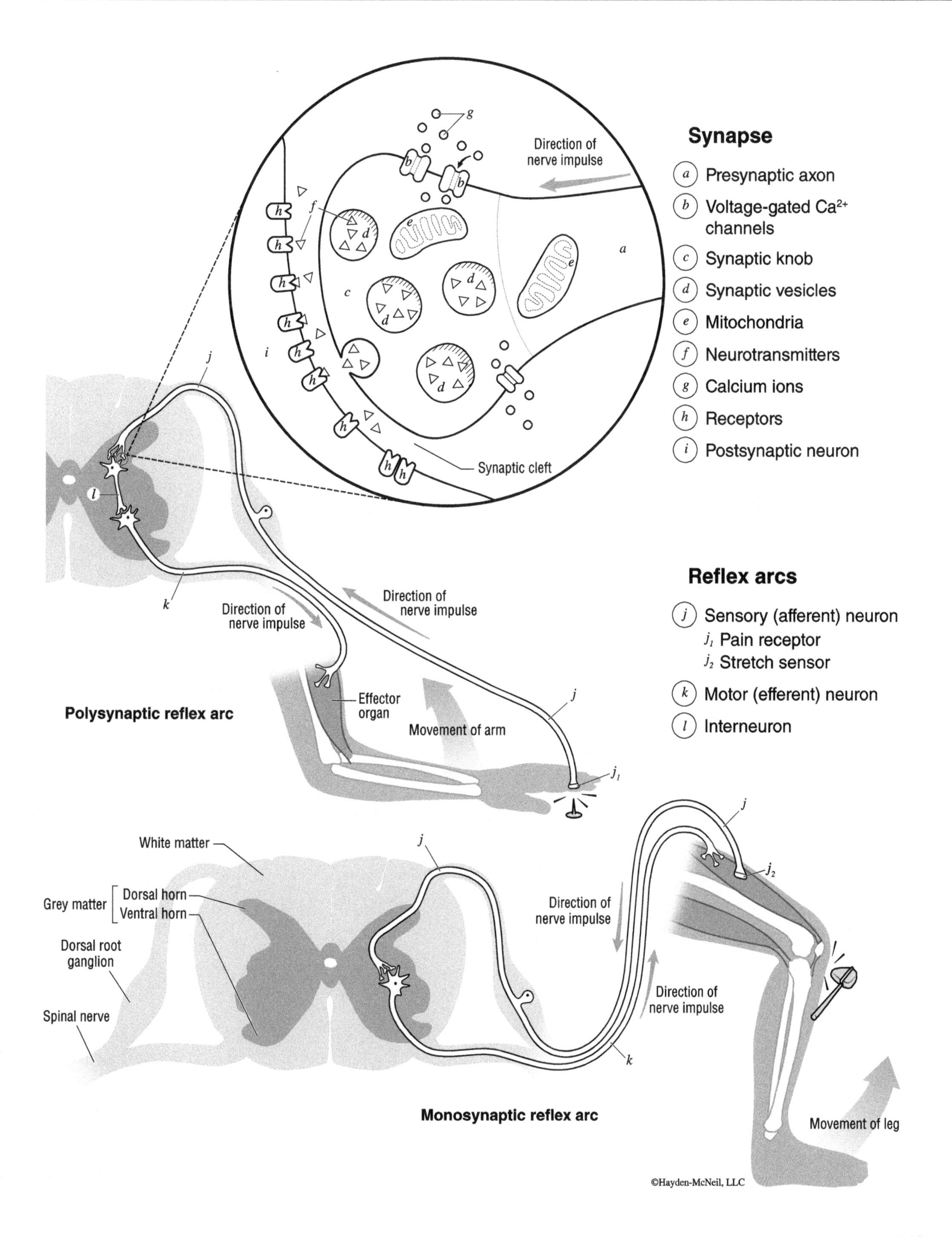
Direction of nerve impulse
Synaptic cleft
Synapse
a Presynaptic axon
b Voltage-gated Ca2+ channels
c Synaptic knob
d Synaptic vesicles
e Mitochondria
f Neurotransmitters
g Calcium ions
h Receptors
i Postsynaptic neuron
Reflex arcs
j Sensory (afferent) neuron
j1 Pain receptor
j2 Stretch sensor
k Motor (efferent) neuron
l Interneuron
Direction of nerve impulse
Direction of nerve impulse
Effector organ
Movement of arm
Polysynaptic reflex arc
White matter
Grey matter
Dorsal horn
Ventral horn
Dorsal root ganglion
Spinal nerve
Direction of nerve impulse
Direction of nerve impulse
Monosynaptic reflex arc
Movement of leg

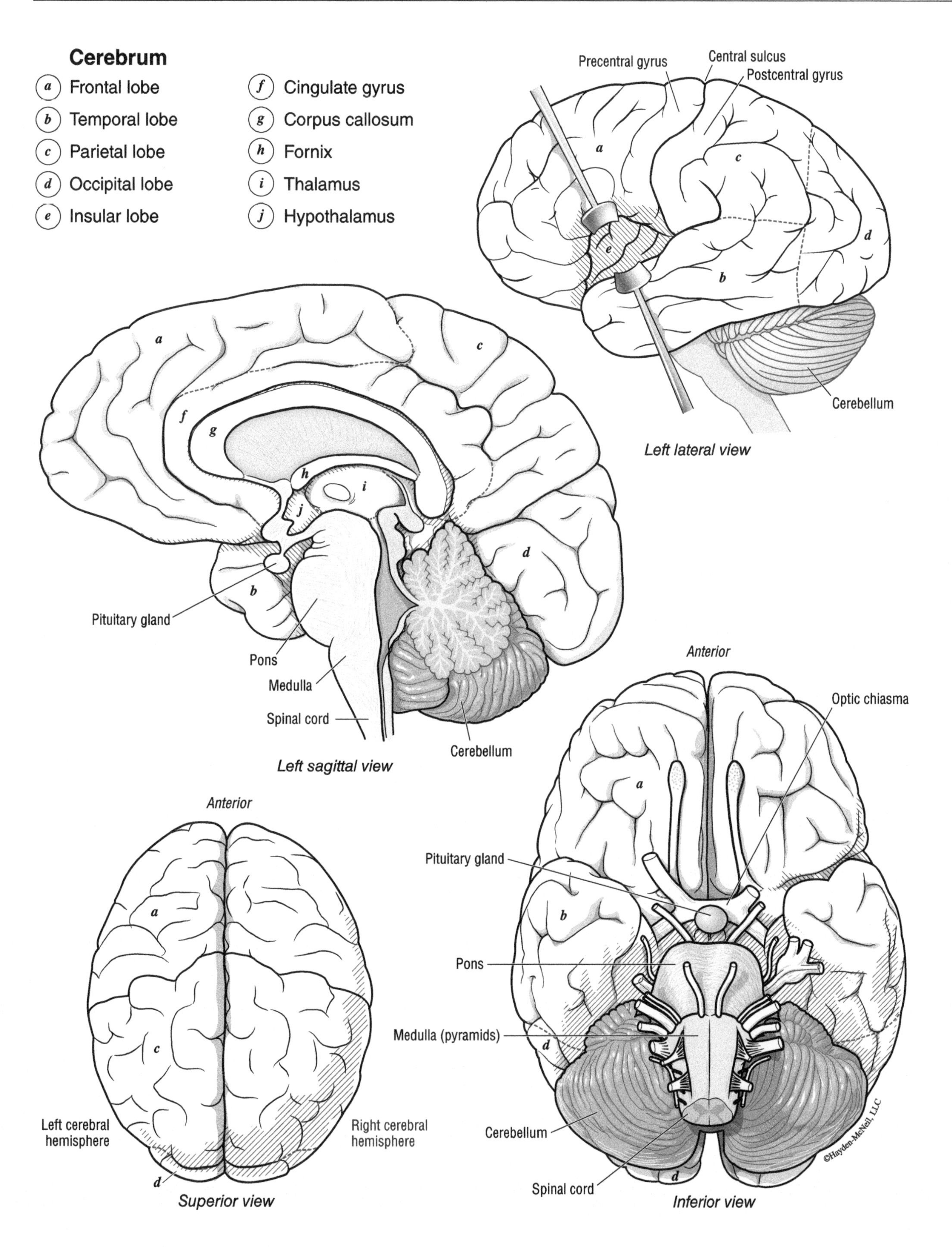
Cerebrum
a Frontal lobe
b Temporal lobe
c Parietal lobe
d Occipital lobe
e Insular lobe
f Cingulate gyrus
g Corpus callosum
h Fornix
i Thalamus
j Hypothalamus
Precentral gyrus
Central sulcus
Postcentral gyrus
Cerebellum
Left lateral view
Pituitary gland
Pons
Medulla
Spinal cord
Cerebellum
Left sagittal view
Anterior
Optic chiasma
Anterior
Pituitary gland
Pons
Medulla (pyramids)
Left cerebral hemisphere
Right cerebral hemisphere
Cerebellum
Spinal cord
Superior view
Inferior view
©Hayden-McNeil, LLC

Cerebellum

- (a) Anterior lobe
  - a_1 Quadrangular lobule
- (b) Posterior lobe
  - b_1 Simple lobule
  - b_2 Superior semilunar lobule
  - b_3 Inferior semilunar lobule
- (c) Arbor vitae *Use a light color.*
- (d) Cerebellar cortex *Use a dark color.*

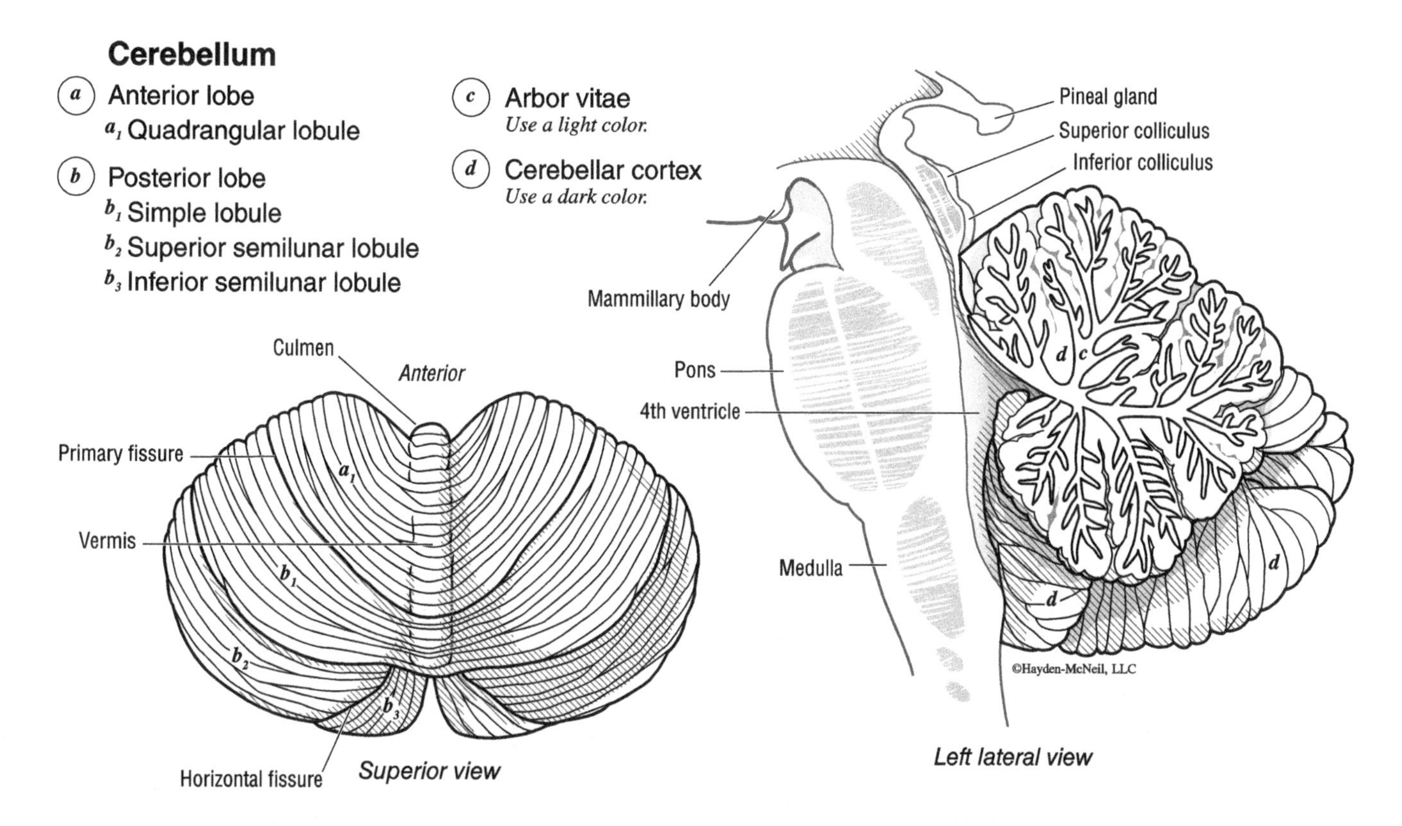

Brainstem and midbrain *Do not color the cranial nerves (CN) or the cervical spinal nerve.*

- (e) Insula
- (f) Thalamus
- (g) Pineal body
- (h) Quadrigeminal plate
- (i) Mammillary bodies
- (j) Pons
- (k) Cerebral peduncles
- (l) Cerebellar peduncles
- (m) Medulla oblongata
- (n) Choroid plexus
- (o) Fornix
- (p) Spinal cord

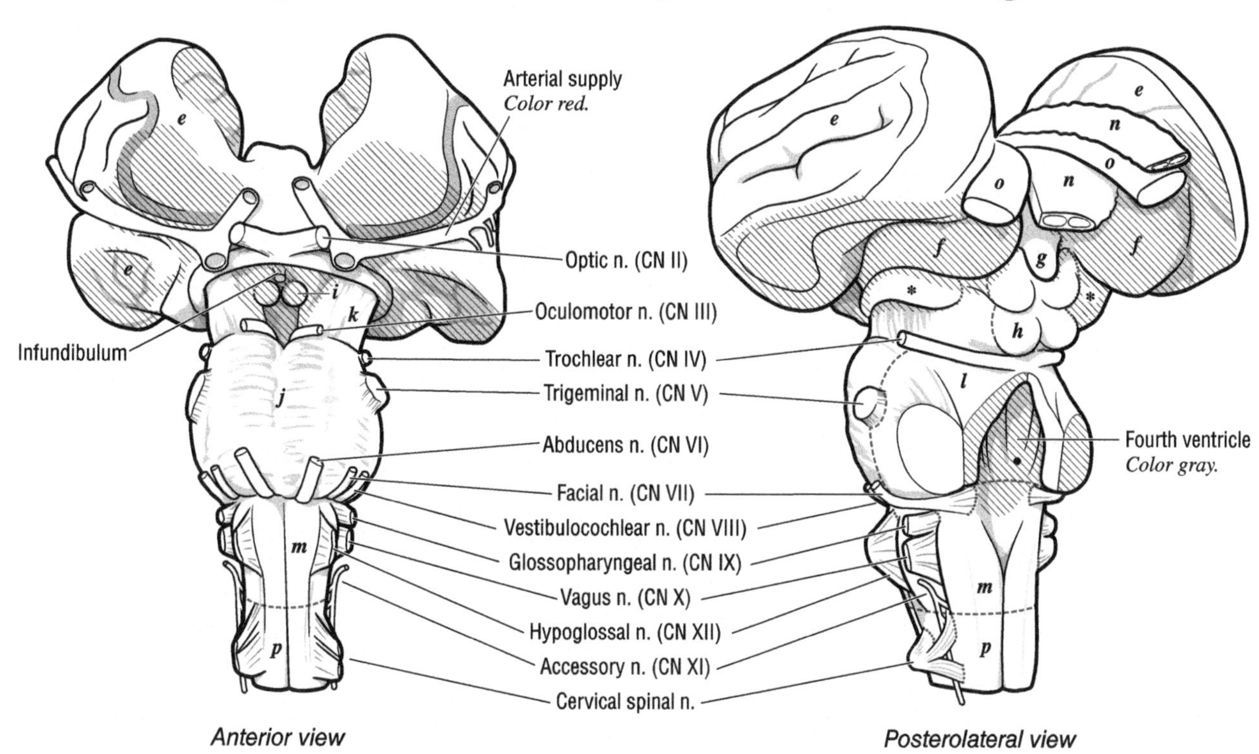

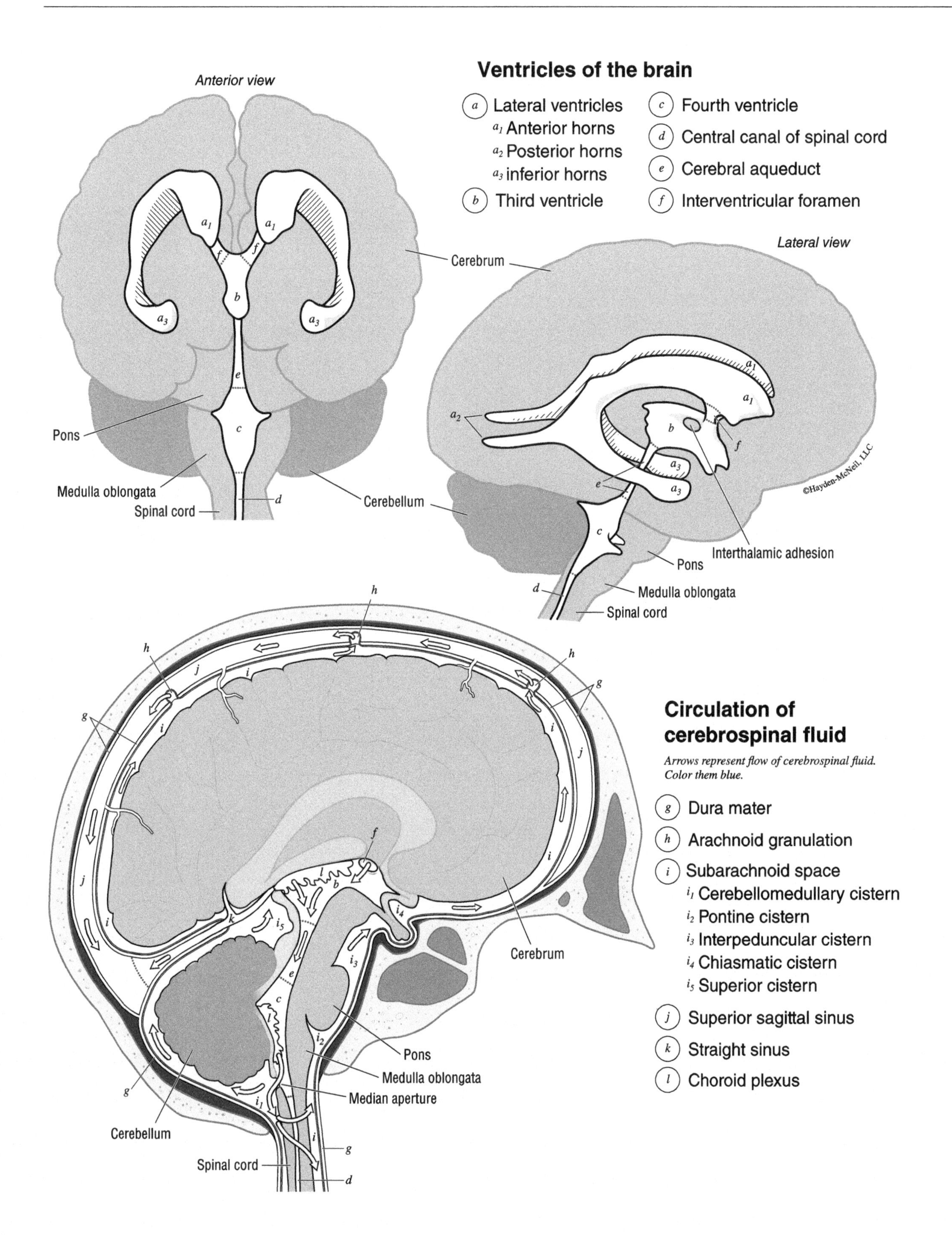
Anterior view
Ventricles of the brain
a Lateral ventricles
a1 Anterior horns
a2 Posterior horns
a3 inferior horns
b Third ventricle
c Fourth ventricle
d Central canal of spinal cord
e Cerebral aqueduct
f Interventricular foramen
Lateral view
Cerebrum
Pons
Medulla oblongata
Spinal cord
Cerebellum
Interthalamic adhesion
©Hayden-McNeil, LLC
Circulation of cerebrospinal fluid
Arrows represent flow of cerebrospinal fluid. Color them blue.
g Dura mater
h Arachnoid granulation
i Subarachnoid space
i1 Cerebellomedullary cistern
i2 Pontine cistern
i3 Interpeduncular cistern
i4 Chiasmatic cistern
i5 Superior cistern
j Superior sagittal sinus
k Straight sinus
l Choroid plexus
Cerebrum
Pons
Medulla oblongata
Median aperture
Cerebellum
Spinal cord

Fiber tracts

- (a) Short association fibers
- (b) Superior longitudinal fasciculus
- (c) Perpendicular fasciculus
- (d) Inferior longitudinal fasciculus
- (e) Uncinate fasciculus

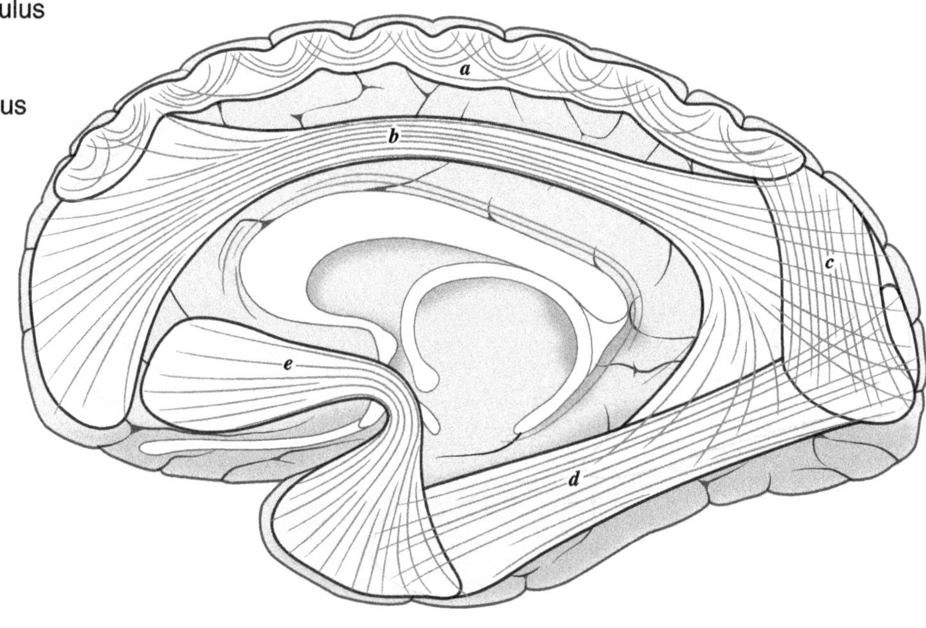

Left medial view

Cerebral basal ganglia

- (f) Caudate nucleus
- (g) Putamen
- (h) Globus pallidus
- (i) Red nucleus

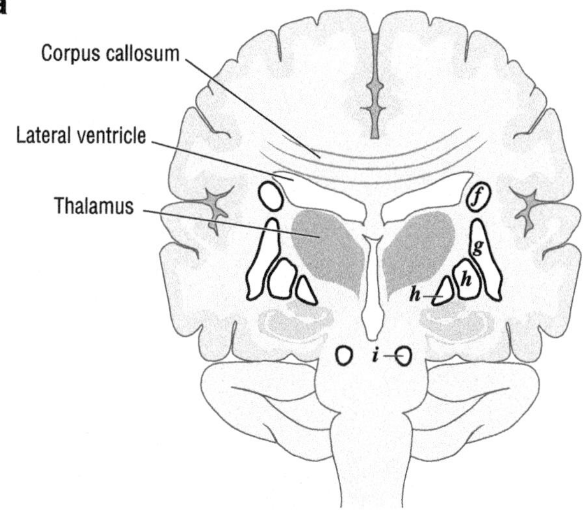

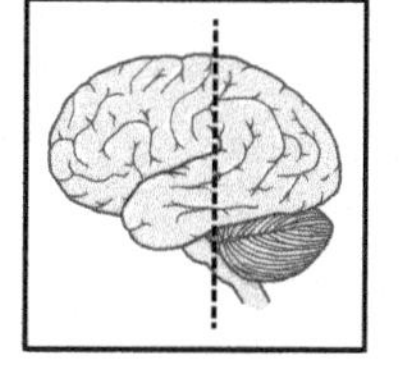

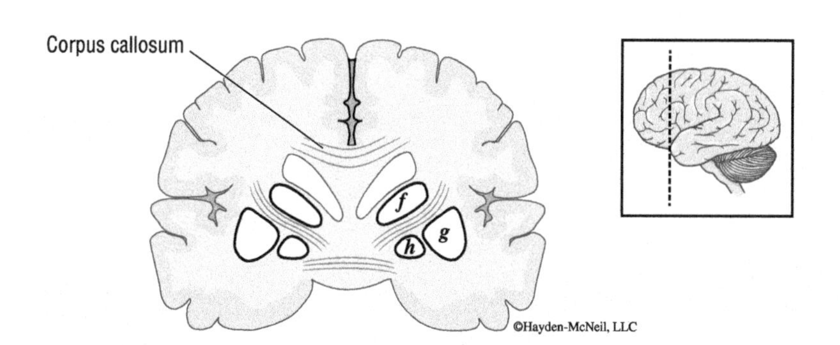

Limbic system

- (a) Thalamus
- (b) Hypothalamus
- (c) Mammillary body
- (d) Corpus callosum
- (e) Fornix
- (f) Pineal gland
- (g) Amygdala
- (h) Hippocampus

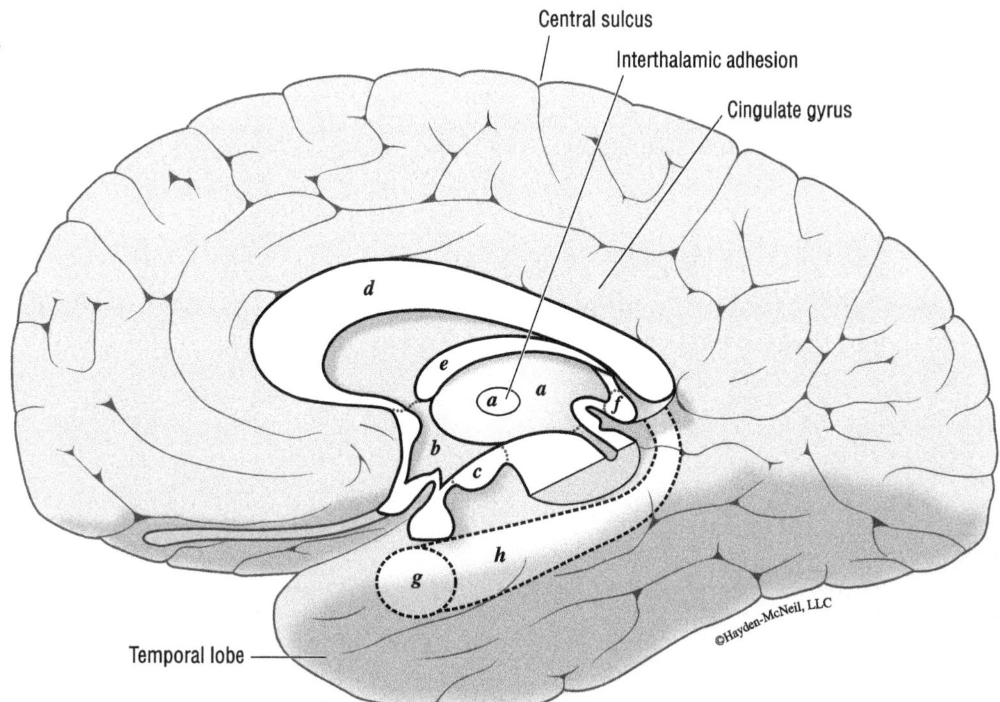

Left sagittal view

Hypothalamic nuclei

- (i) Preoptic area
- (j) Anterior (supraoptic) region
- (k) Middle (tuberal) region
- (l) Posterior (mammillary) region

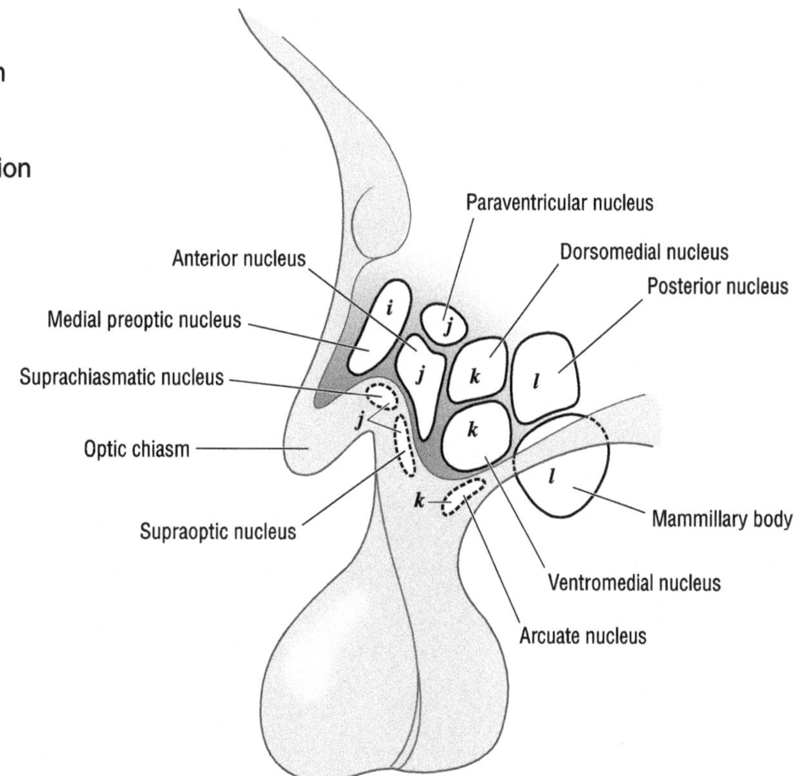

Brain meninges

Color the veins marked with (•) blue.

- (a) Scalp
- (b) Cranial bone
- (c) Dura mater
- (d) Arachnoid mater
  - d_1 Arachnoid villus
- (e) Pia mater
- (f) Cerebral cortex

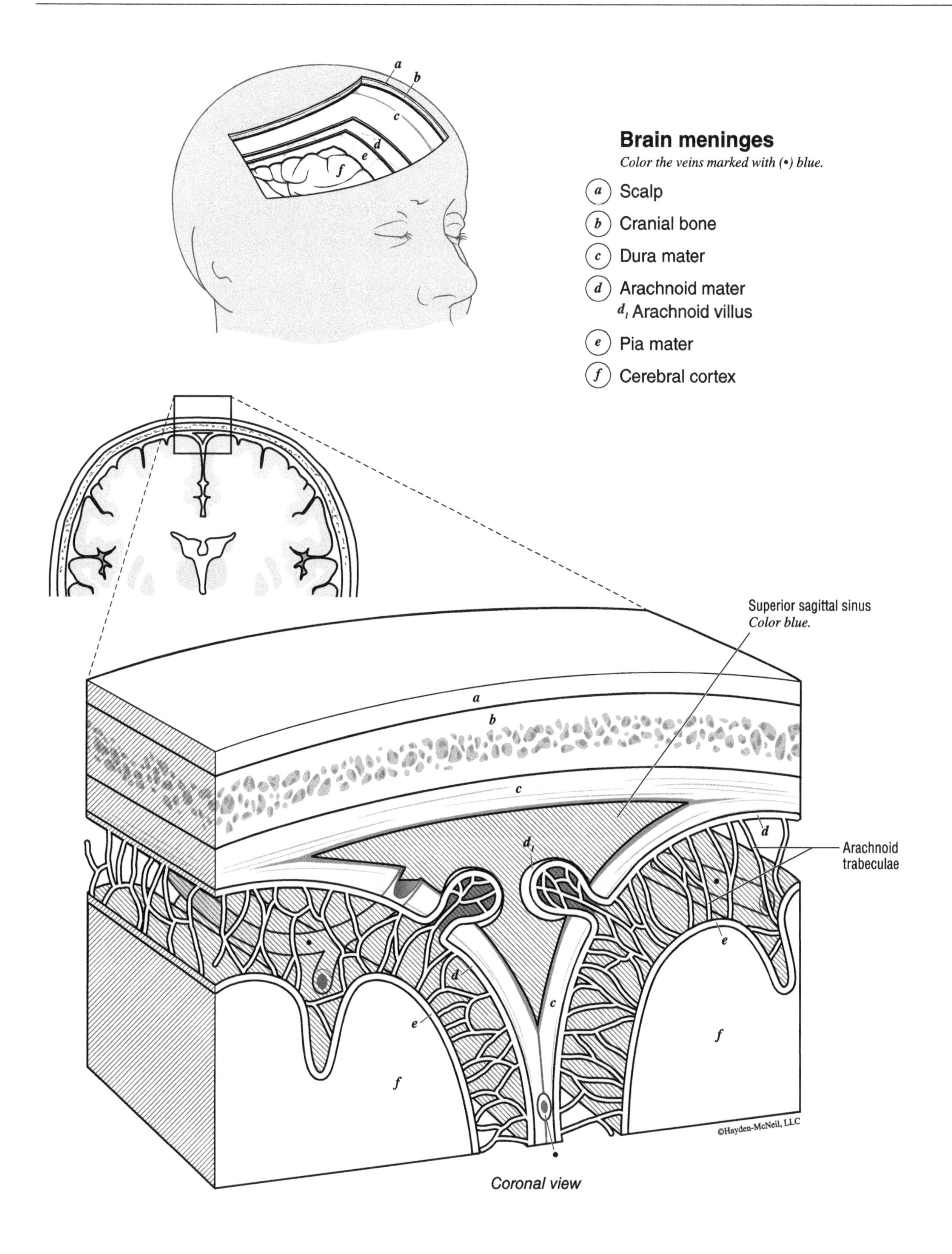

Coronal view

Precentral and postcentral regions

(a) Somatosensory cortex

(b) Primary motor cortex

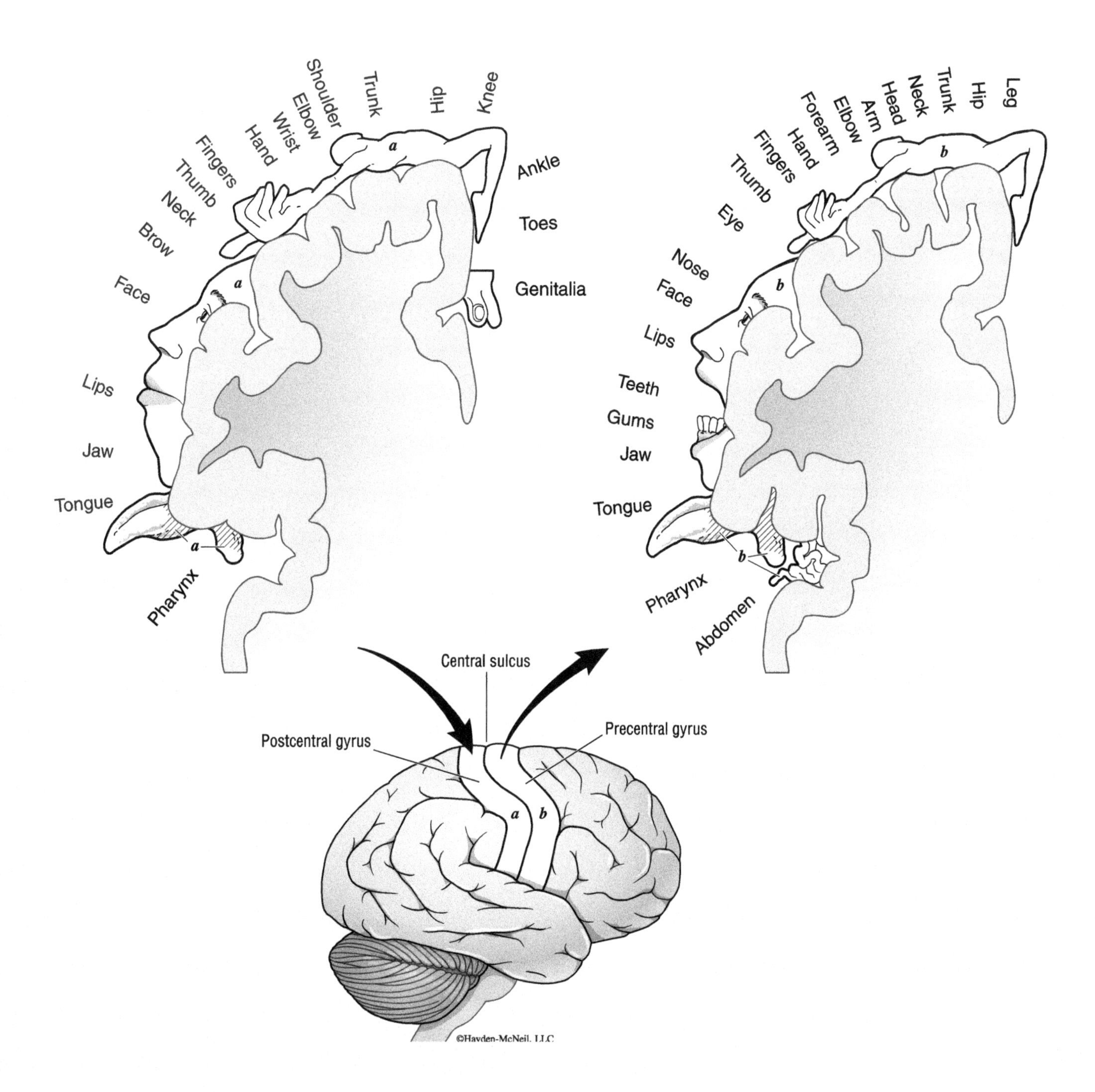

Cranial nerves

Color the Roman numeral for each cranial nerve, then color the corresponding nerve in the inferior brain view, and finally color the structure(s) innervated by that nerve. Use two unique colors for the sensory and motor arrows to the right, and then color the arrows throughout the drawing according to your key.

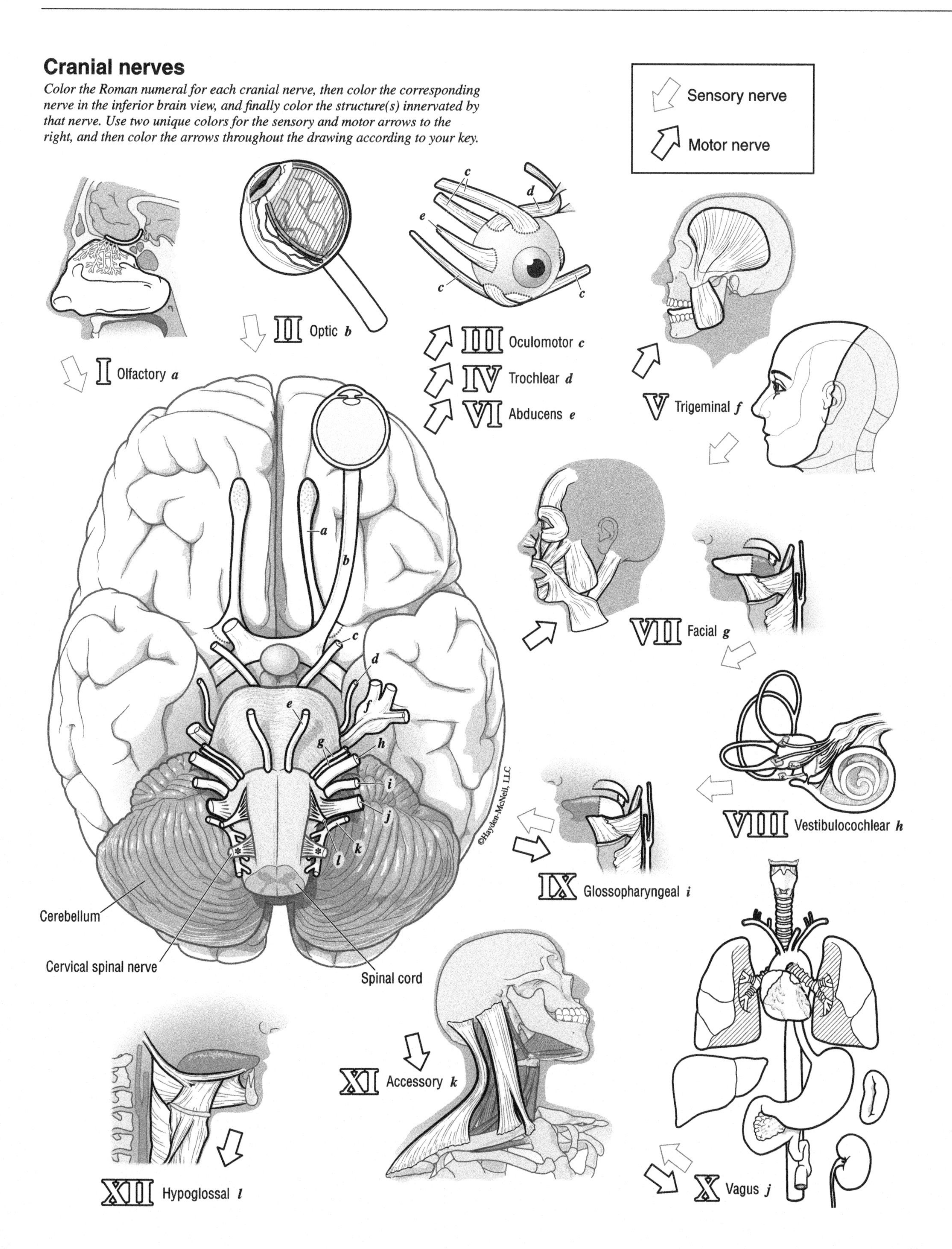

Spinal cord

(a) Cervical segments

(b) Thoracic segments

(c) Lumbar, sacral, coccygeal segments

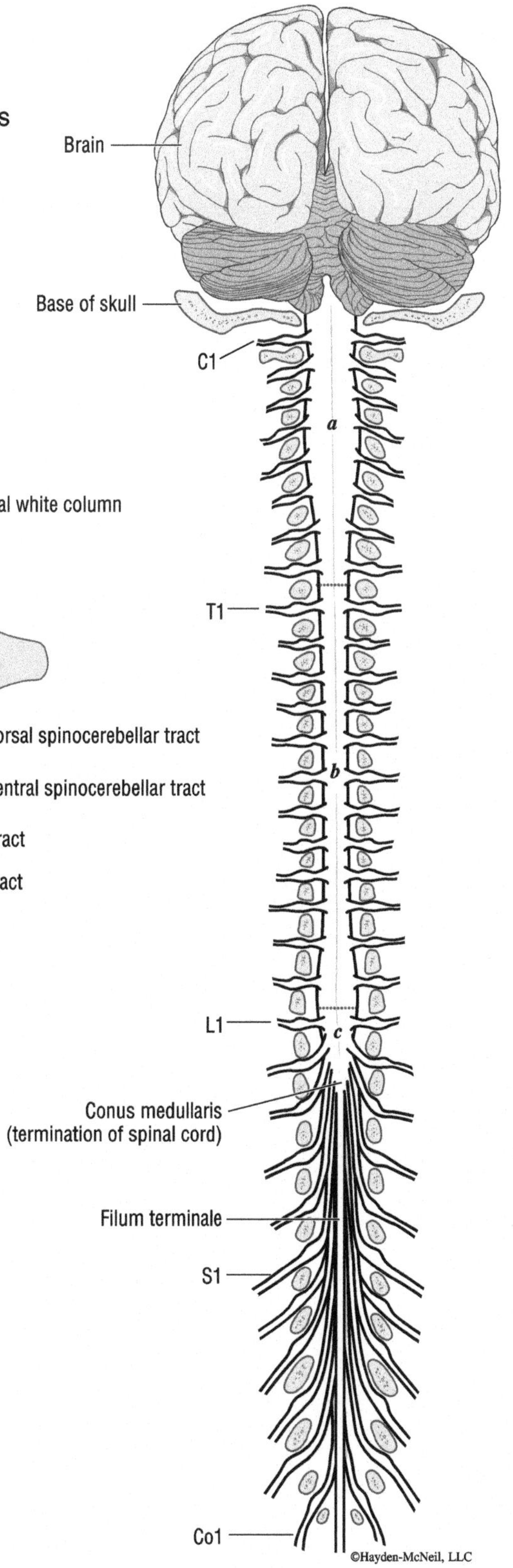

Brain
Base of skull
C1
a
T1
b
L1
c
Conus medullaris (termination of spinal cord)
Filum terminale
S1
Co1

©Hayden-McNeil, LLC

Spinal cord tracts

Choose one color for the descending tracts and arrow (thick black outline) and another for the ascending tracks and arrow (thin gray outline).

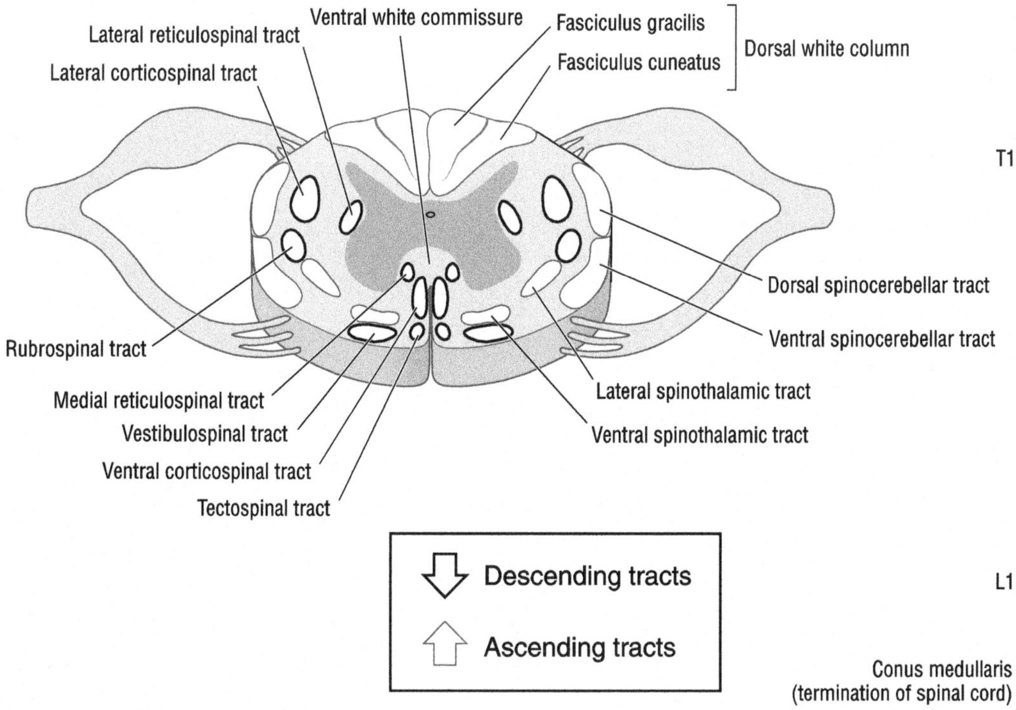

Lateral reticulospinal tract
Ventral white commissure
Fasciculus gracilis
Fasciculus cuneatus
Dorsal white column
Lateral corticospinal tract
Dorsal spinocerebellar tract
Ventral spinocerebellar tract
Rubrospinal tract
Lateral spinothalamic tract
Medial reticulospinal tract
Ventral spinothalamic tract
Vestibulospinal tract
Ventral corticospinal tract
Tectospinal tract

Descending tracts

Ascending tracts

Spinal cord and meninges

Color the spinal and vertebral vessels red and blue.

- (a) Gray matter *Use a dark color.*
- (b) White matter *Use a light color.*
- (c) Pia mater
- (d) Arachnoid mater
- (e) Dura mater
- (f) Ventral nerve root
- (g) Dorsal nerve root

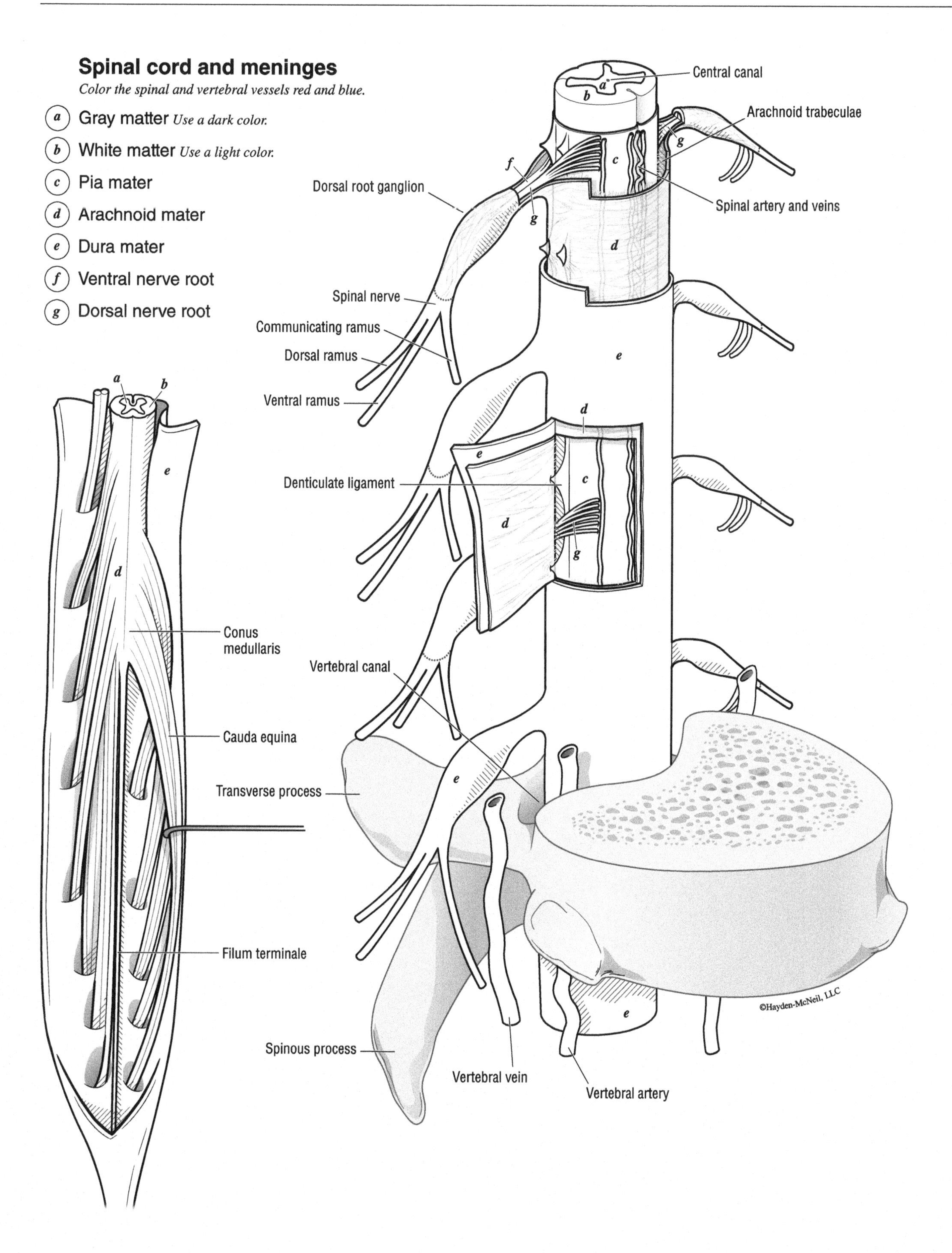

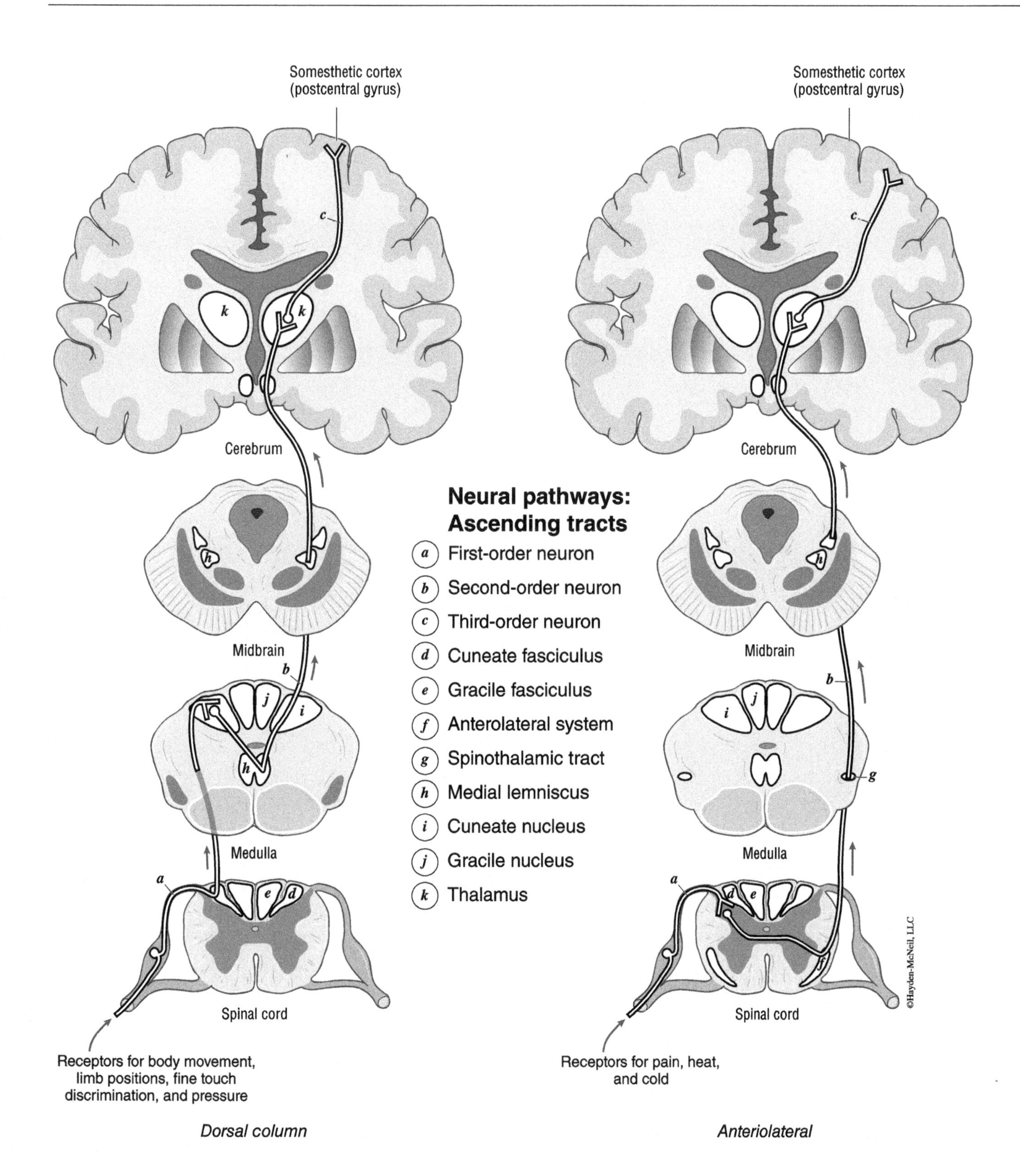

Somesthetic cortex
(postcentral gyrus)
c
k
k
Cerebrum
h
Midbrain
b
j
i
h
Medulla
a
e
d
Spinal cord
Receptors for body movement,
limb positions, fine touch
discrimination, and pressure
Dorsal column
Neural pathways:
Ascending tracts
a First-order neuron
b Second-order neuron
c Third-order neuron
d Cuneate fasciculus
e Gracile fasciculus
f Anterolateral system
g Spinothalamic tract
h Medial lemniscus
i Cuneate nucleus
j Gracile nucleus
k Thalamus
Somesthetic cortex
(postcentral gyrus)
c
h
Cerebrum
h
Midbrain
b
i
j
g
Medulla
a
d
e
f
Spinal cord
Receptors for pain, heat,
and cold
Anteriolateral
©Hayden-McNeil, LLC

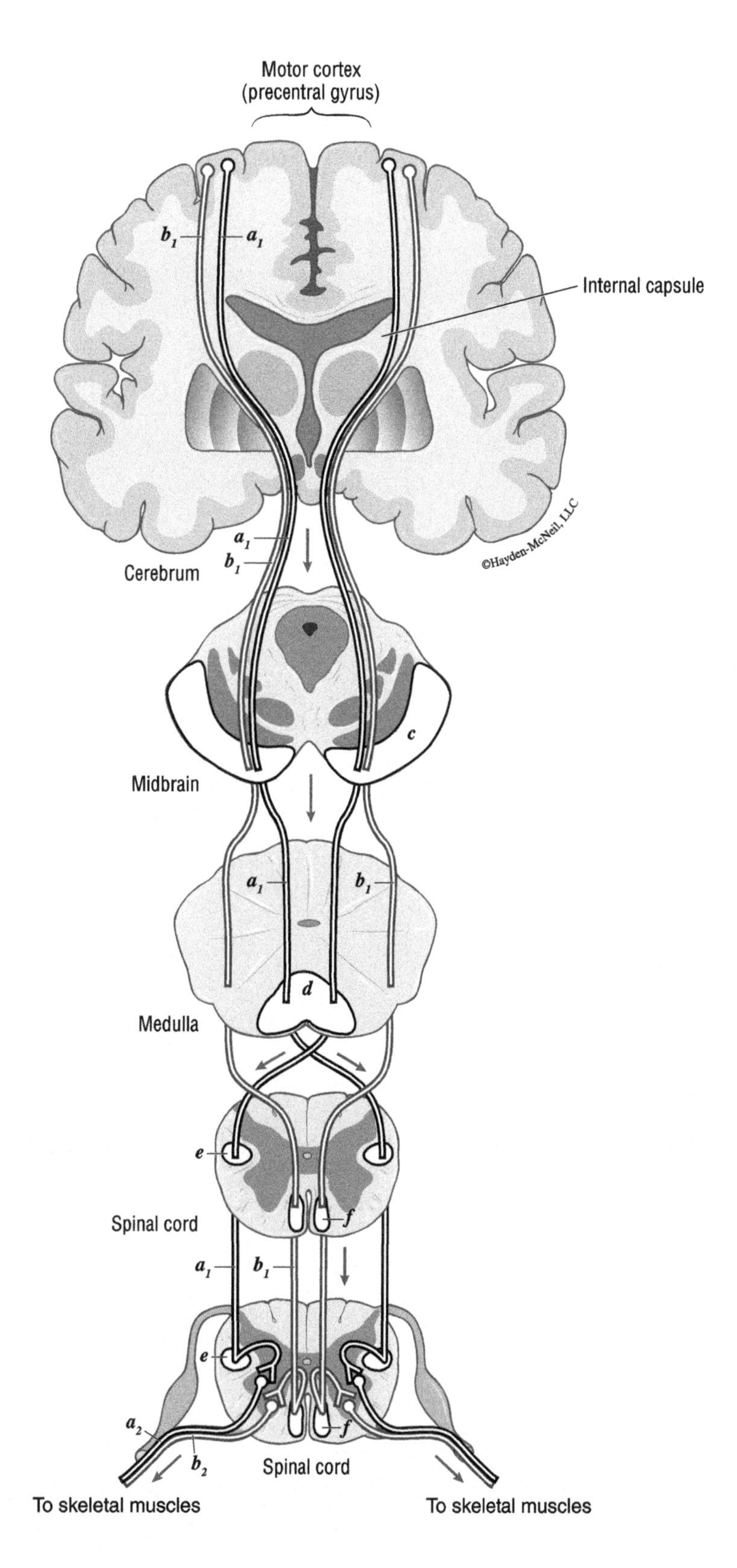

Neural pathways: Descending tracts

- (a) Motor neuron (A)
  - a_1 Upper motor neuron
  - a_2 Lower motor neuron
- (b) Motor neuron (B)
  - b_1 Upper motor neuron
  - b_2 Lower motor neuron
- (c) Cerebral peduncle
- (d) Medullary pyramid
- (e) Lateral corticospinal tract
- (f) Ventral corticospinal tract

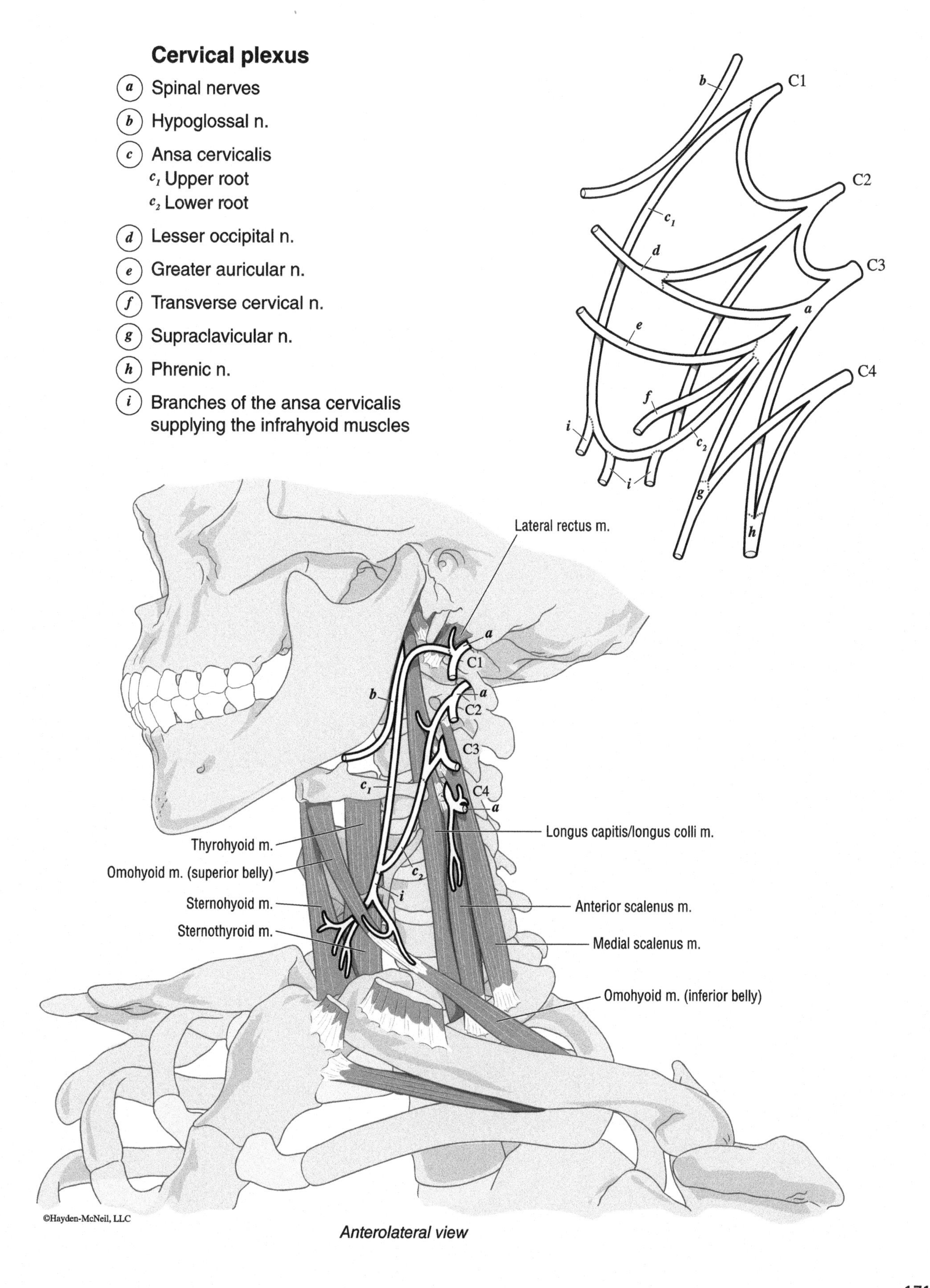

Cervical plexus
a Spinal nerves
b Hypoglossal n.
c Ansa cervicalis
c1 Upper root
c2 Lower root
d Lesser occipital n.
e Greater auricular n.
f Transverse cervical n.
g Supraclavicular n.
h Phrenic n.
i Branches of the ansa cervicalis supplying the infrahyoid muscles
C1
C2
C3
C4
Lateral rectus m.
Longus capitis/longus colli m.
Thyrohyoid m.
Omohyoid m. (superior belly)
Sternohyoid m.
Sternothyroid m.
Anterior scalenus m.
Medial scalenus m.
Omohyoid m. (inferior belly)

Anterolateral view

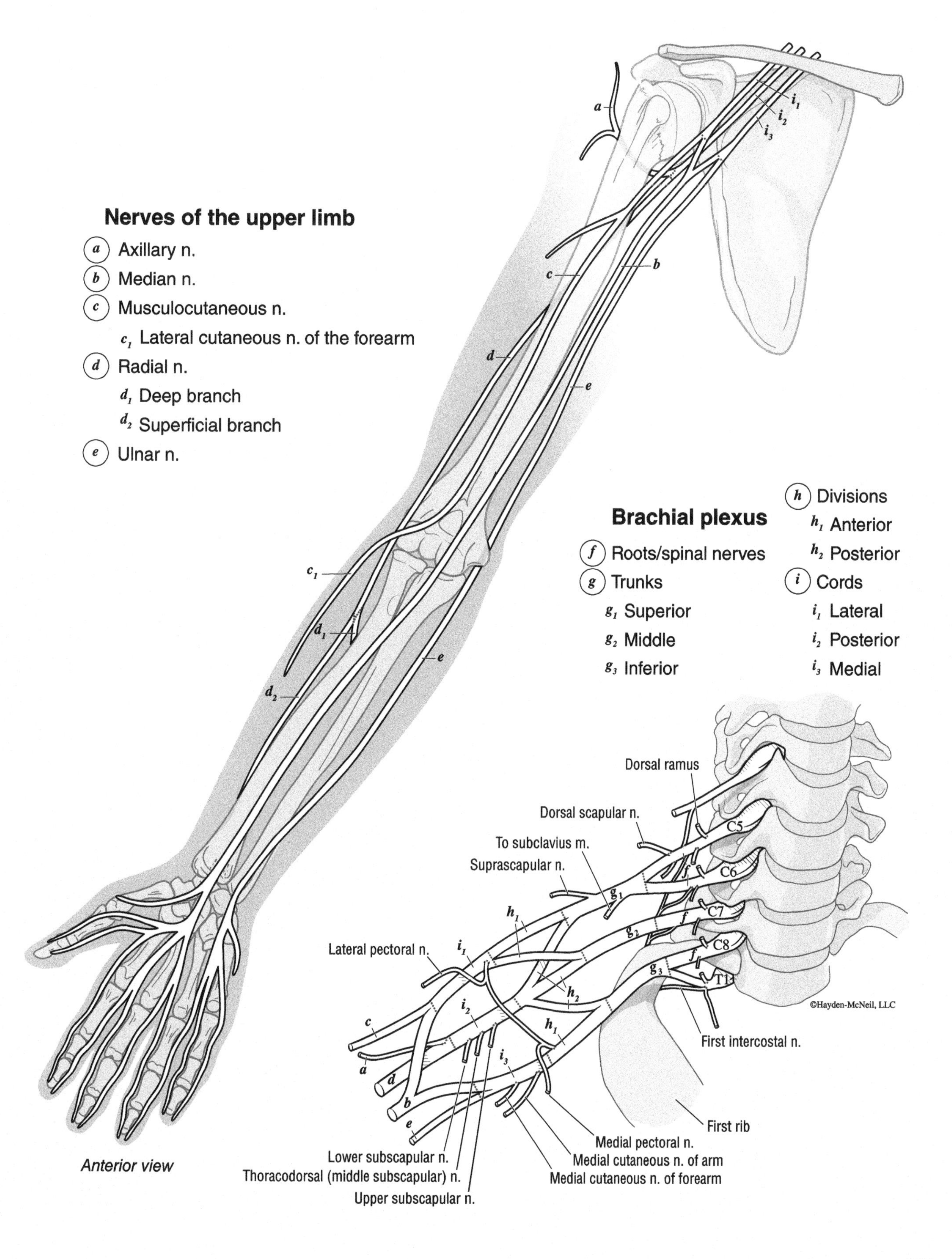
Nerves of the upper limb
a Axillary n.
b Median n.
c Musculocutaneous n.
c1 Lateral cutaneous n. of the forearm
d Radial n.
d1 Deep branch
d2 Superficial branch
e Ulnar n.
Brachial plexus
f Roots/spinal nerves
g Trunks
g1 Superior
g2 Middle
g3 Inferior
h Divisions
h1 Anterior
h2 Posterior
i Cords
i1 Lateral
i2 Posterior
i3 Medial
Dorsal ramus
Dorsal scapular n.
To subclavius m.
Suprascapular n.
C5
C6
C7
C8
T1
Lateral pectoral n.
First intercostal n.
First rib
Medial pectoral n.
Medial cutaneous n. of arm
Medial cutaneous n. of forearm
Lower subscapular n.
Thoracodorsal (middle subscapular) n.
Upper subscapular n.
Anterior view
©Hayden-McNeil, LLC

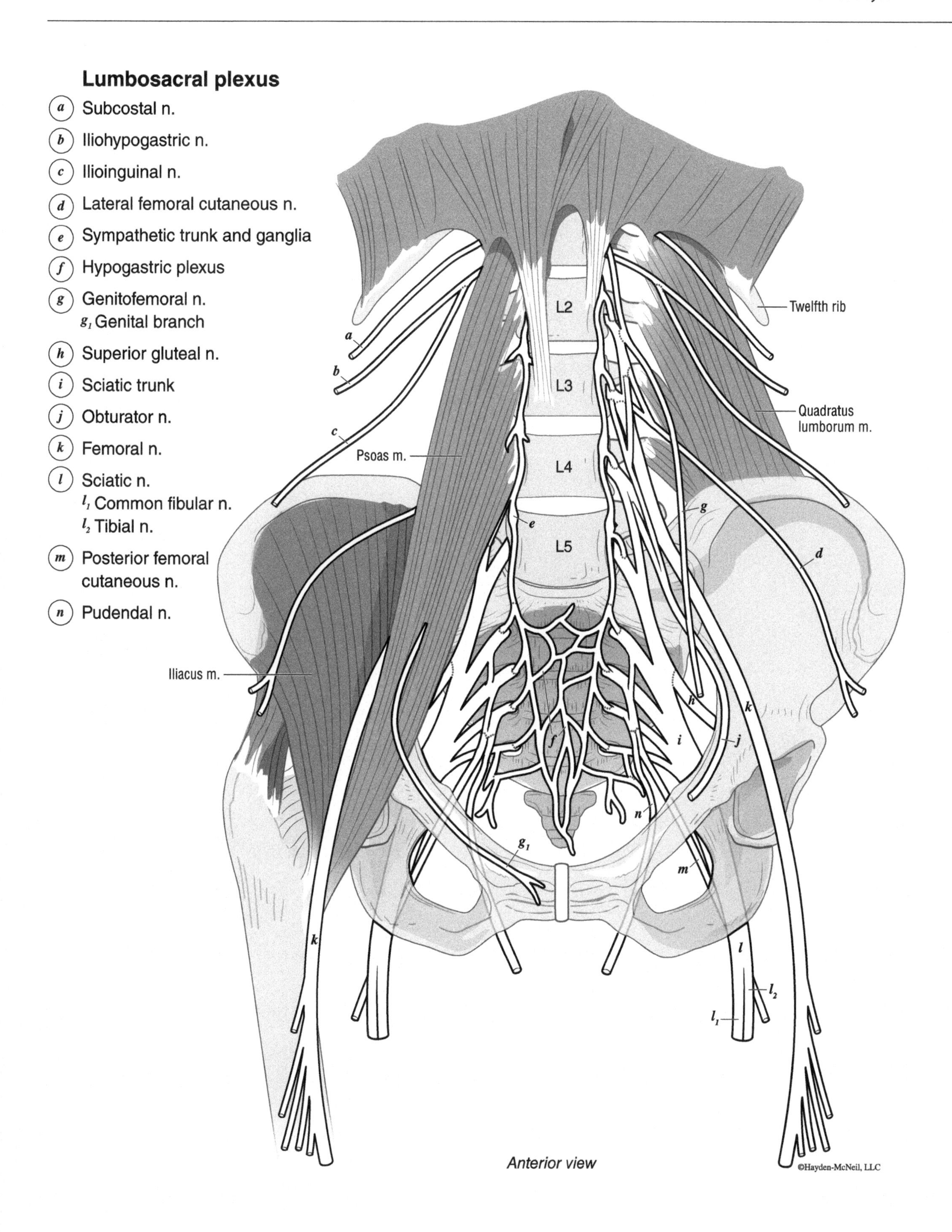

Anterior view

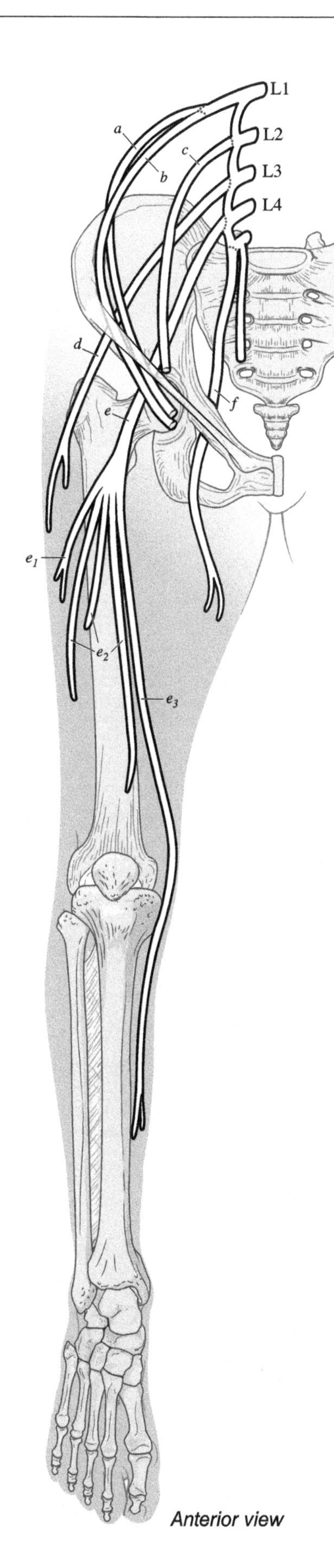

Anterior view

Nerves of the lower limb

(a) Ilioinguinal n

(b) Iliohypogastric n.

(c) Genitofemoral n.

(d) Lateral femoral cutaneous n.

(e) Femoral n.
- e_1 Anterior cutaneous branches
- e_2 Motor branches to extensor muscles
- e_3 Saphenous n.

(f) Obturator n.

(g) Sciatic n.

(h) Common fibular n.
- h_1 Deep fibular n.
- h_2 Superficial fibular n.

(i) Tibial n.
- i_1 Plantar branches

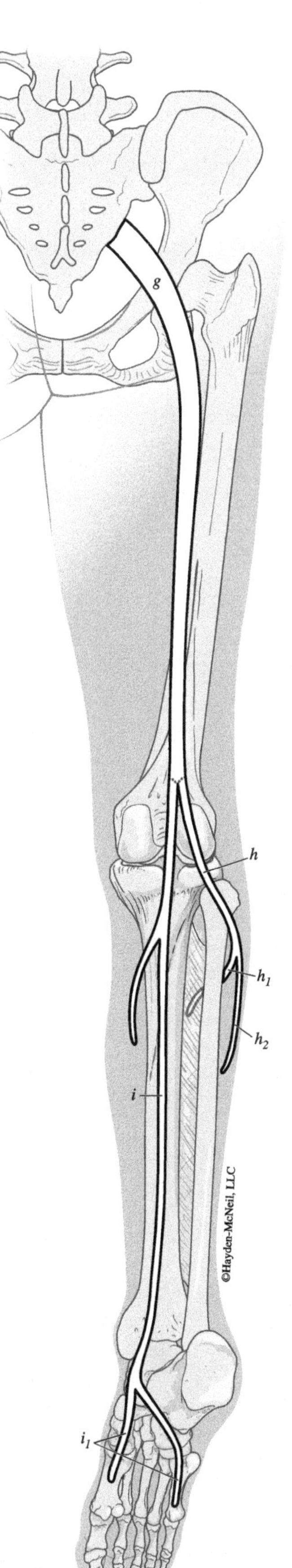

Posterior view

Dermatomes

Dermatomes are symmetrical; use the same colors on the left and right sides of the body.

- (a) Facial dermatomes (served by cranial nerve V)
- (b) Cervical dermatomes (C2–C8)
- (c) Thoracic dermatomes (T1–T12)
- (d) Lumbar dermatomes (L1–L5)
- (e) Sacral dermatomes (S1–S5)

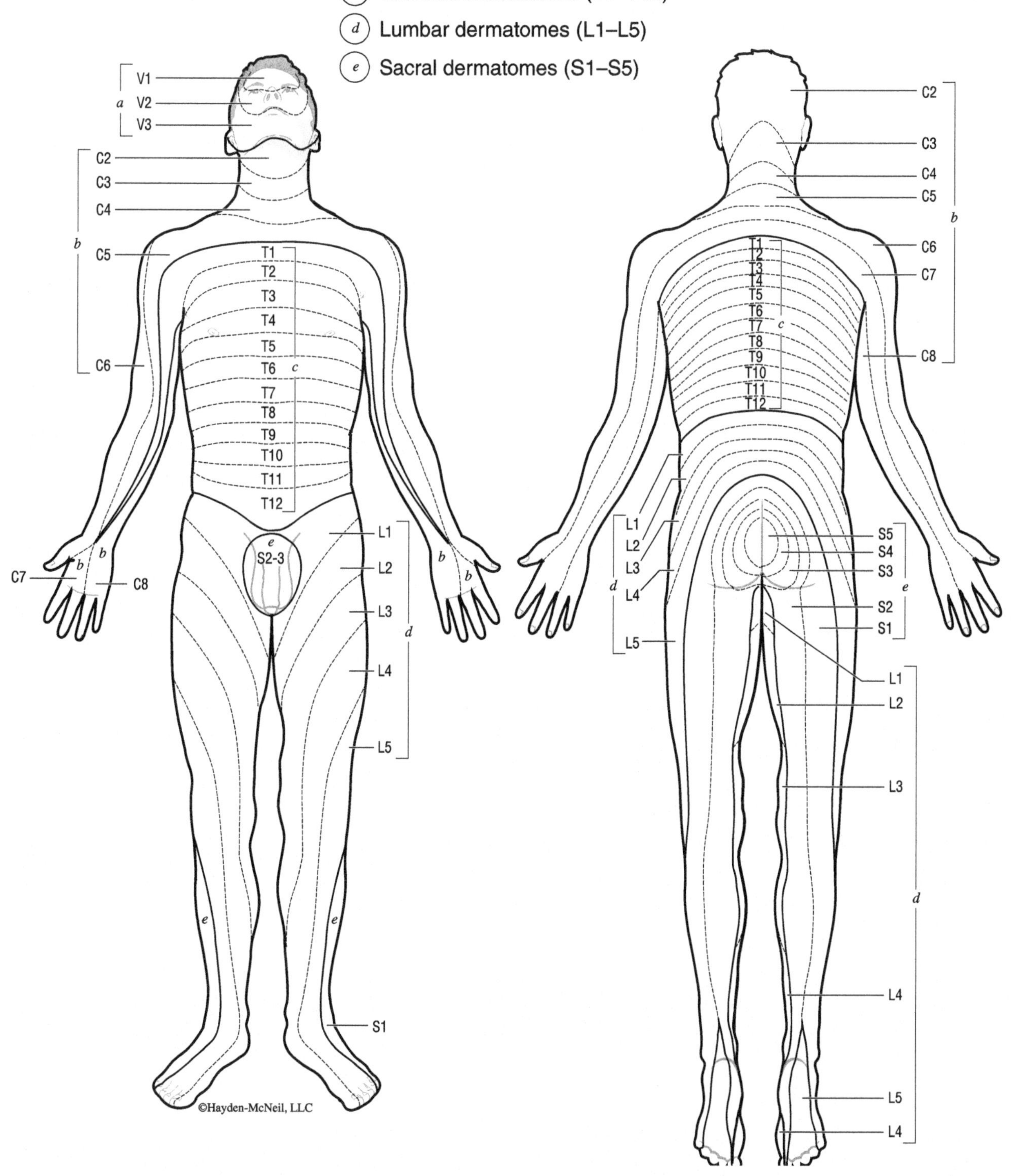

Autonomic nervous system

Use one color for the sympathetic system and the organs it affects (on the left side of the drawing). Use a different color for the parasympathetic system and the organs it affects (on the right side of the drawing). Use green for the increase (dark) arrows and red for the decrease (light) arrows.

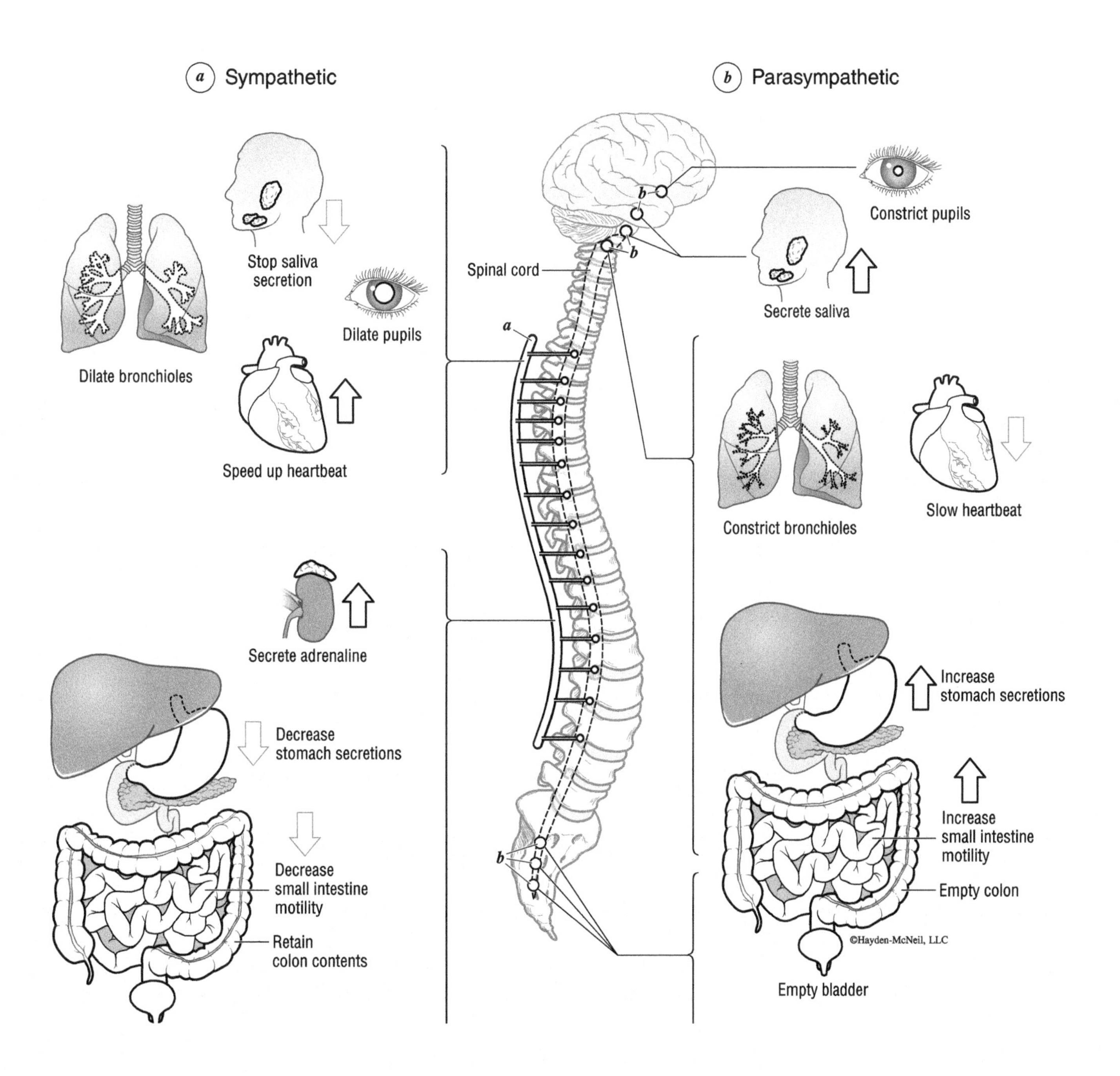

9

The Special Senses

Eye

Retina

Eye muscles

Optic pathway

Ear

Inner ear equilibrium structures

Olfactory reception

Taste reception

Taste buds

Join us on a **sensory system** safari through the wonderous world of perception. We will be cruising through the wild ecosystems of sight, sound, smell, taste, and touch, with each page providing a peek into the practicalities of our senses. While you color, we encourage you to take note of your surroundings. You will be tuning in to the satisfying scratch of your coloring utensils and the gentle rustle of the paper, the feeling of the pencil in your hand, and watching as colors spill from your imagination onto the page. So, buckle up and let's "see" what our senses "hold" for us in the habitats of sensory reception and interpretation.

Eye

Color the retinal blood vessels red and blue.

- (a) Conjunctiva
- (b) Superior and inferior rectus muscles
- (c) Cornea
- (d) Anterior chamber *(part of anterior cavity)*
- (e) Posterior chamber *(part of anterior cavity)*
- (f) Iris
- (g) Lens
- (h) Suspensory ligaments
- (i) Ciliary muscle
- (j) Sclera
- (k) Choroid
- (l) Retina
- (m) Posterior cavity
- (n) Optic nerve

Retina

- (o) Ganglion cell
- (p) Amacrine cell
- (q) Bipolar cell
- (r) Horizontal cell
- (s) Rod
- (t) Cone
- (u) Pigmented epithelium
- (v) Basement membrane

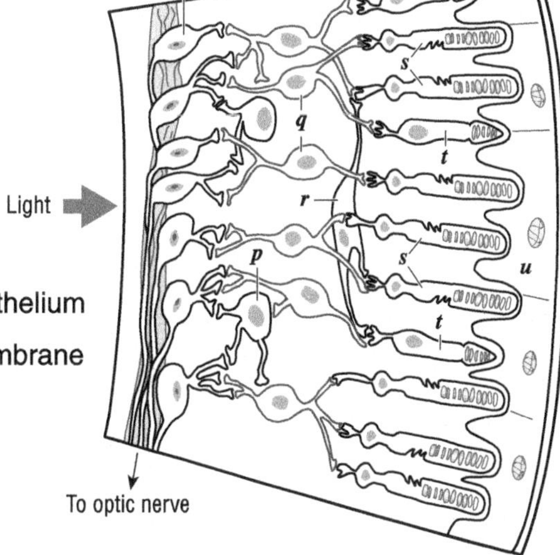

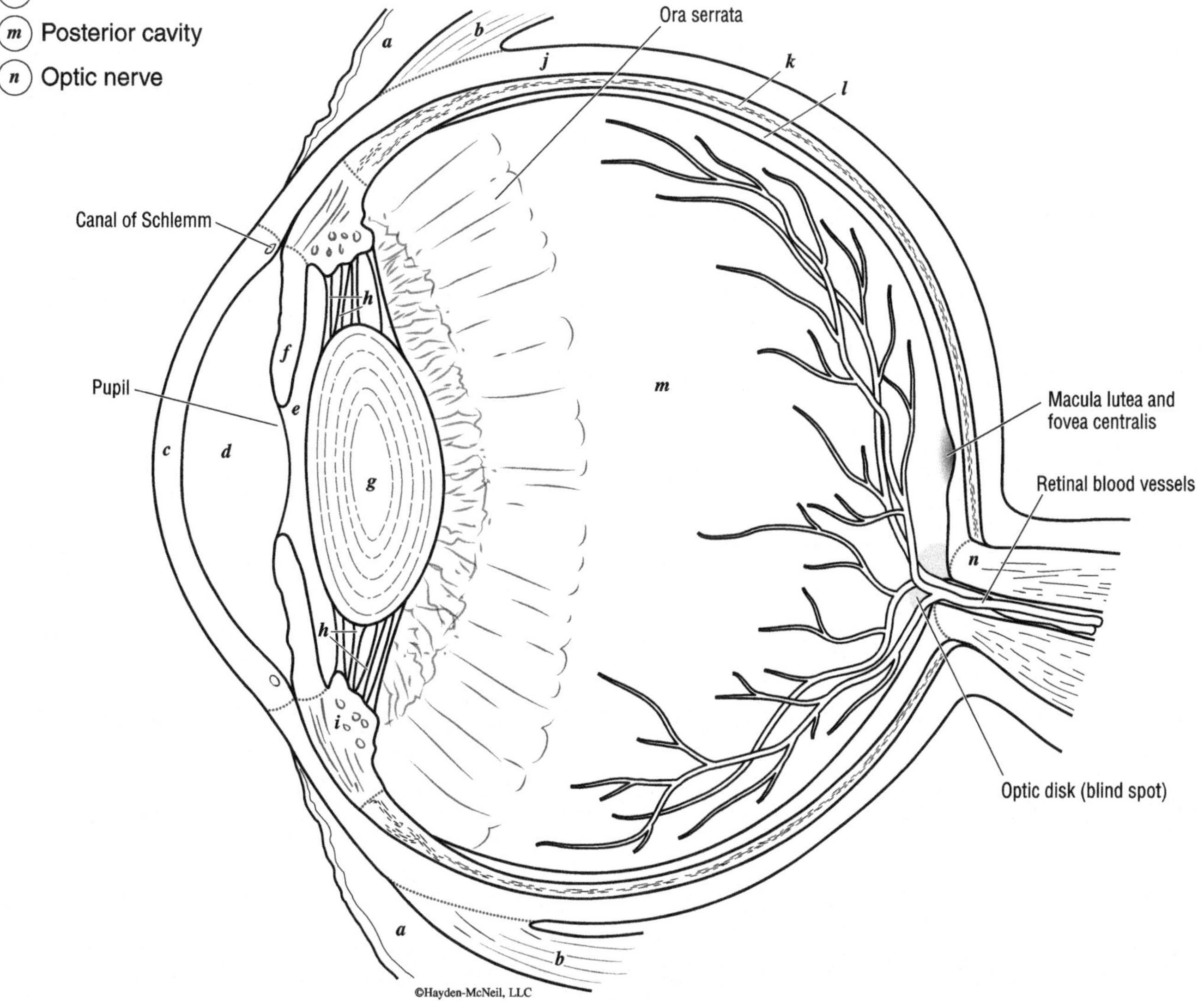

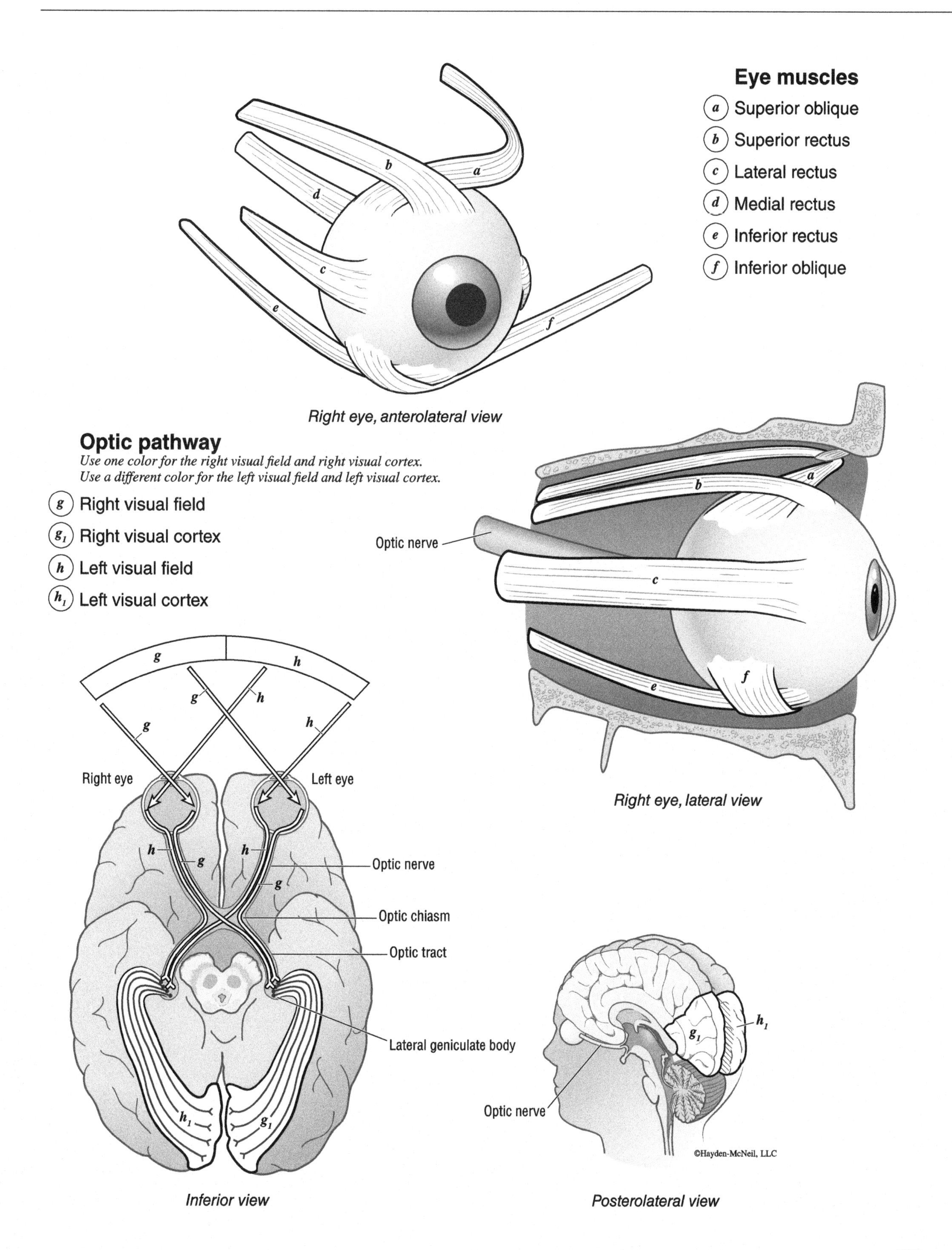
Eye muscles
a Superior oblique
b Superior rectus
c Lateral rectus
d Medial rectus
e Inferior rectus
f Inferior oblique
a
b
c
d
e
f
Right eye, anterolateral view
Optic pathway
Use one color for the right visual field and right visual cortex.
Use a different color for the left visual field and left visual cortex.
g Right visual field
g1 Right visual cortex
h Left visual field
h1 Left visual cortex
Optic nerve
Right eye, lateral view
g
h
Right eye
Left eye
Optic nerve
Optic chiasm
Optic tract
Lateral geniculate body
h1
g1
Inferior view
Optic nerve
Posterolateral view
©Hayden-McNeil, LLC

Ear

- (a) External acoustic meatus
- (b) Tympanic membrane
- (c) Malleus
- (d) Incus
- (e) Stapes
- (f) Semicircular canals
- (g) Vestibule
- (h) Cochlea
- (i) Auditory tube
- (j) Vestibular nerve
- (k) Cochlear nerve
- (l) Basilar membrane
- (m) Tectorial membrane
- (n) Spiral organ (of Corti)
- (o) Vestibular (Reissner's) membrane
- (p) Scala media (cochlear duct)
- (q) Scala vestibuli
- (r) Scala tympani

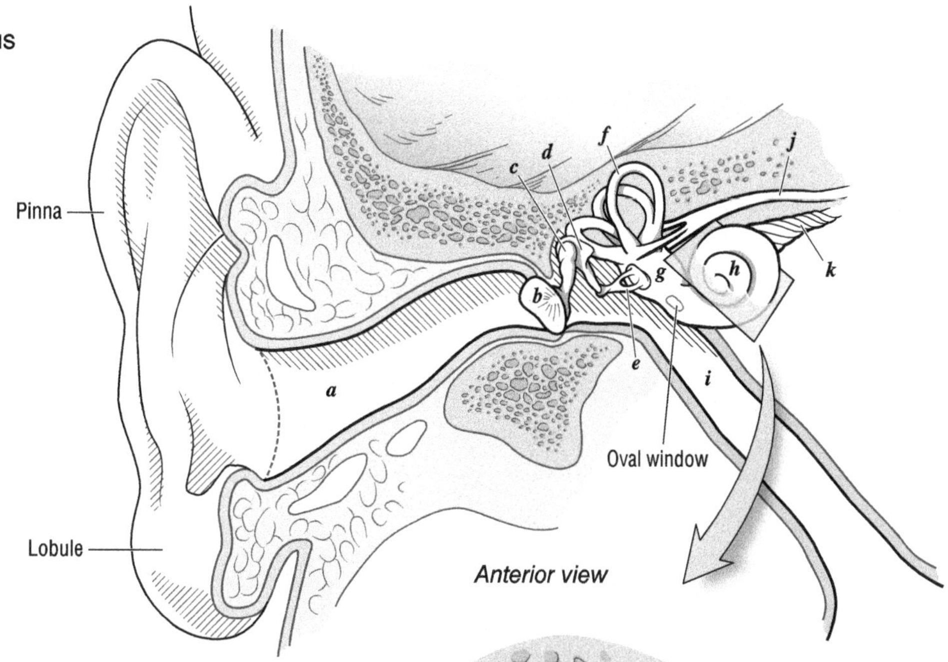

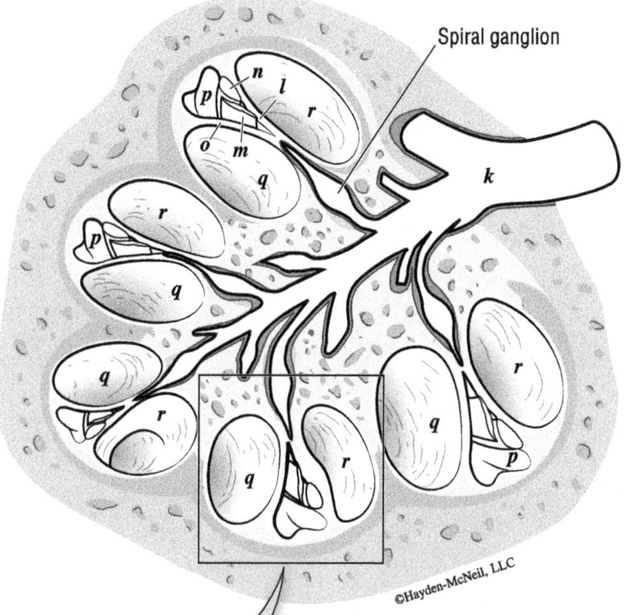

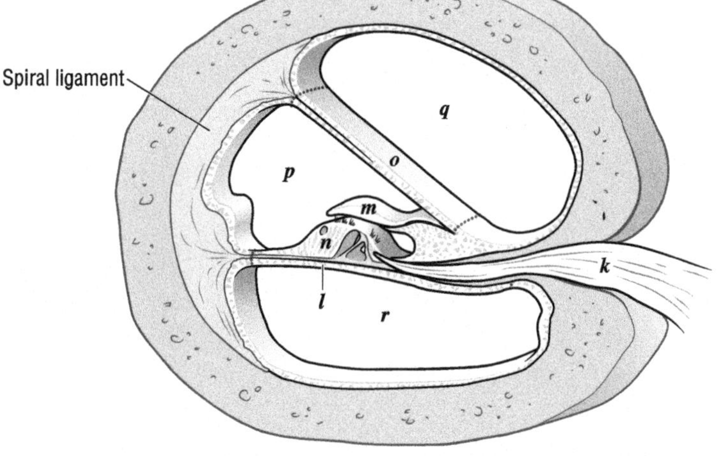

Inner ear equilibrium structures

- (a) Cochlea
- (b) Utricle
- (c) Saccule
- (d) Macula
- (e) Semicircular canals
- (f) Endolymph
- (g) Otolithic membrane
- (h) Otoliths
- (i) Hair cells
- (j) Crista ampullaris
- (k) Cupula

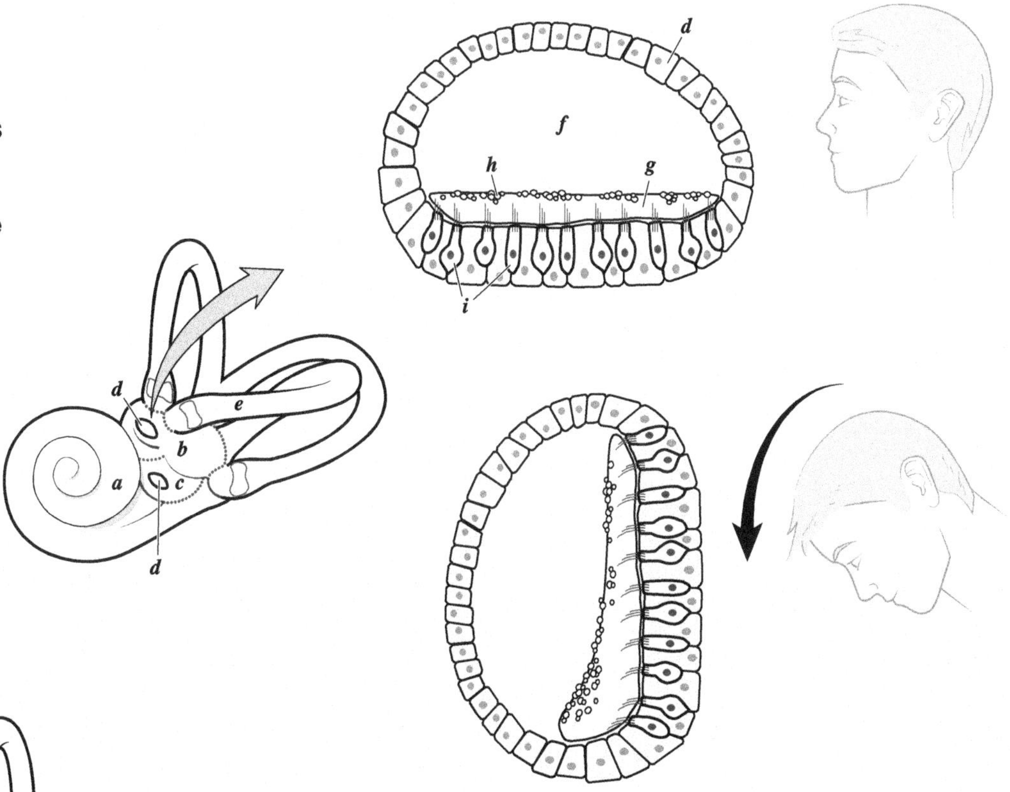

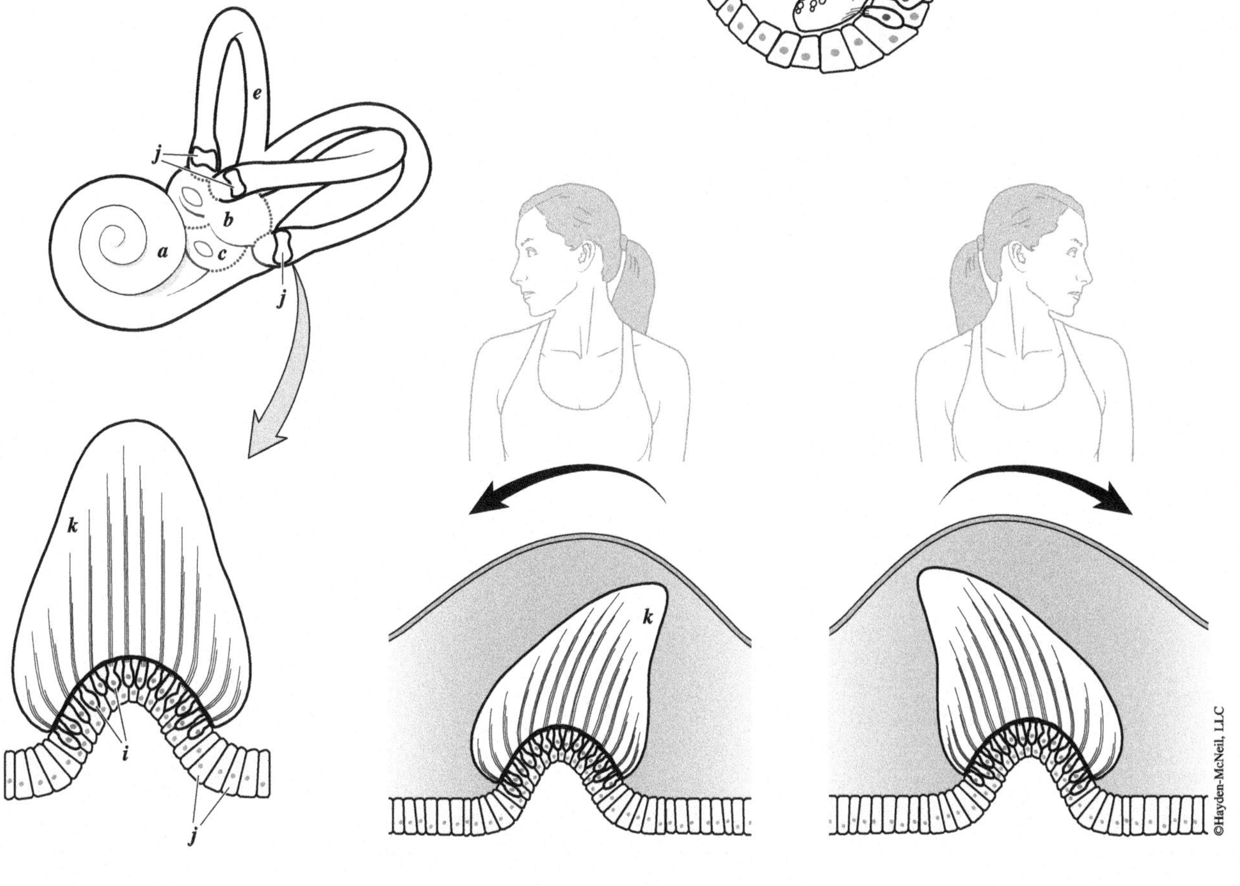

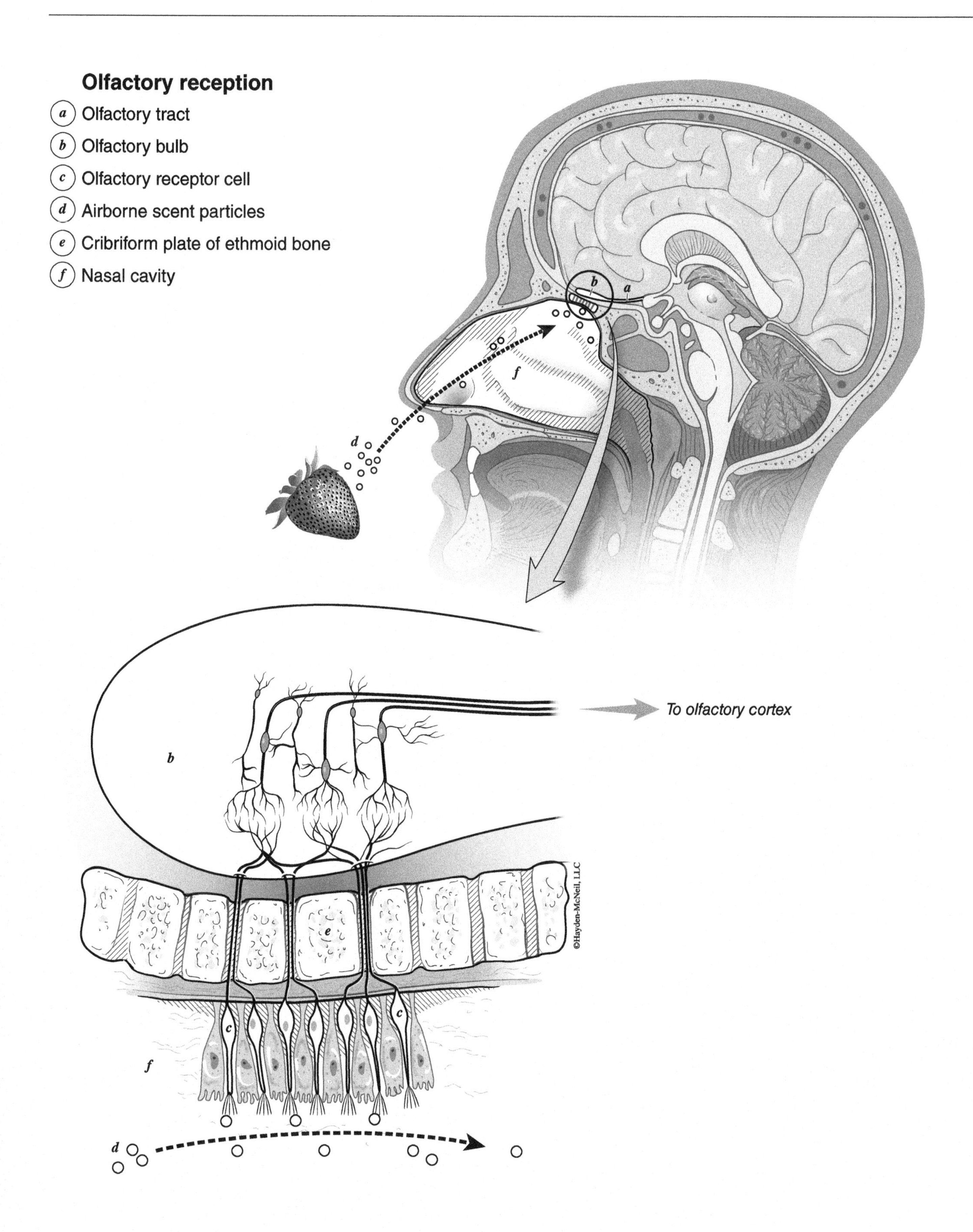
Olfactory reception
a Olfactory tract
b Olfactory bulb
c Olfactory receptor cell
d Airborne scent particles
e Cribriform plate of ethmoid bone
f Nasal cavity
b
a
f
d
To olfactory cortex
b
e
c
c
f
d
©Hayden-McNeil, LLC

Taste reception

- (a) Circumvallate (vallate) papilla
- (b) Foliate papilla
- (c) Fungiform papilla
- (d) Taste bud
- (e) Sensory nerve fiber
- (f) Epithelial cells
- (g) Taste cell
- (h) Supporting cell

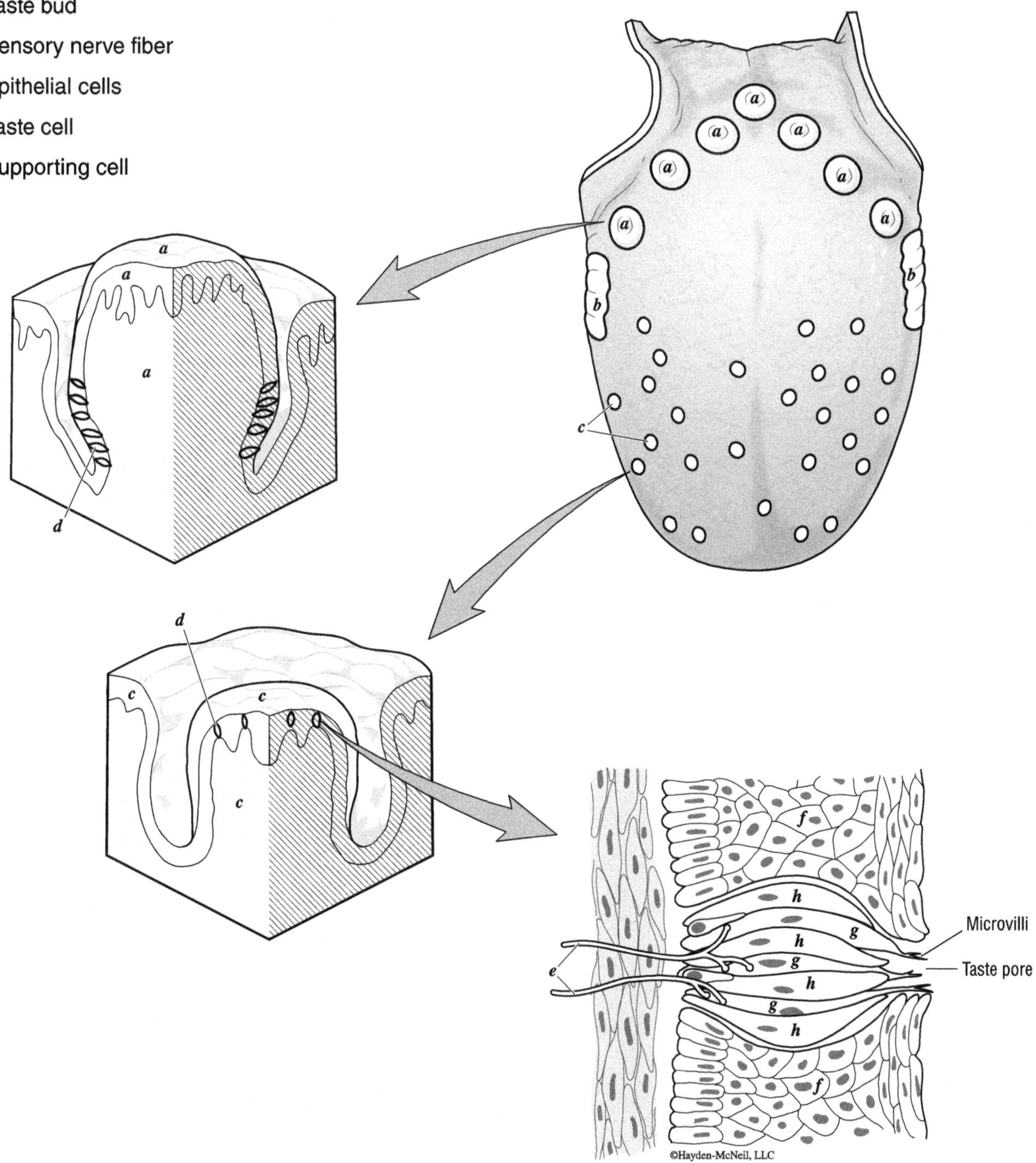

Taste buds

(a) Circumvallate (vallate) papilla
(b) Folate papilla
(c) Fungiform papilla
(d) Filiform papilla
(e) Regions of taste

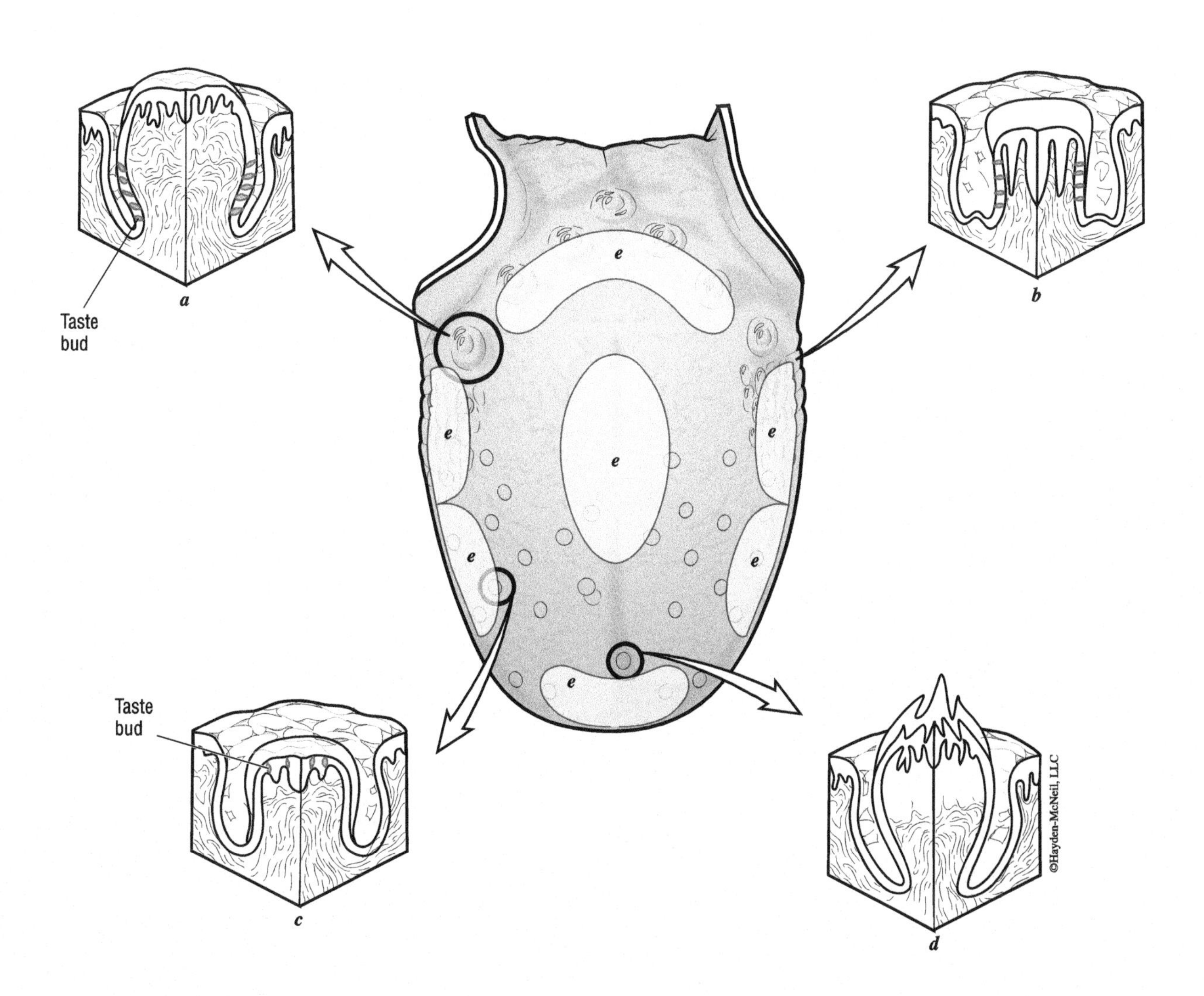

10

The Endocrine System

Endocrine glands
Pituitary gland
Thyroid gland
Parathyroid glands
Adrenal gland
Hormone secretions
Pancreas
Islet of Langerhans

For this section, imagine a quiet, abundant garden, where the **endocrine system** acts as the master gardener, tending to the delicate balance of life's processes. You will be wandering through verdant paths lined with the glands of the body, each a unique bloom releasing the nectar that will sustain us—hormones. While coloring these pages, you will be breathing life into the pancreatic blossoms that regulate sugar, the adrenal blooms that respond to stress, and the pituitary bud that oversees growth and reproduction. Let us help cultivate color in this glorious greenhouse of hormonal harmony!

Endocrine glands

- (*a*) Hypophysis (pituitary)
- (*b*) Pineal
- (*c*) Thyroid
- (*d*) Parathyroids
- (*e*) Thymus
- (*f*) Adrenals
- (*g*) Pancreas
- (*h*) Ovaries
- (*i*) Testes

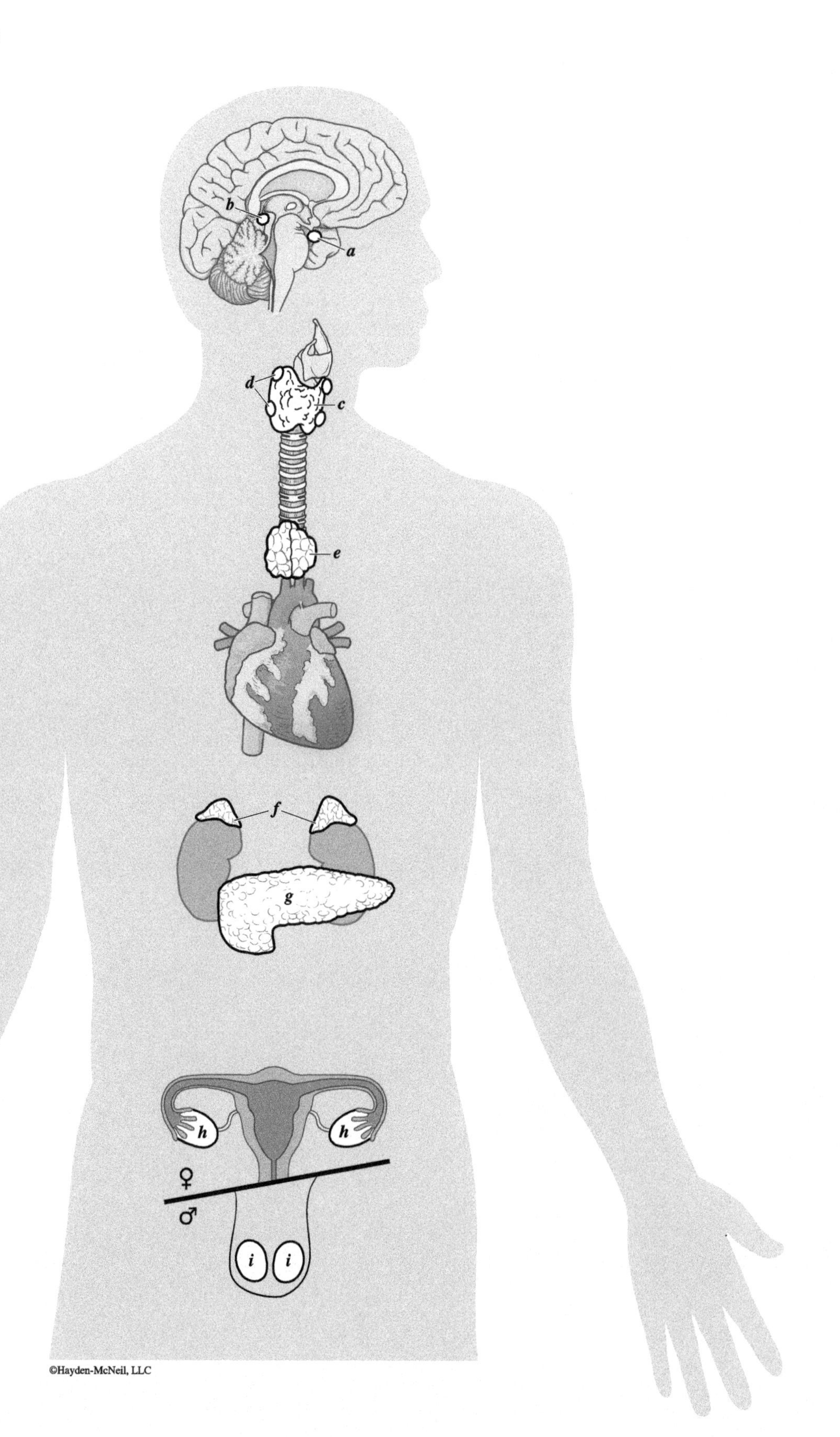

Pituitary gland

Color the hypophyseal arteries red, the portal system purple, and the hypophyseal veins blue. Color the neurons yellow.

- (a) Infundibulum (pituitary stalk)
- (b) Anterior pituitary
- (c) Posterior pituitary
- (d) Hormone-secreting cells
- (e) Superior hypophyseal artery
- (f) Inferior hypophyseal artery
- (g) Portal system
  - g_1 Capillary bed
  - g_2 Venule
  - g_3 Sinusoids
- (h) Hypophyseal veins
- (i) Capillary network
- (j) Neurons

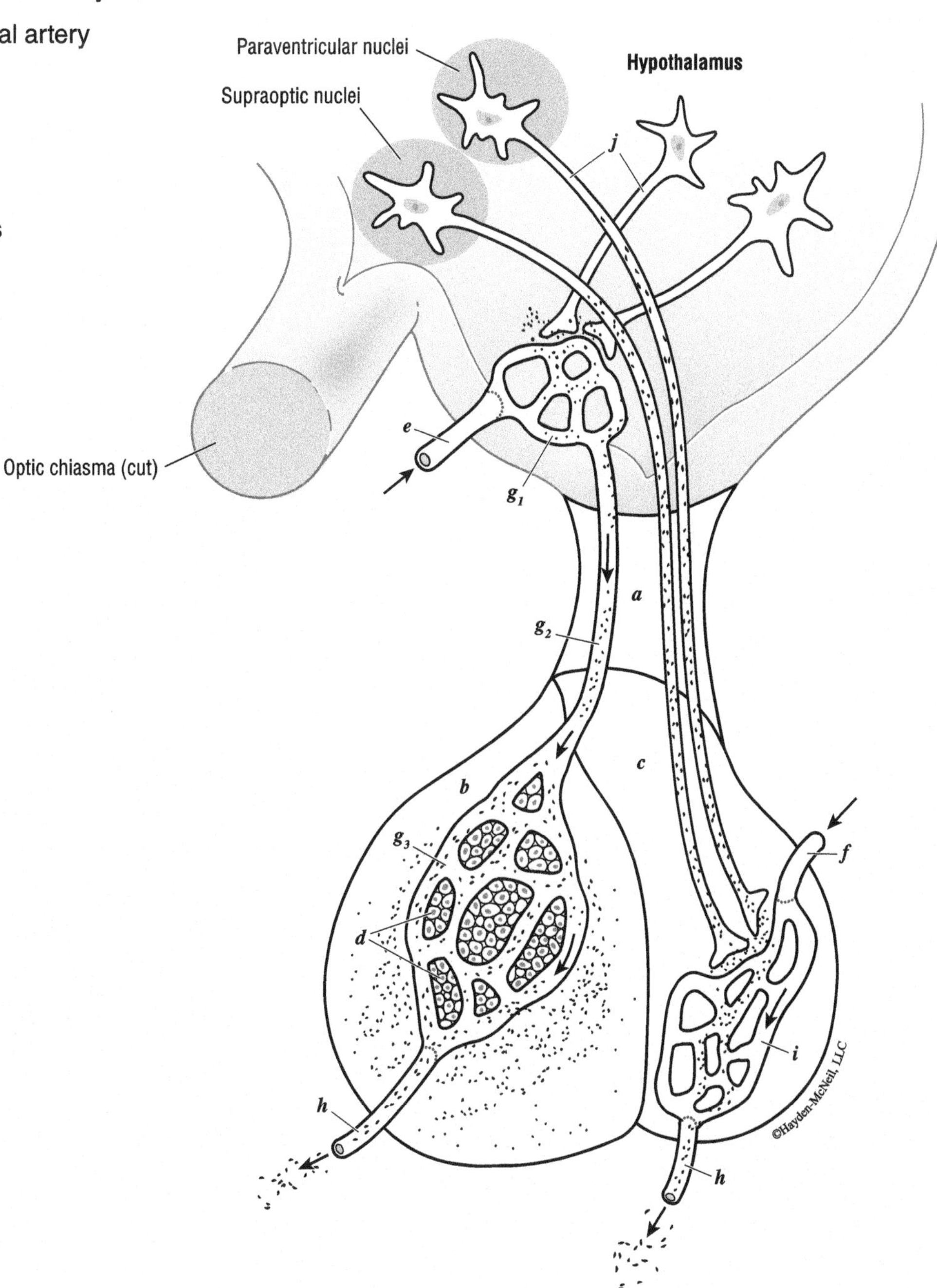

Thyroid gland

- (a) Thyroid gland
  - a_1 Right lobe
  - a_2 Isthmus
  - a_3 Left lobe
- (b) Follicular cells
- (c) Follicles filled with thyroglobulin
- (d) C cells

Parathyroid glands

- (e) Parathyroid glands
- (f) Chief cells
- (g) Oxyphil cells
- (h) Blood vessels
- (i) Connective tissue

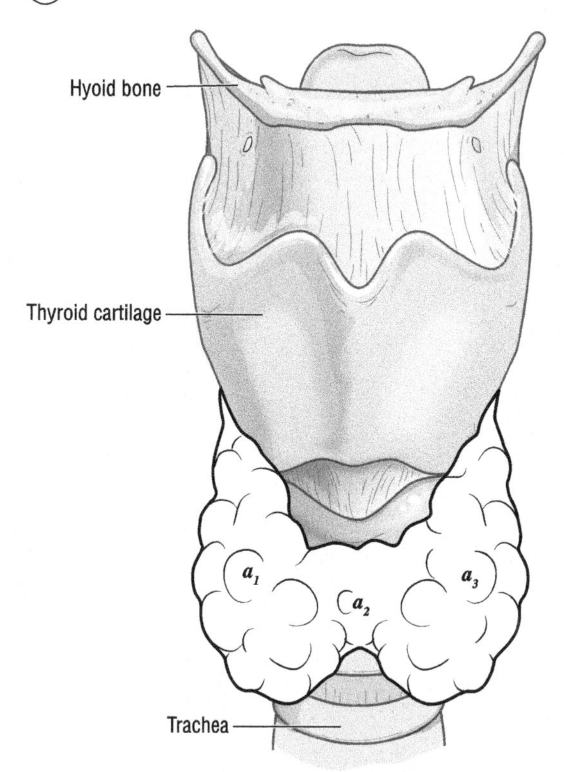

Anterior view

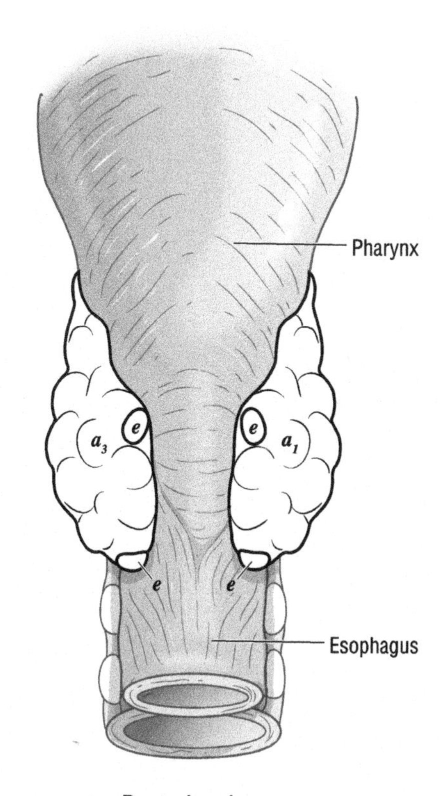

Posterior view

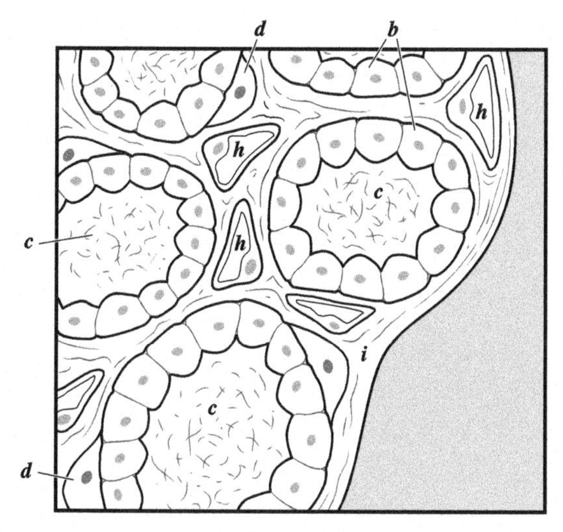

Transverse section

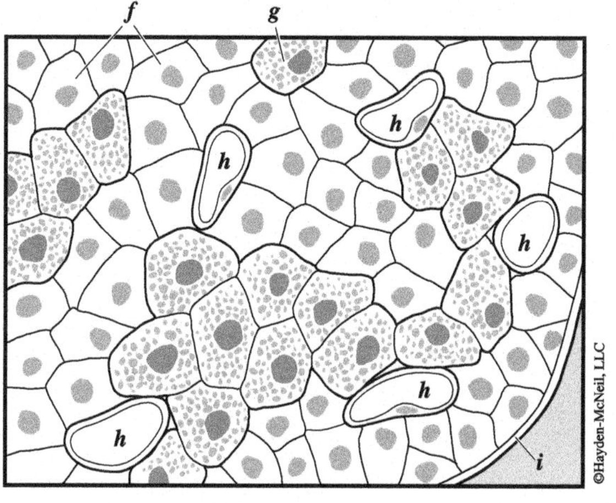

Transverse section

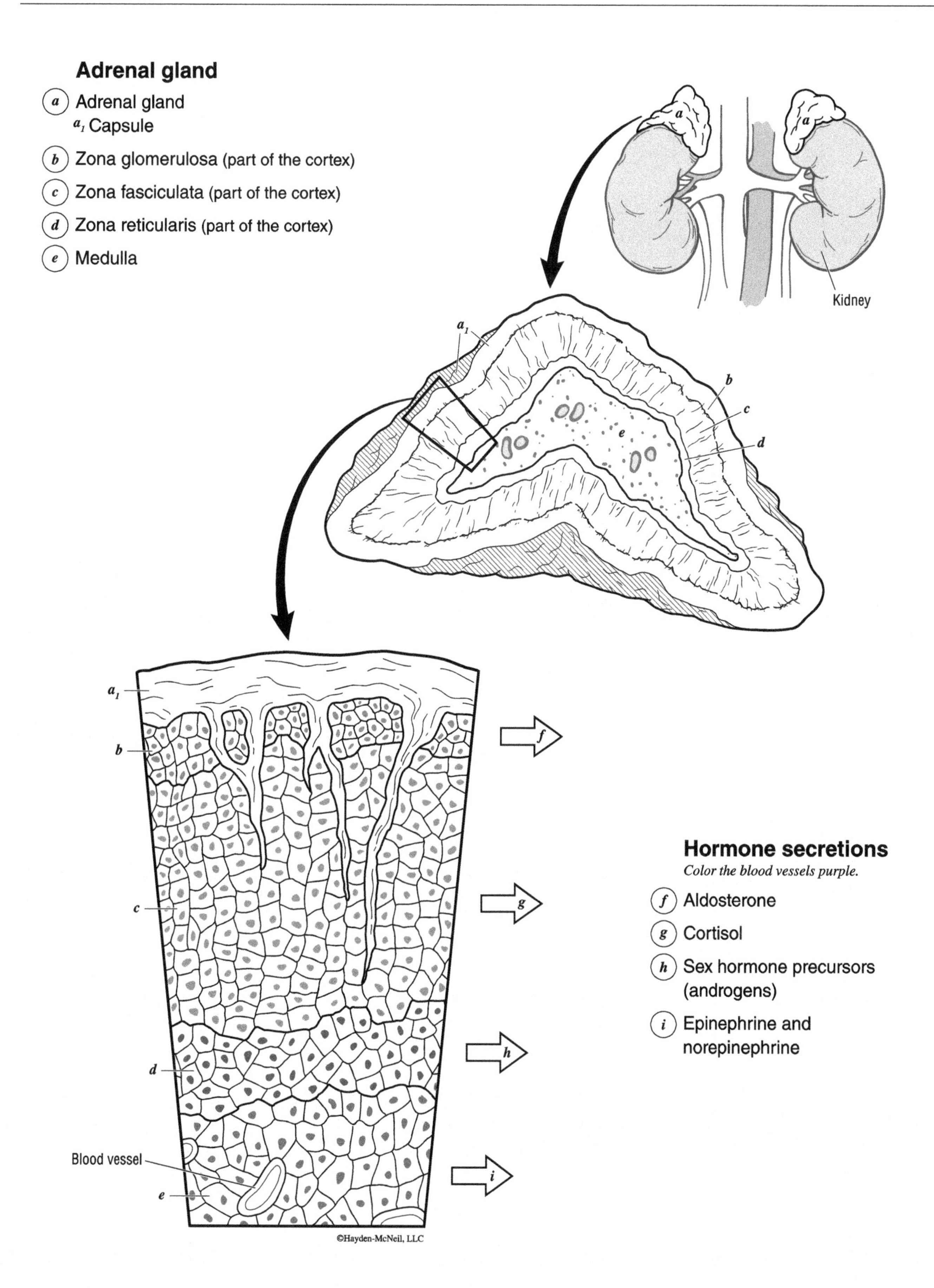
Adrenal gland
a Adrenal gland
a1 Capsule
b Zona glomerulosa (part of the cortex)
c Zona fasciculata (part of the cortex)
d Zona reticularis (part of the cortex)
e Medulla
a
a
Kidney
a1
b
c
d
e
a1
b
c
d
Blood vessel
e
f
g
h
i
Hormone secretions
Color the blood vessels purple.
f Aldosterone
g Cortisol
h Sex hormone precursors (androgens)
i Epinephrine and norepinephrine
©Hayden-McNeil, LLC

Pancreas

- (a) Pancreas
  - a_1 Head
  - a_2 Uncinate process
  - a_3 Body
  - a_4 Tail
- (b) Pancreatic duct
- (c) Duodenal papilla
- (d) Splenic vein
- (e) Acinar cells (exocrine pancreatic secretions)

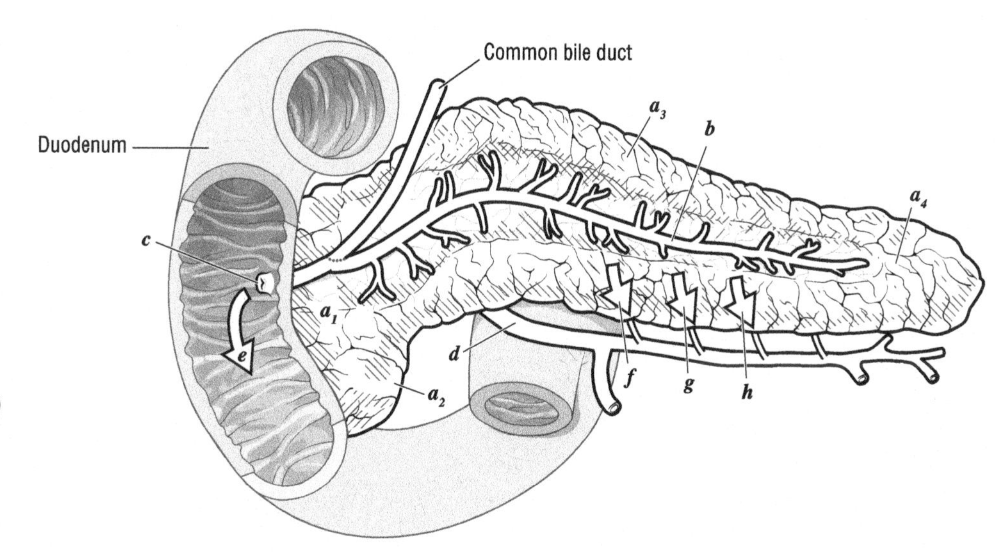

Anterior view

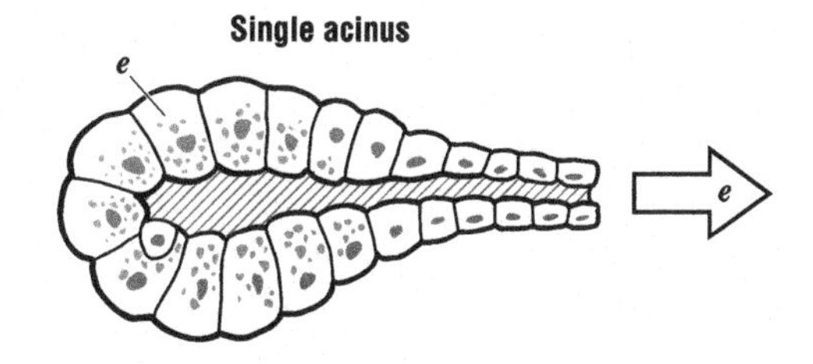

Longitudinal section

Islet of Langerhans (Endocrine pancreatic secretions)

Color the blood vessels purple.

- (f) Alpha cells (glucagon)
- (g) Beta cells (insulin)
- (h) Delta cells (somatostatin)

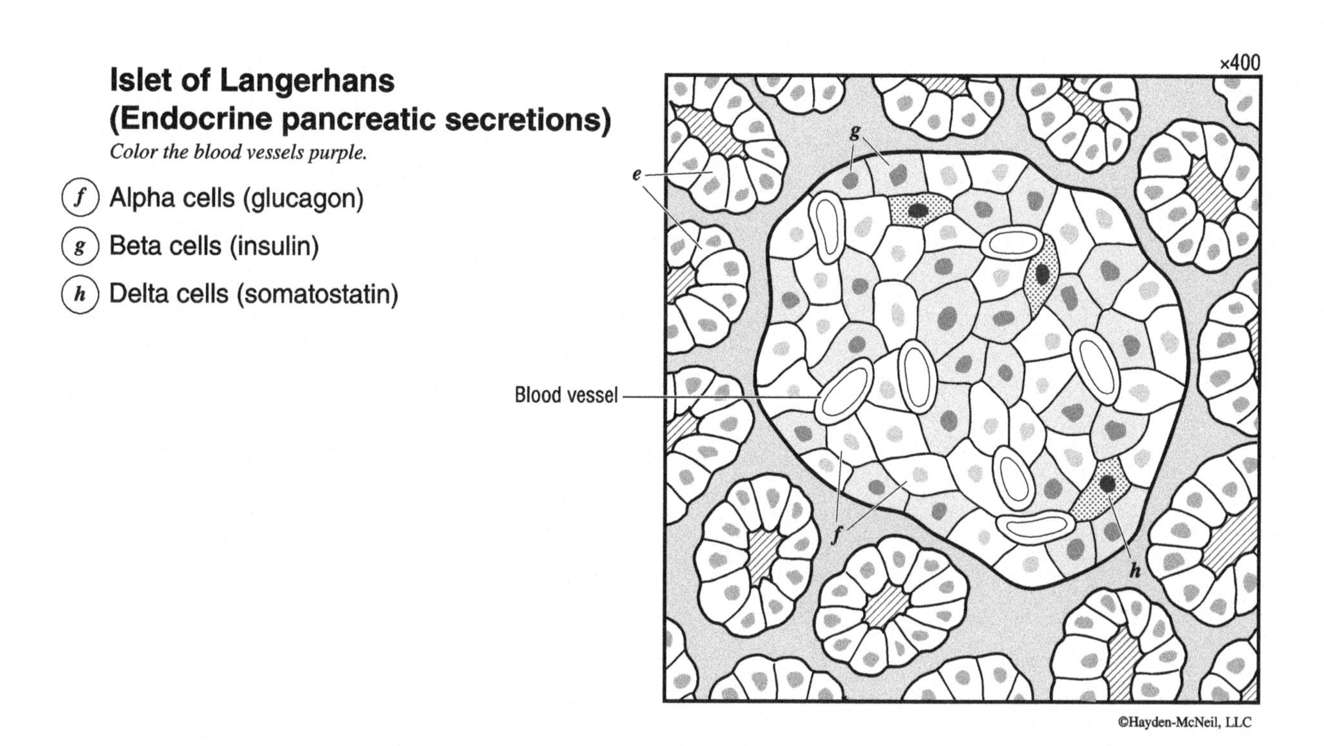

11

The Cardiovascular System

Blood and blood cells
Mediastinum and exterior heart
Interior heart and blood flow
Conduction system of the heart
Artery and vein structure
Capillary structure
Head and neck arteries
Head and neck veins
Cranial dural sinuses
Arteries of the brain
Circle of Willis
Arteries of the upper limb
Veins of the upper limb
Arteries of the lower limb
Veins of the lower limb
Aorta and its branches
Hepatic portal system
Pelvic arteries

Within the upcoming sections, you are invited to experience the **cardiovascular system** symphony, where every muscle plays a vital role in the rhythm of movement and life. Imagine yourself sitting in the orchestra pit, watching the intuitive direction of this concert of contractions, where every beat contributes to the collegial flow within our bodies. The heart acts as a vital conductor, providing the rhythm, support, and direction, while the blood vessels work in perfect harmony to bring blood to every organ and tissue and nutrients to every cell. This chapter invites you to experience the song that often goes unnoticed. Stop and listen—can you hear the music?

Blood and blood cells

Do not color areas marked with (*).

(a) Erythrocytes (red blood cells) *Color red.*

Leukocytes (white blood cells) *Color cytoplasm light blue.*

(b) Eosinophil *Color nucleus purple, cytoplasmic granules red.*

(c) Basophil *Color nucleus purple, cytoplasmic granules dark blue.*

(d) Neutrophil *Color nucleus purple.*

(e) Lymphocyte *Color nucleus light purple.*

(f) Monocyte *Color nucleus light purple.*

(g) Platelets *Color blue.*

(h) Plasma *Color light tan.*

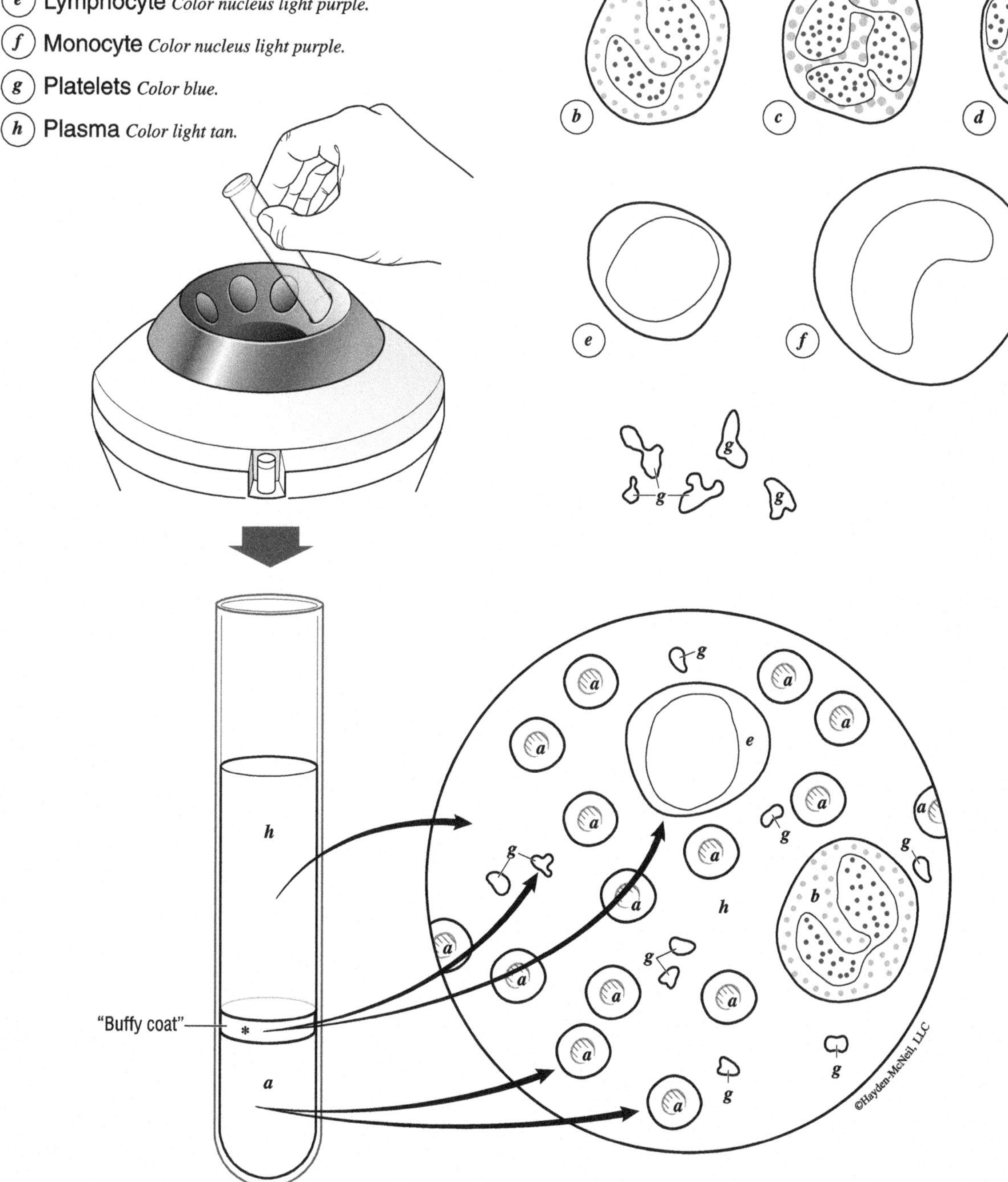

Centrifuged blood sample

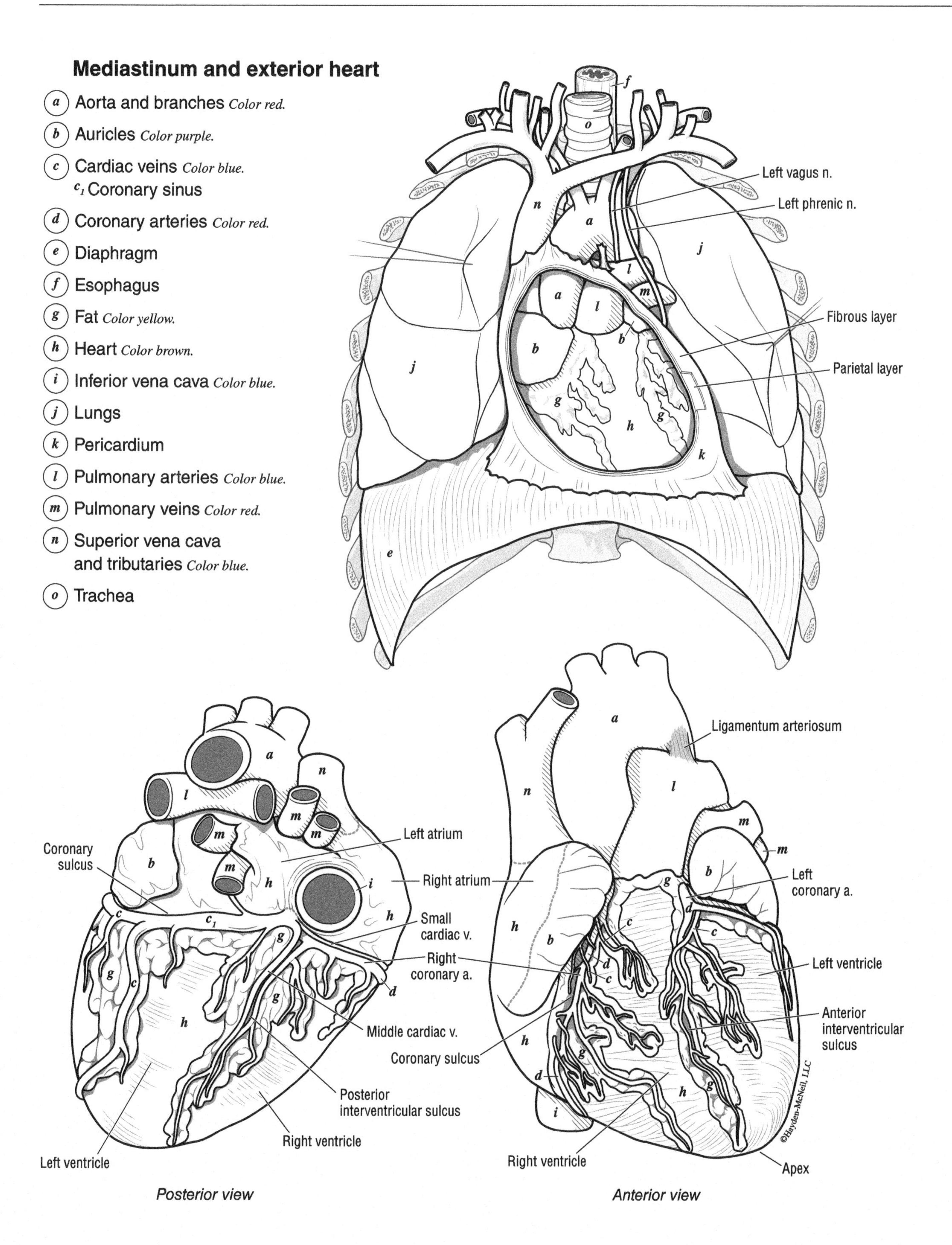
Mediastinum and exterior heart
a Aorta and branches Color red.
b Auricles Color purple.
c Cardiac veins Color blue.
c_1 Coronary sinus
d Coronary arteries Color red.
e Diaphragm
f Esophagus
g Fat Color yellow.
h Heart Color brown.
i Inferior vena cava Color blue.
j Lungs
k Pericardium
l Pulmonary arteries Color blue.
m Pulmonary veins Color red.
n Superior vena cava and tributaries Color blue.
o Trachea
Left vagus n.
Left phrenic n.
Fibrous layer
Parietal layer
Coronary sulcus
Left atrium
Right atrium
Small cardiac v.
Right coronary a.
Middle cardiac v.
Posterior interventricular sulcus
Right ventricle
Left ventricle
Posterior view
Ligamentum arteriosum
Left coronary a.
Left ventricle
Anterior interventricular sulcus
Coronary sulcus
Right ventricle
Apex
Anterior view
©Hayden-McNeil, LLC

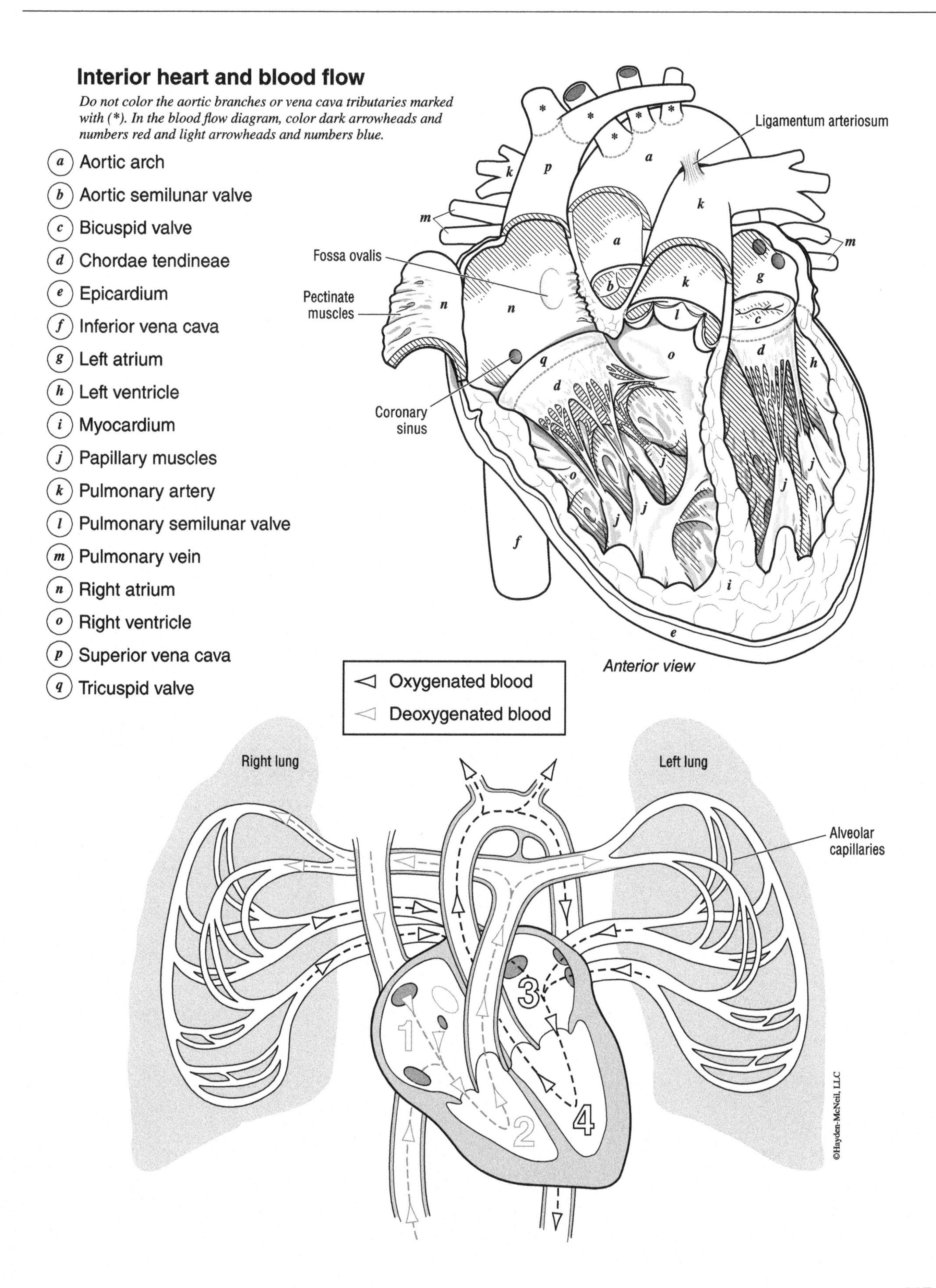

Interior heart and blood flow
Do not color the aortic branches or vena cava tributaries marked with (*). In the blood flow diagram, color dark arrowheads and numbers red and light arrowheads and numbers blue.
a Aortic arch
b Aortic semilunar valve
c Bicuspid valve
d Chordae tendineae
e Epicardium
f Inferior vena cava
g Left atrium
h Left ventricle
i Myocardium
j Papillary muscles
k Pulmonary artery
l Pulmonary semilunar valve
m Pulmonary vein
n Right atrium
o Right ventricle
p Superior vena cava
q Tricuspid valve
Ligamentum arteriosum
Fossa ovalis
Pectinate muscles
Coronary sinus
Anterior view
Oxygenated blood
Deoxygenated blood
Right lung
Left lung
Alveolar capillaries
1
2
3
4
©Hayden-McNeil, LLC

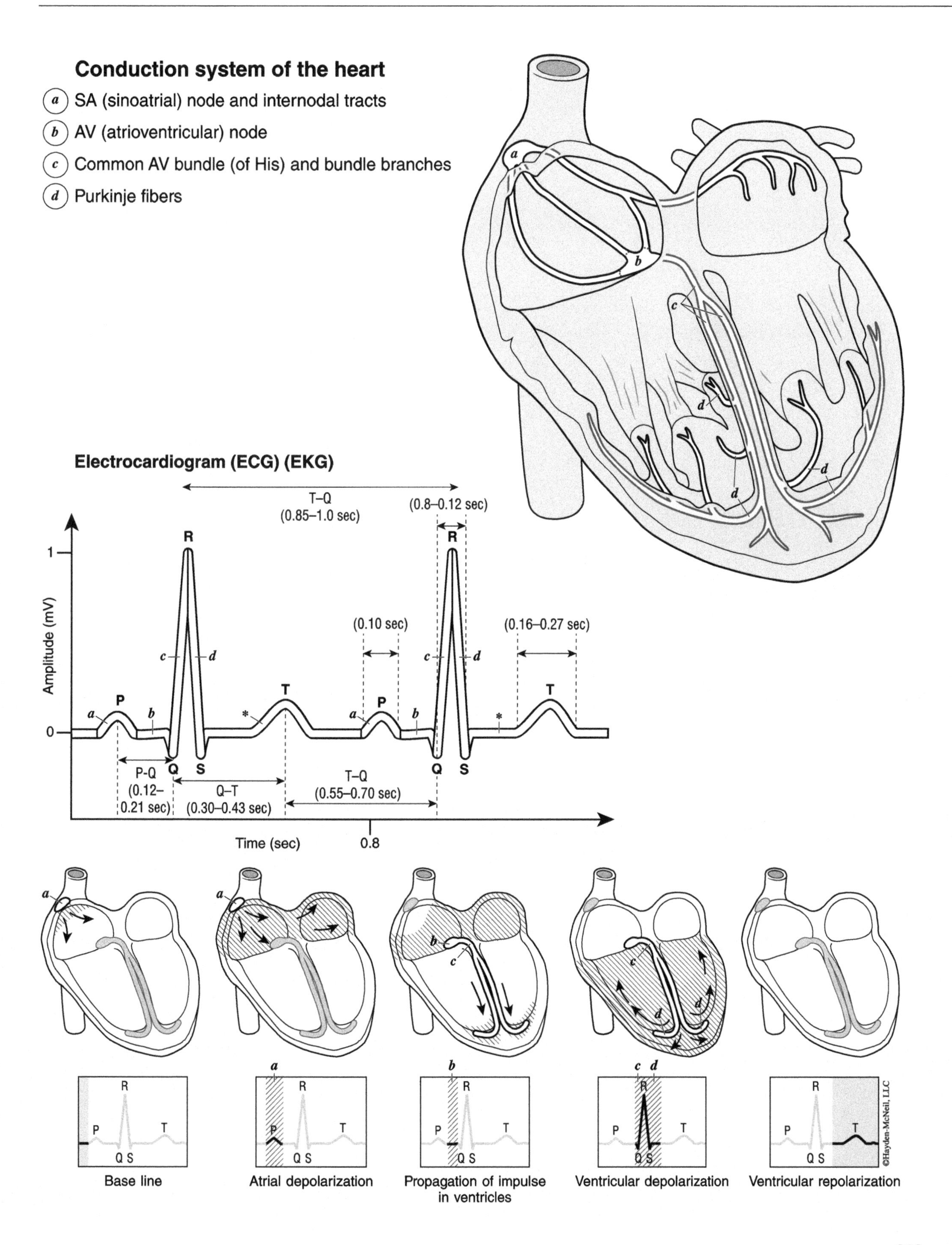
Conduction system of the heart
a SA (sinoatrial) node and internodal tracts
b AV (atrioventricular) node
c Common AV bundle (of His) and bundle branches
d Purkinje fibers
a
b
c
d
d
d
Electrocardiogram (ECG) (EKG)
T–Q
(0.85–1.0 sec)
(0.8–0.12 sec)
R
R
1
Amplitude (mV)
(0.10 sec)
(0.16–0.27 sec)
c
d
c
d
P
T
P
T
a
b
*
a
b
*
0
Q
S
Q
S
P-Q
(0.12–
0.21 sec)
Q–T
(0.30–0.43 sec)
T–Q
(0.55–0.70 sec)
Time (sec)
0.8
a
a
b
c
c
d
d
a
b
c d
R
P
T
Q S
Base line
Atrial depolarization
Propagation of impulse in ventricles
Ventricular depolarization
Ventricular repolarization
©Hayden-McNeil, LLC

Artery and vein structure

Tunica interna

- (a) Endothelium
- (b) Subendothelial layer
- (c) Internal elastic membrane

Tunica media

- (d) Smooth muscle

Tunica externa

- (e) External elastic membrane
- (f) Connective tissue

Artery

Vein

Valve

©Hayden-McNeil, LLC

Capillary structure

- (g) Basal lamina
- (a) Endothelium
- (h) Red blood cell

Intercellular cleft

Continuous capillary

Intercellular cleft

Fenestrations

Fenestrated capillary

Sinusoids

Large intercellular cleft

Sinusoidal capillary

Head and neck arteries

- (a) Axillary a.
- (b) Basilar a.
- (c) Brachiocephalic trunk
- (d) Costocervical trunk
- (e) External carotid a.
- (f) Facial a.
- (g) Inferior thyroid a.
- (h) Internal carotid a. (cut)
- (i) Internal thoracic a.
- (j) Left common carotid a.
- (k) Left subclavian a.
- (l) Lingual a.
- (m) Maxillary a.
- (n) Occipital a.
- (o) Ophthalmic a.
- (p) Right common carotid a.
- (q) Right subclavian a.
- (r) Superficial temporal a.
- (s) Superior thyroid a.
- (t) Thyrocervical trunk
- (u) Transverse facial a.
- (v) Vertebral a.

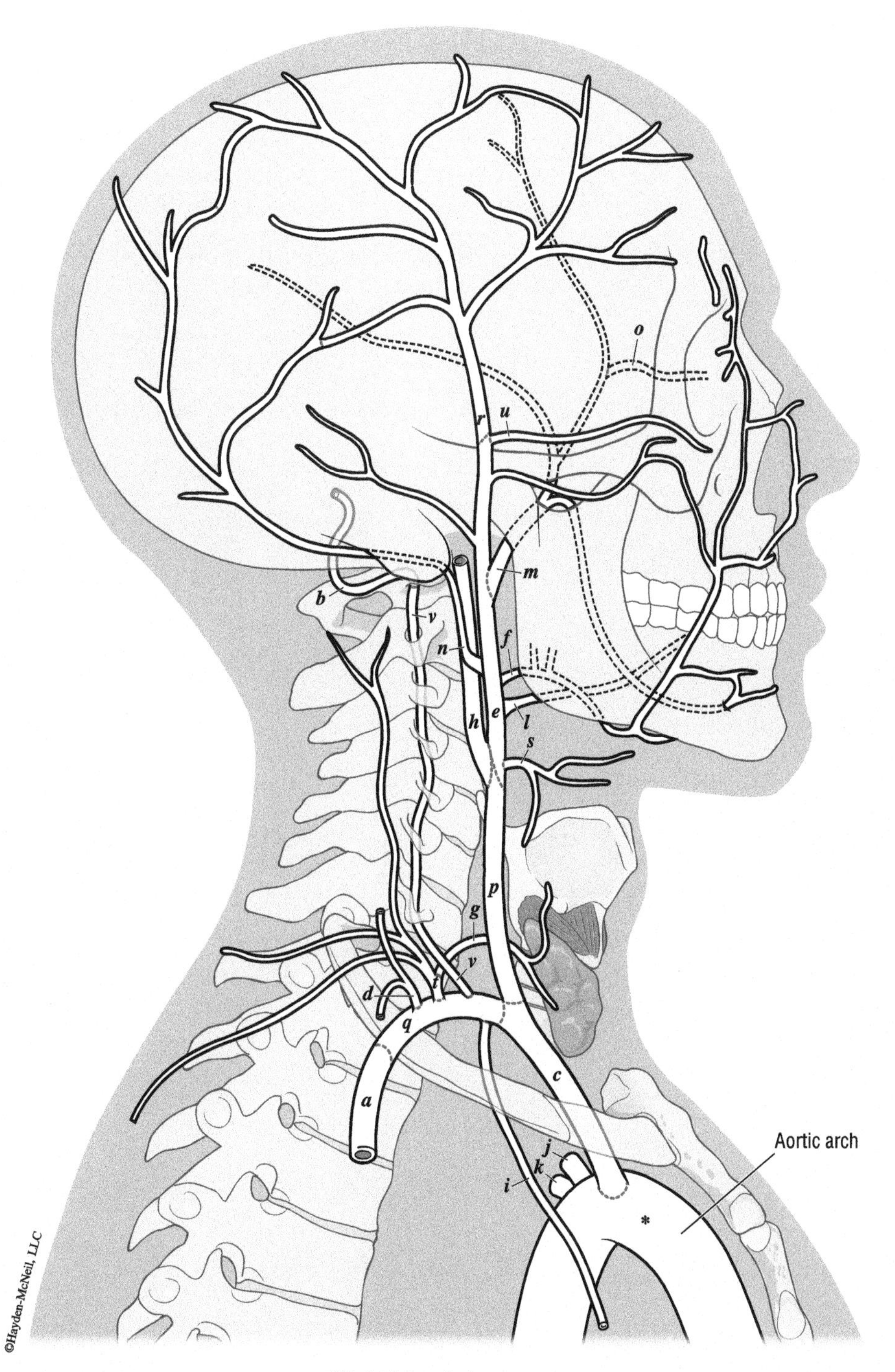

Right lateral view

Head and neck veins

- (a) Common facial v.
- (b) External jugular v.
- (c) Facial v.
- (d) Inferior alveolar v.
- (e) Internal jugular v.
- (f) Left brachiocephalic v.
- (g) Middle thyroid v.
- (h) Occipital v.
- (i) Ophthalmic v.
- (j) Posterior auricular v.
- (k) Retromandibular v.
- (l) Right brachiocephalic v.
- (m) Right subclavian v.
- (n) Superficial temporal v.
- (o) Superior thyroid v.
- (p) Vertebral v.

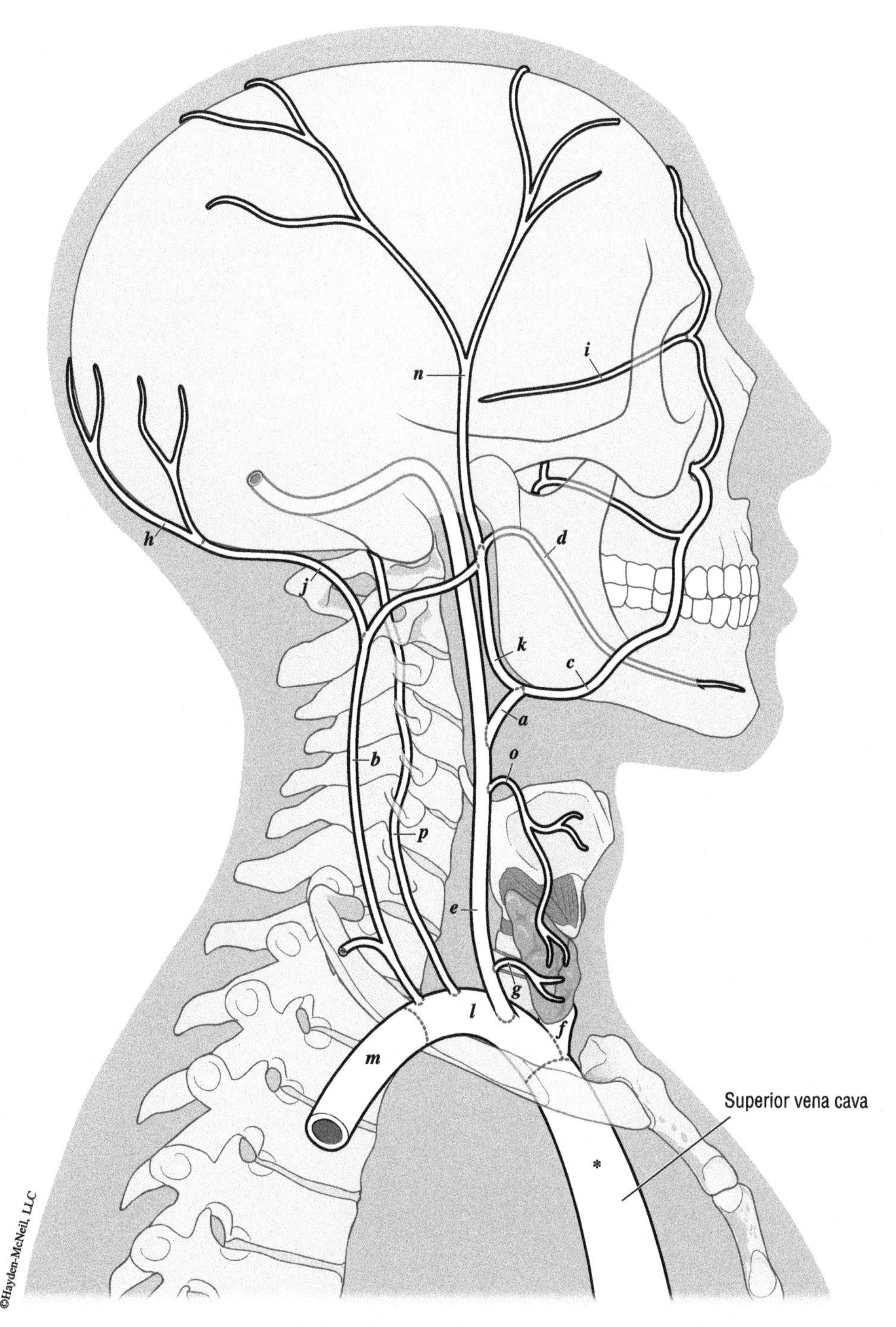

Right lateral view

Cranial dural sinuses

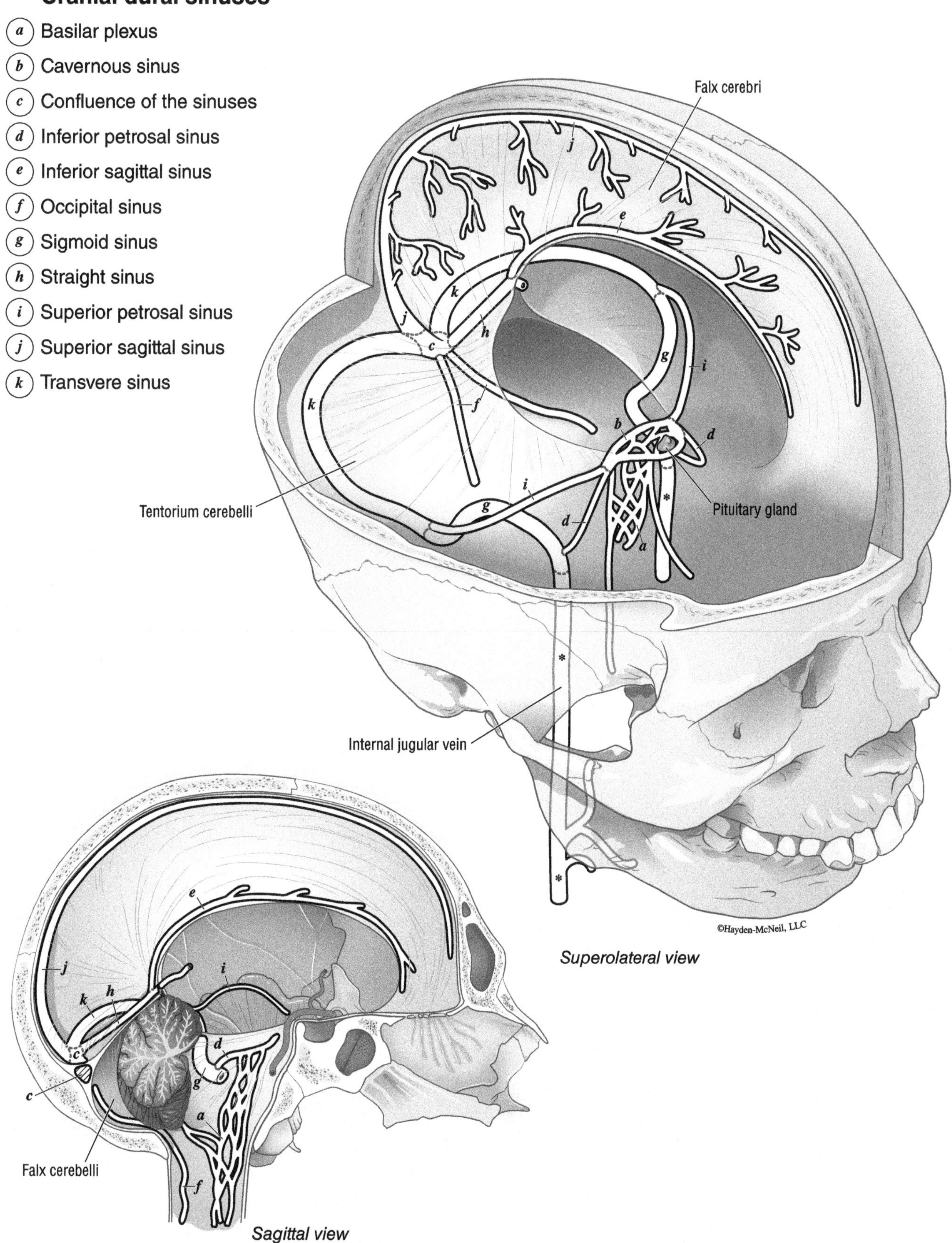
(a) Basilar plexus
(b) Cavernous sinus
(c) Confluence of the sinuses
(d) Inferior petrosal sinus
(e) Inferior sagittal sinus
(f) Occipital sinus
(g) Sigmoid sinus
(h) Straight sinus
(i) Superior petrosal sinus
(j) Superior sagittal sinus
(k) Transvere sinus
Falx cerebri
Tentorium cerebelli
Pituitary gland
Internal jugular vein
©Hayden-McNeil, LLC
Superolateral view
Falx cerebelli
Sagittal view

Arteries of the brain

- (a) Anterior inferior cerebellar a.
- (b) Basilar a.
- (c) Callosomarginal a.
- (d) Internal carotid a.
- (e) Middle cerebral a. (MCA)
  - (f) Inferior terminal branches
  - (g) Superior terminal branches
  - (h) Lateral frontobasal a.
- (i) Pericallosal a.
- (j) Posterior inferior cerebellar a.
- (k) Superior cerebellar a. (SCA)
- (l) Vertebral a.

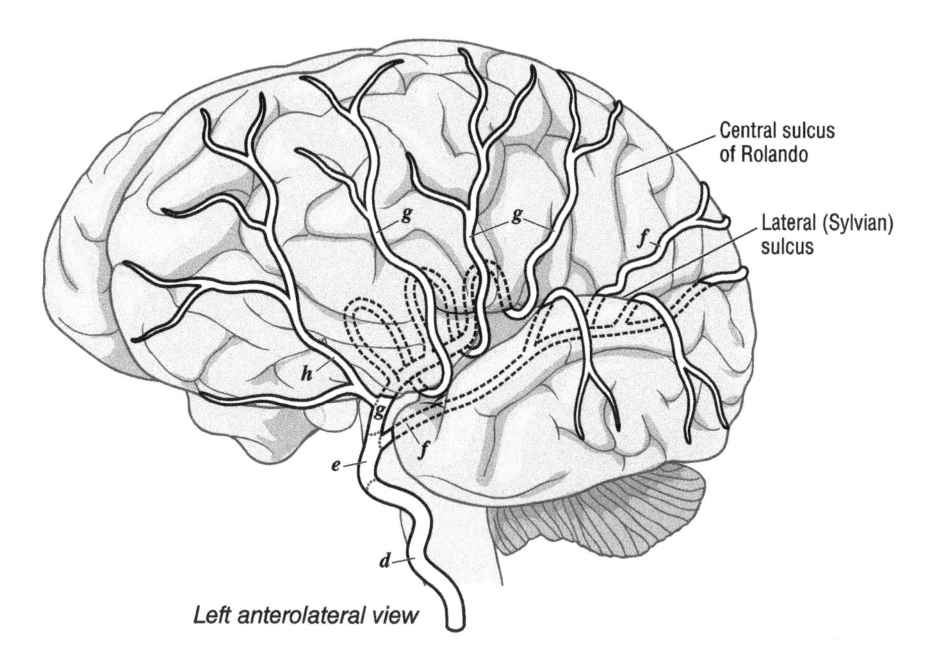

Left anterolateral view

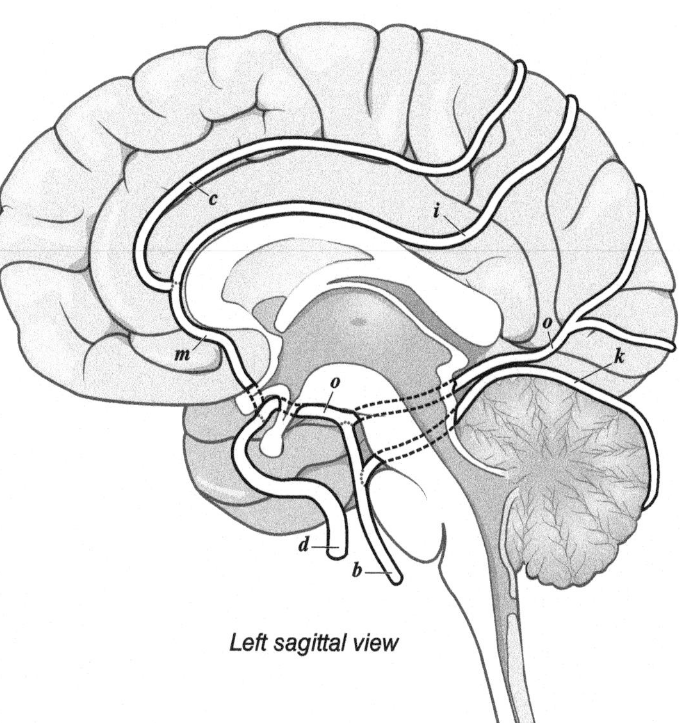

Left sagittal view

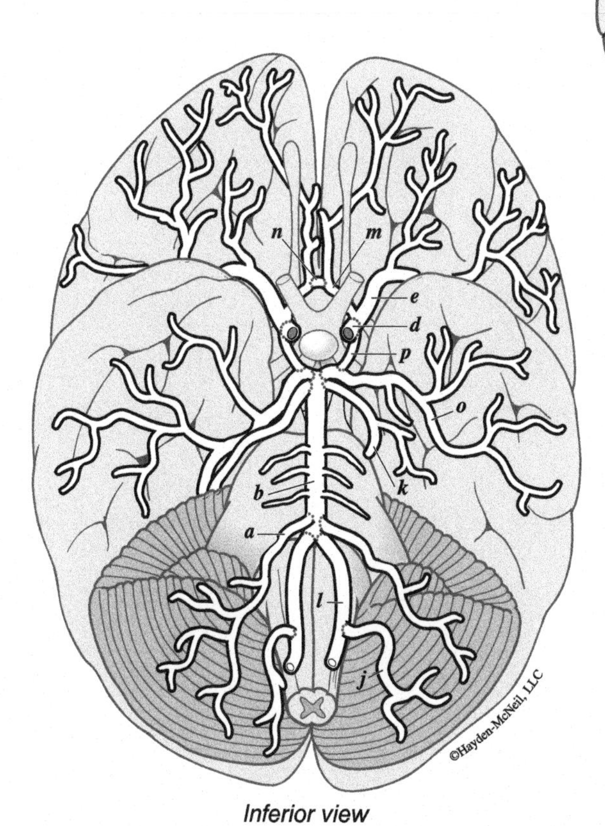

Inferior view

Circle of Willis

- (m) Anterior cerebral a. (ACA)
- (n) Anterior communicating a.
- (o) Posterior cerebral a. (PCA)
- (p) Posterior communicating a.

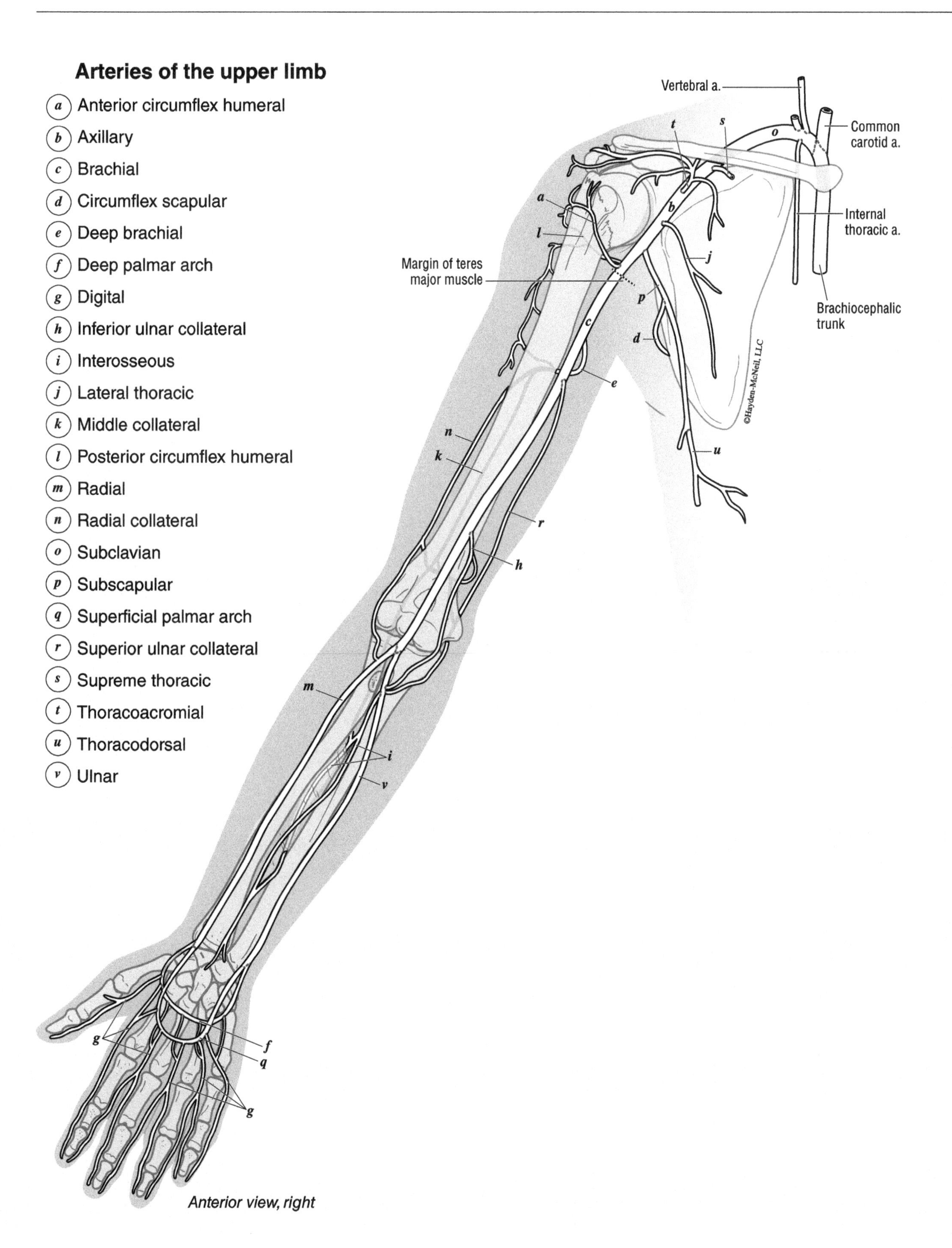
Arteries of the upper limb
a Anterior circumflex humeral
b Axillary
c Brachial
d Circumflex scapular
e Deep brachial
f Deep palmar arch
g Digital
h Inferior ulnar collateral
i Interosseous
j Lateral thoracic
k Middle collateral
l Posterior circumflex humeral
m Radial
n Radial collateral
o Subclavian
p Subscapular
q Superficial palmar arch
r Superior ulnar collateral
s Supreme thoracic
t Thoracoacromial
u Thoracodorsal
v Ulnar
Vertebral a.
Common carotid a.
Internal thoracic a.
Brachiocephalic trunk
Margin of teres major muscle
©Hayden-McNeil, LLC
Anterior view, right

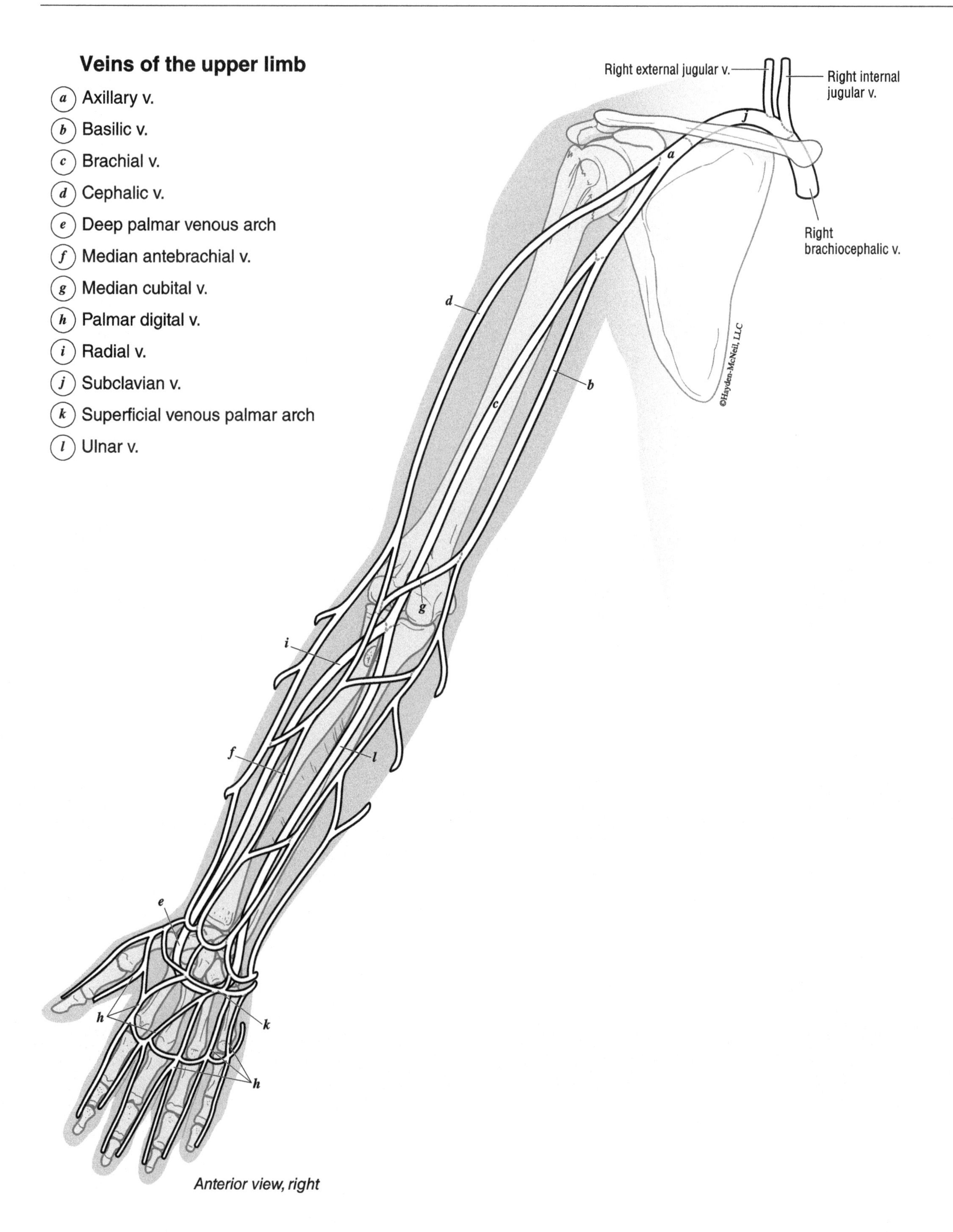

Anterior view, right

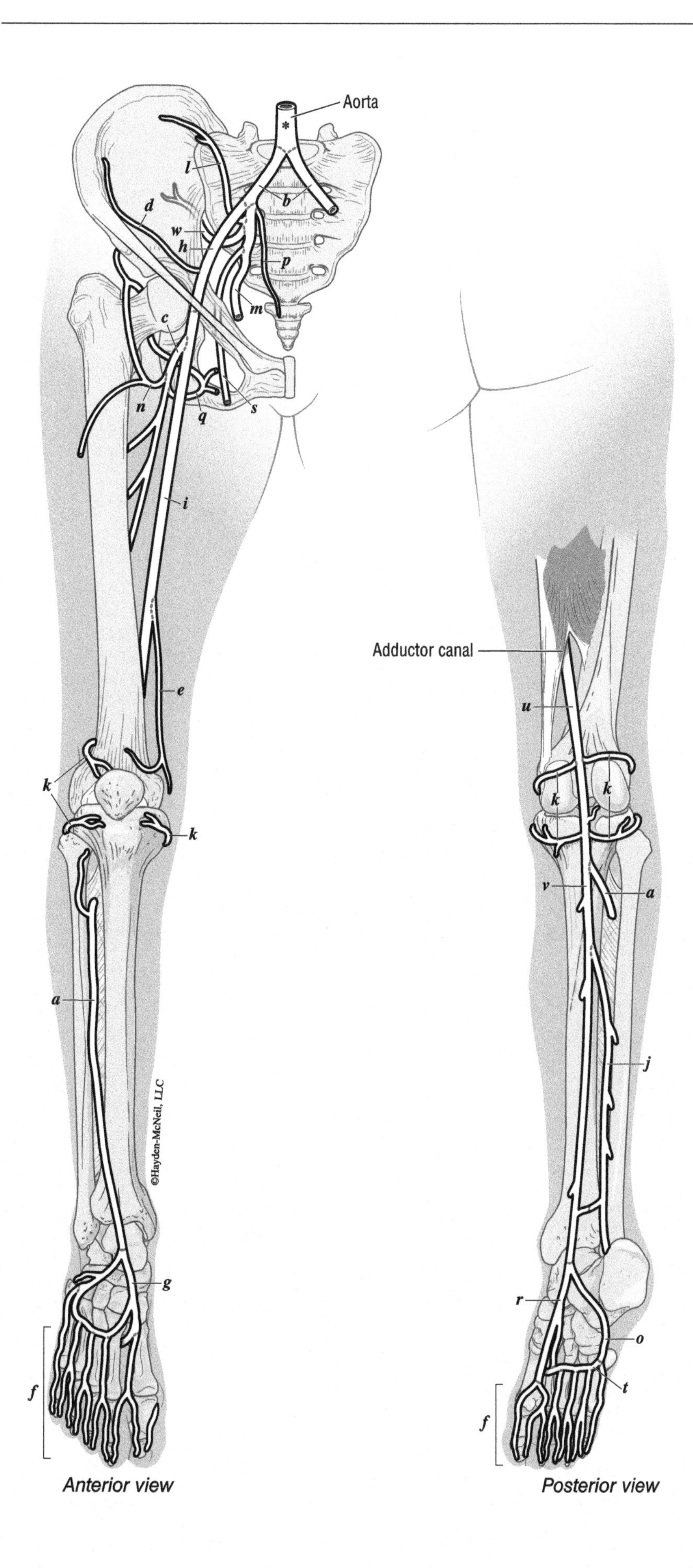

Anterior view *Posterior view*

Arteries of the lower limb

- (a) Anterior tibial a.
- (b) Common iliac a.
- (c) Deep a. of the thigh
- (d) Deep circumflex iliac a.
- (e) Descending genicular a.
- (f) Digital a.
- (g) Dorsalis pedis a.
- (h) External iliac a.
- (i) Femoral a.
- (j) Fibular a.
- (k) Genicular a.
- (l) Iliolumbar a.
- (m) Internal iliac a.
- (n) Lateral circumflex femoral a.
- (o) Lateral plantar a.
- (p) Lateral sacral a.
- (q) Medial circumflex femoral a.
- (r) Medial plantar a.
- (s) Obturator a.
- (t) Plantar arch
- (u) Popliteal a.
- (v) Posterior tibial a.
- (w) Superior gluteal a.

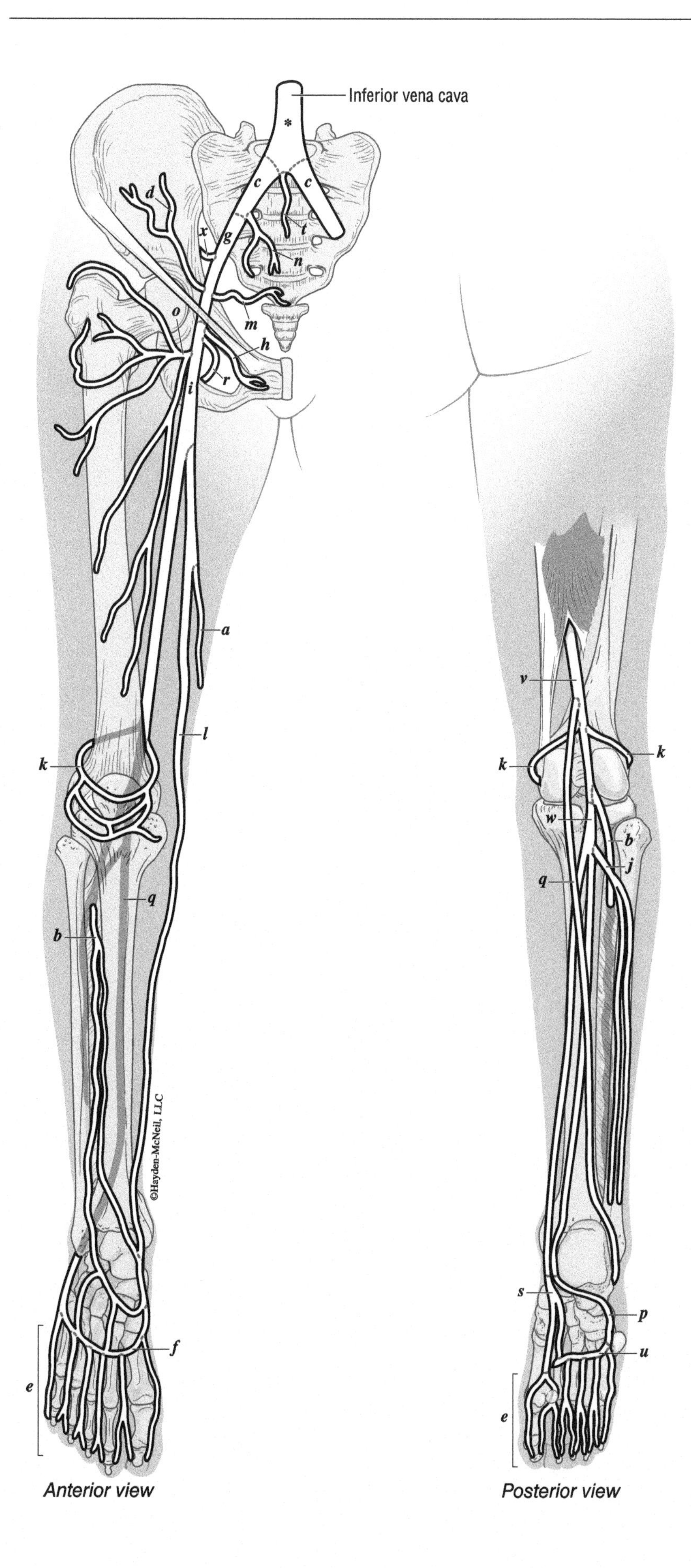

Anterior view

Posterior view

Veins of the lower limb

- (a) Accessory saphenous v.
- (b) Anterior tibial v.
- (c) Common iliac v.
- (d) Deep circumflex iliac v.
- (e) Digital v.
- (f) Dorsal venous arch
- (g) External iliac v.
- (h) External pudendal v.
- (i) Femoral v.
- (j) Fibular v.
- (k) Genicular v.
- (l) Great saphenous v.
- (m) Inferior epigastric v.
- (n) Internal iliac v.
- (o) Lateral circumflex femoral v.
- (p) Lateral plantar v.
- (q) Lesser saphenous v.
- (r) Medial circumflex femoral v.
- (s) Medial plantar v.
- (t) Median sacral v.
- (u) Plantar venous arch
- (v) Popliteal v.
- (w) Posterior tibial v.
- (x) Superior gluteal v.

Aorta and its branches

- (a) Aorta
  - a_1 Ascending aorta
  - a_2 Aortic arch
  - a_3 Descending aorta
- (b) Brachiocephalic trunk
- (c) Right subclavian a.
- (d) Right common carotid a.
- (e) Left common carotid a.
- (f) Left subclavian a.
- (g) Coronary a.
  - g_1 Right coronary a.
  - g_2 Left coronary a.
- (h) Celiac trunk
- (i) Left gastric a.
- (j) Common hepatic a.
- (k) Right gastric a.
- (l) Splenic a.
- (m) Gastroduodenal a.
- (n) Gastroomental a.
- (o) Superior mesenteric a.
- (p) Renal a.
- (q) Inferior mesenteric a.
- (r) Gonadal a.
- (s) Common iliac a.
- (t) External iliac a.
- (u) Internal iliac a.

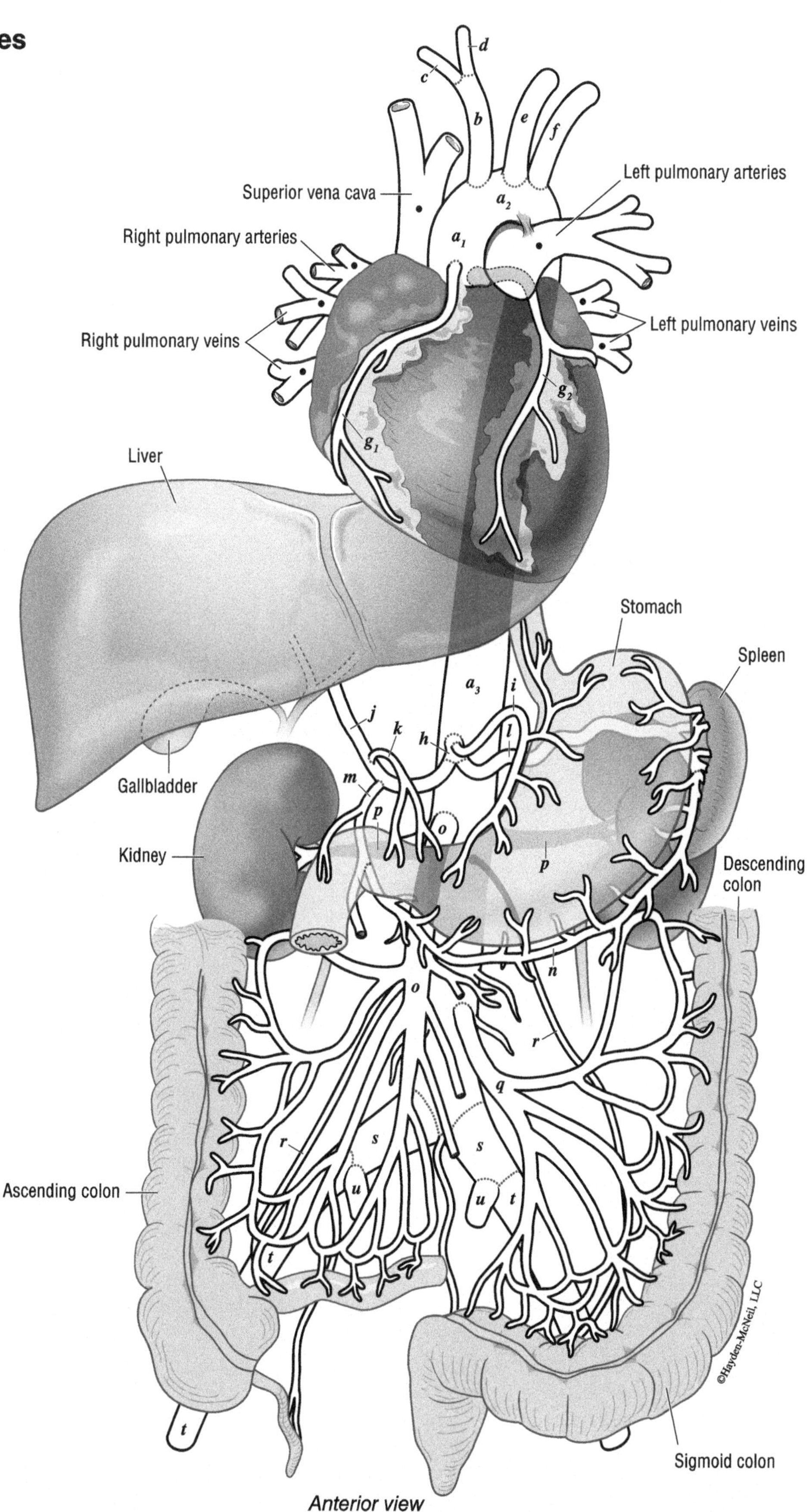

Anterior view

Hepatic portal system

Color the right and left common iliac veins (•) light gray.

- (a) Inferior vena cava
- (b) Hepatic v.
- (c) Hepatic portal v.
- (d) Cystic v.
- (e) Esophageal v.
- (f) Gastric v.
- (g) Splenic v.
- (h) Inferior mesenteric v.
- (i) Superior mesenteric v.
- (j) Gastroomental v.
- (k) Pancreaticoduodenal v.
- (l) Ileocolic v.
- (m) Right colic v.
- (n) Middle colic v.
- (o) Jejunal and ileal v.
- (p) Left colic v.
- (q) Sigmoid v.
- (r) Superior rectal v.

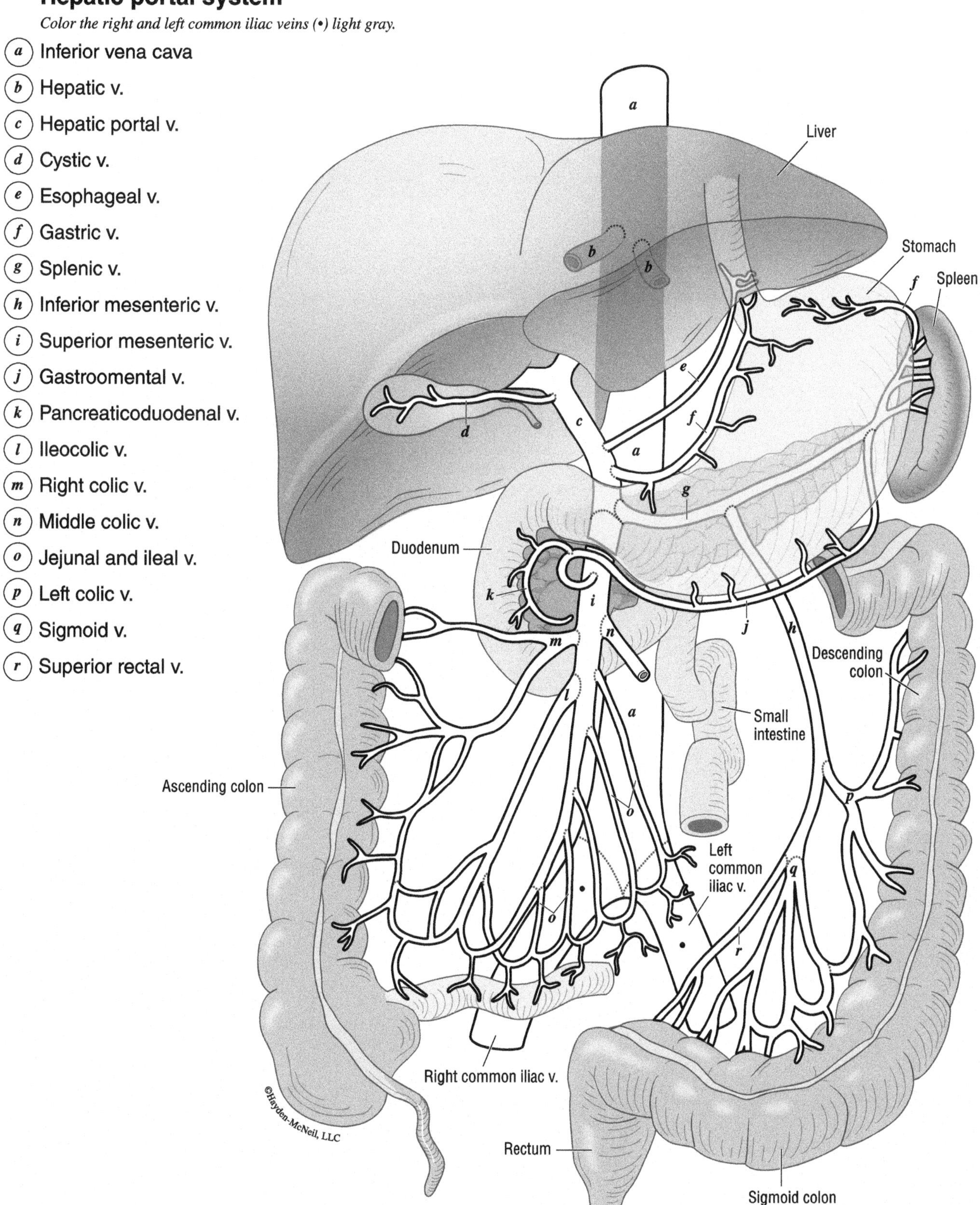

Pelvic arteries

- (a) Common iliac a.
- (b) Internal iliac a.
- (c) External iliac a.
- (d) Iliolumbar a.
- (e) Lateral sacral a.
- (f) Superior gluteal a.
- (g) Inferior gluteal a.
- (h) Deep circumflex iliac a.
- (i) Inferior epigastric a.
- (j) Inferior vesical a.
  - j_1 Prostatic branch
- (k) Middle rectal a.
- (l) Obturator a.
- (m) Umbilical a.
  - m_1 Patent part
  - m_2 Occluded part
- (n) Superior vesical a.
- (o) Uterine a.
- (p) Vaginal a.

Iliacus m.
Sacrum
Inguinal ligament
Infrapiriform foramen
Urinary bladder
Prostate
Rectum

Left anterolateral view, male

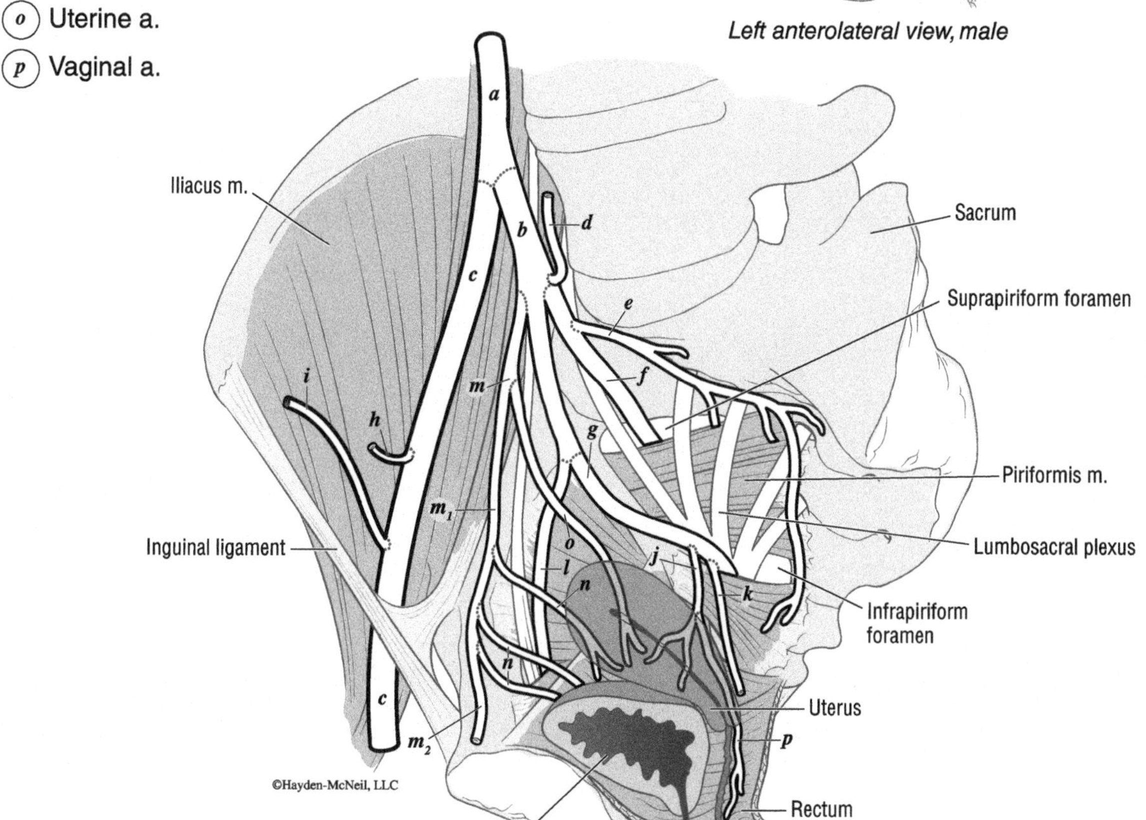

Left anterolateral view, female

12

The Lymphatic System

Lymphatic system
Lymph node and vessel
Lymphocyte development

Prepare yourself, adventurer, to embark on a noble quest through the kingdom of our body's natural defense system: the **lymphatic system**! In this chapter, you will be exploring the majestic realm of the lymphatic system, where lymphocytes act as valiant knights, the bastions of the body, guarding our body against harmful pathogens and marshaling the very best immune responses. With every step, you will be venturing deeper into the dragon's lair and getting to know the intrepid heroes of our body's innermost stronghold.

Lymphatic system

- (a) Tonsils
- (b) Thymus gland
- (c) Spleen
- (d) Peyer's patches
- (e) Lymphoid nodules of intestine
- (f) Bone marrow
- (g) Lymph nodes
  - g_1 Cervical nodes
  - g_2 Axillary nodes
  - g_3 Iliac nodes
  - g_4 Inguinal nodes
- (h) Thoracic duct
- (i) Right lymphatic duct
- (j) Cisterna chyli
- (k) Lymphatic vessels
  - k_1 Jugular trunks
  - k_2 Subclavian trunks
  - k_3 Bronchomediastinal trunks
  - k_4 Descending thoracic trunks
  - k_5 Intestinal trunks
  - k_6 Lumbar trunks
  - k_7 Vessels from upper limb
  - k_8 Vessels from pelvis
  - k_9 Vessels from lower limb

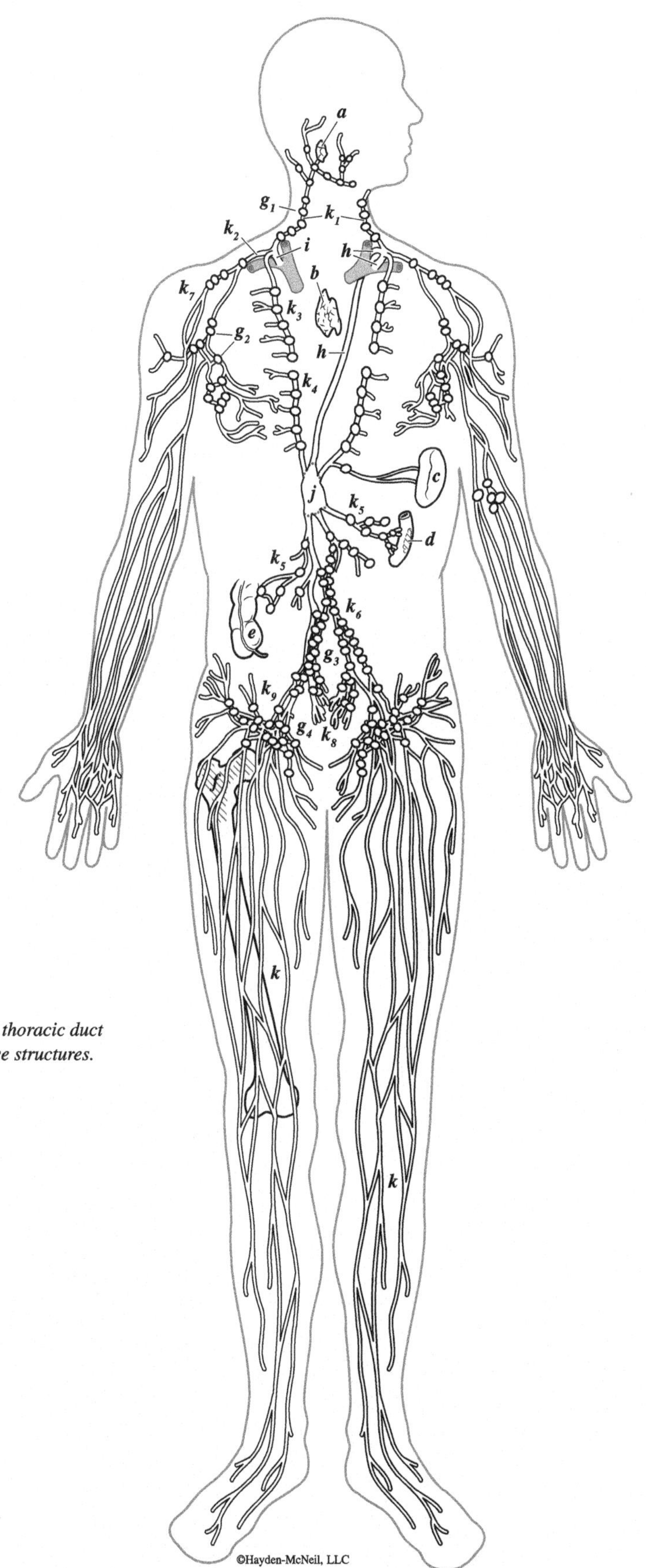

Drainage areas

Use the same colors for the right lymphatic duct and thoracic duct above to distinguish the regions drained through these structures.

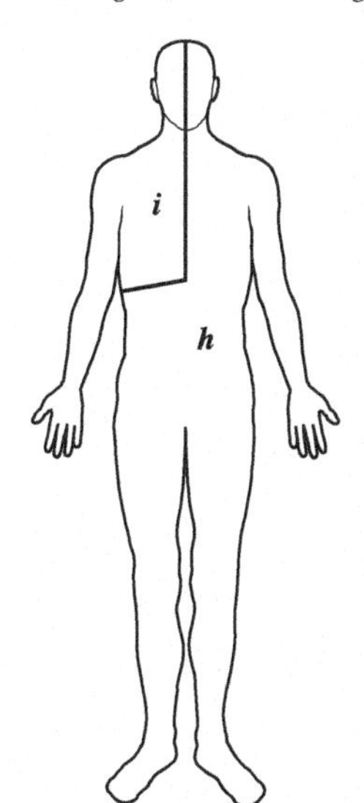

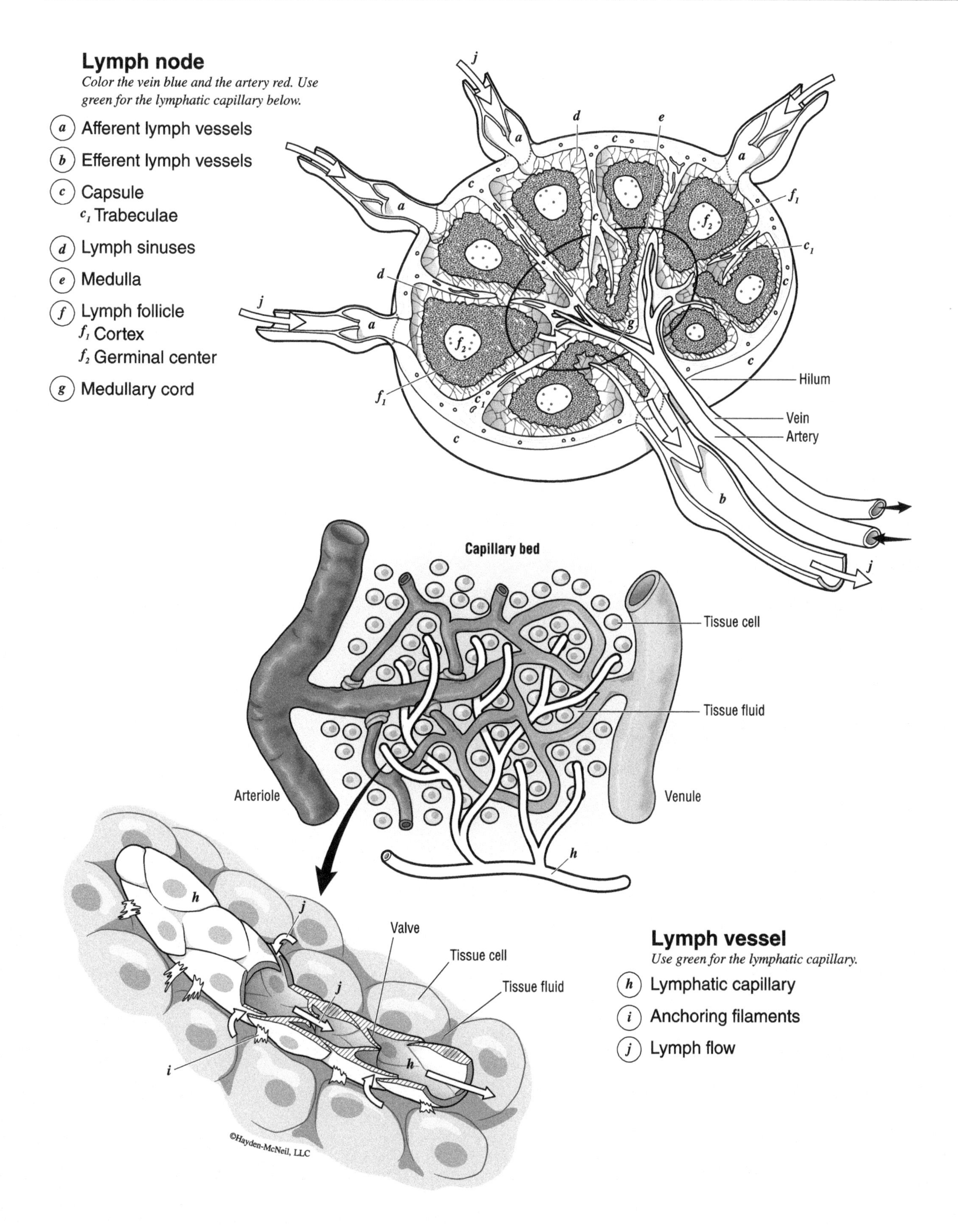

Lymph node
Color the vein blue and the artery red. Use green for the lymphatic capillary below.
a Afferent lymph vessels
b Efferent lymph vessels
c Capsule
c1 Trabeculae
d Lymph sinuses
e Medulla
f Lymph follicle
f1 Cortex
f2 Germinal center
g Medullary cord
Hilum
Vein
Artery
Capillary bed
Tissue cell
Tissue fluid
Arteriole
Venule
Valve
Tissue cell
Tissue fluid
Lymph vessel
Use green for the lymphatic capillary.
h Lymphatic capillary
i Anchoring filaments
j Lymph flow
©Hayden-McNeil, LLC

Lymphocyte development

(a) Thymus gland
(b) Red bone marrow
(c) Lymphoid stem cell
(d) Helper T lymphocyte
(e) Cytotoxic T lymphocyte
(f) B lymphocyte (B cell)
(g) Natural killer cell (NK cell)
(h) Mature T_H cell
(i) Mature T_C cell
(j) Mature B cell
(k) Mature NK cell

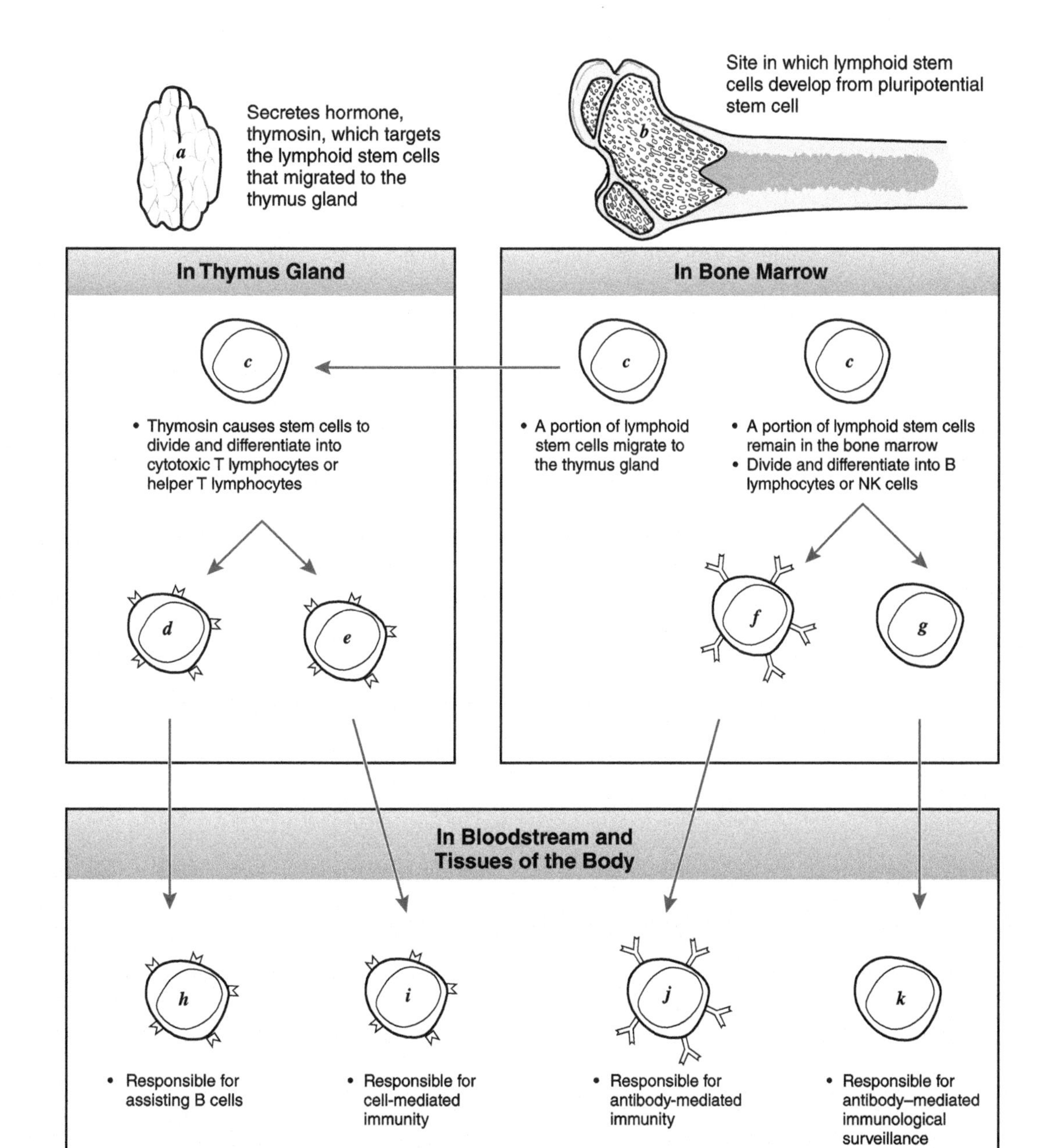

13

The Respiratory System

Respiratory system

External nose

Nasal cavity

Paranasal air sinuses

Pharynx and larynx

Bronchial tree

Lobes and pleurae of the lungs

Alveoli exchange

Respiratory muscles

This chapter of our voyage will introduce you to the breathtaking world of the **respiratory system**. You will be at the helm as our ship's captain, navigating through the mysterious Nasal Passage and the Straits of the Trachea, alluring channels guiding you to the lungs, with twin sails capturing every breath. As you color, you'll map out the journey of oxygen, imagining it as a fleet of invisible ships docking at the alveoli harbors. Get ready for a colorful voyage through the ebbs and flows of our respiratory system, where each wave of color breathes life into every page.

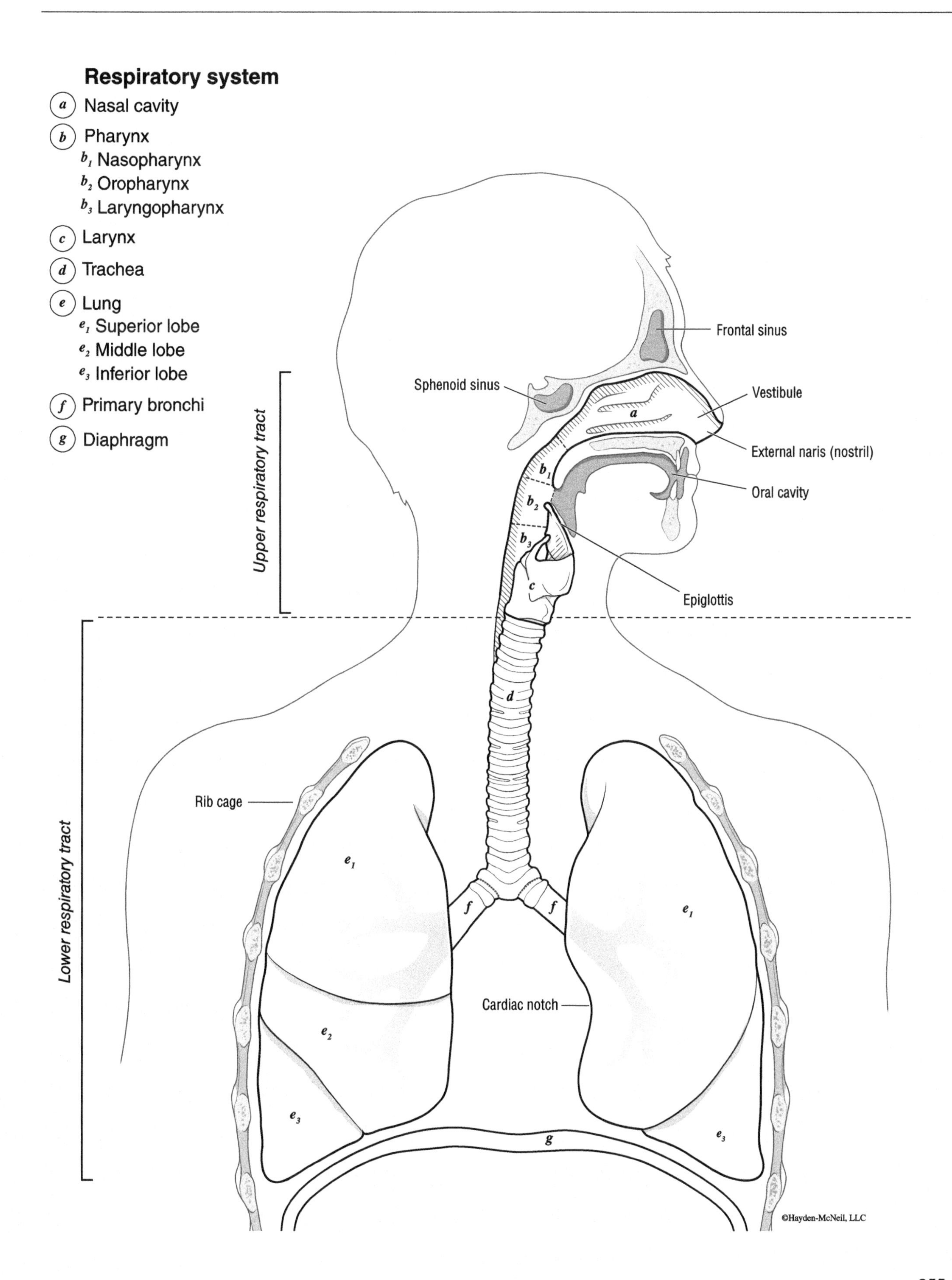

Respiratory system
a Nasal cavity
b Pharynx
b1 Nasopharynx
b2 Oropharynx
b3 Laryngopharynx
c Larynx
d Trachea
e Lung
e1 Superior lobe
e2 Middle lobe
e3 Inferior lobe
f Primary bronchi
g Diaphragm
Upper respiratory tract
Lower respiratory tract
Frontal sinus
Sphenoid sinus
Vestibule
External naris (nostril)
Oral cavity
Epiglottis
Rib cage
Cardiac notch
©Hayden-McNeil, LLC

External nose

- (a) Nasal bones
- (b) Lateral nasal cartilages
- (c) Septal cartilages
- (d) Alar cartilages

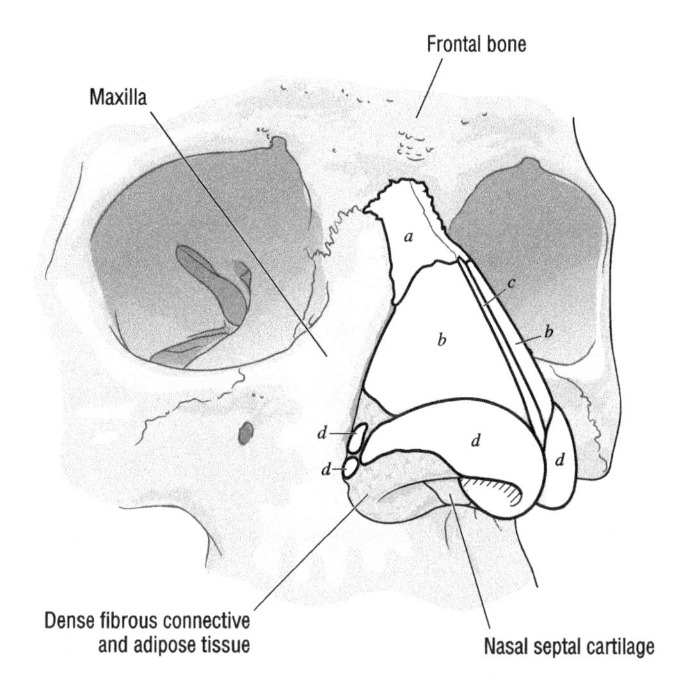

Nasal cavity

- (e) Nasal cavity
- (f) Frontal bone
- (g) Sphenoid bone
- (h) Cribriform plate of ethmoid
- (i) Superior nasal concha
  - i_1 Superior nasal meatus
- (j) Medial nasal concha
  - j_1 Medial nasal meatus
- (k) Inferior nasal concha
  - k_1 Inferior nasal meatus
- (l) Vestibule of nose
- (m) Hard palate
- (n) Soft palate

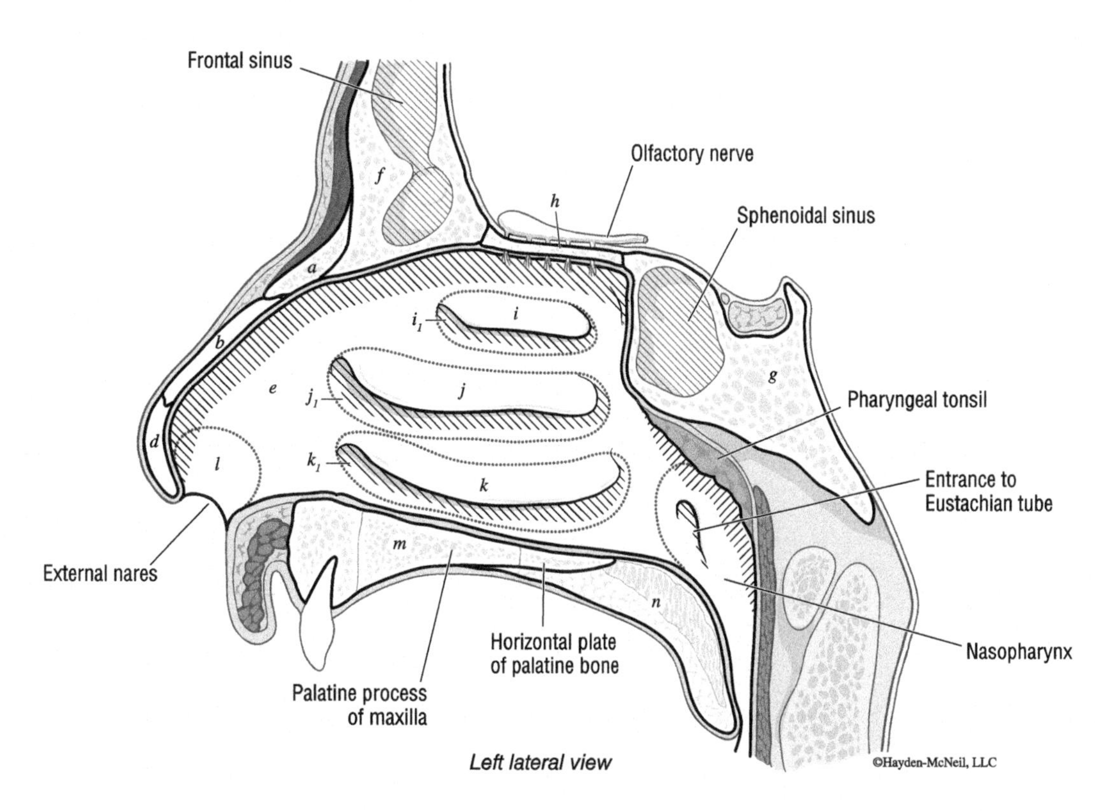

Left lateral view

Paranasal air sinuses

- (a) Frontal sinus
- (b) Ethmoid sinus
- (c) Sphenoid sinus
- (d) Maxillary sinus

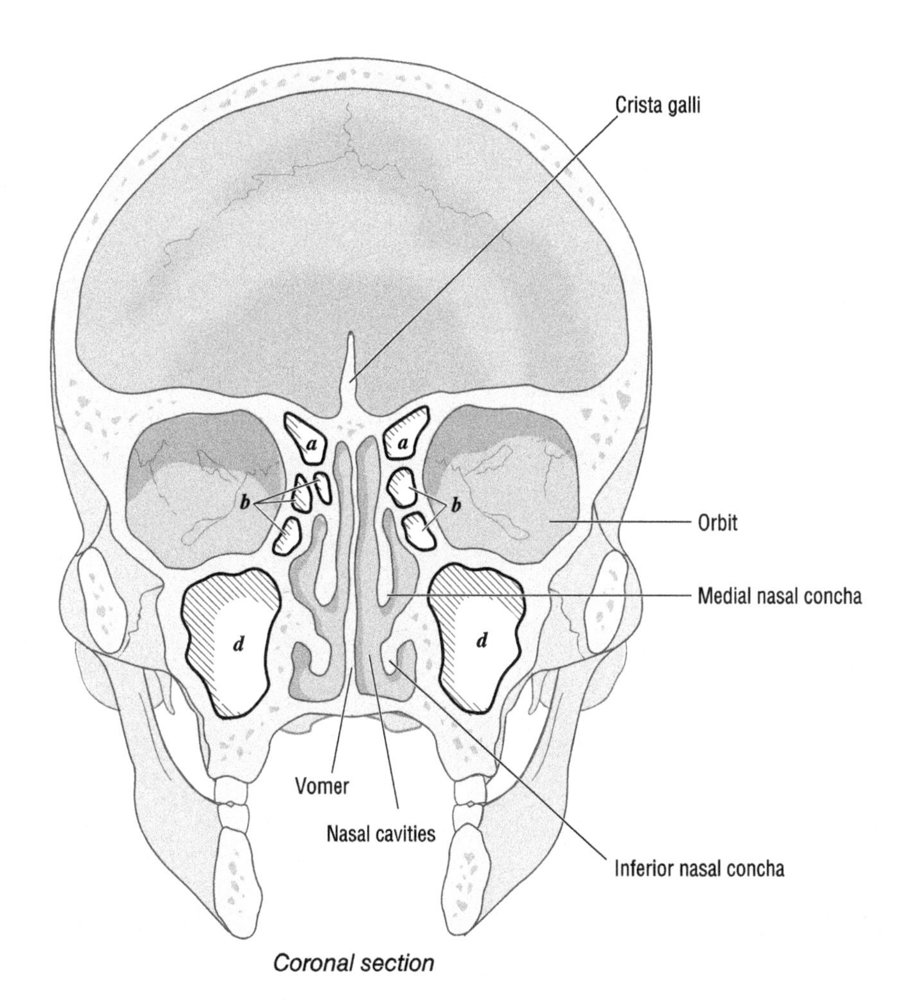

Coronal section

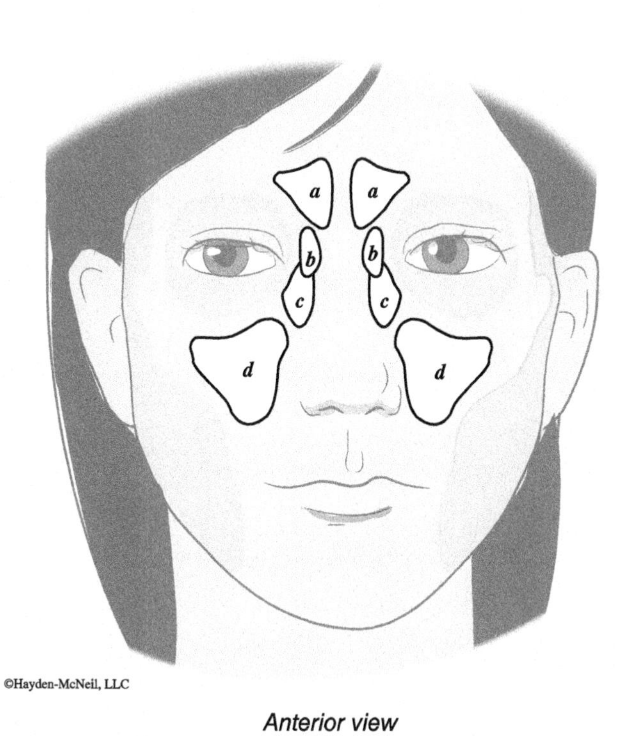

Anterior view

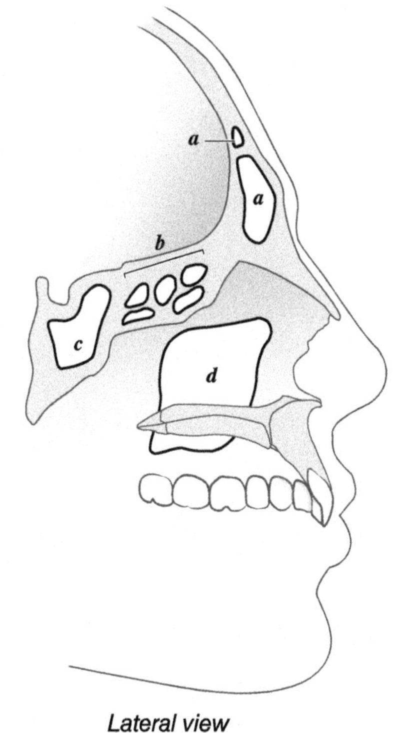

Lateral view

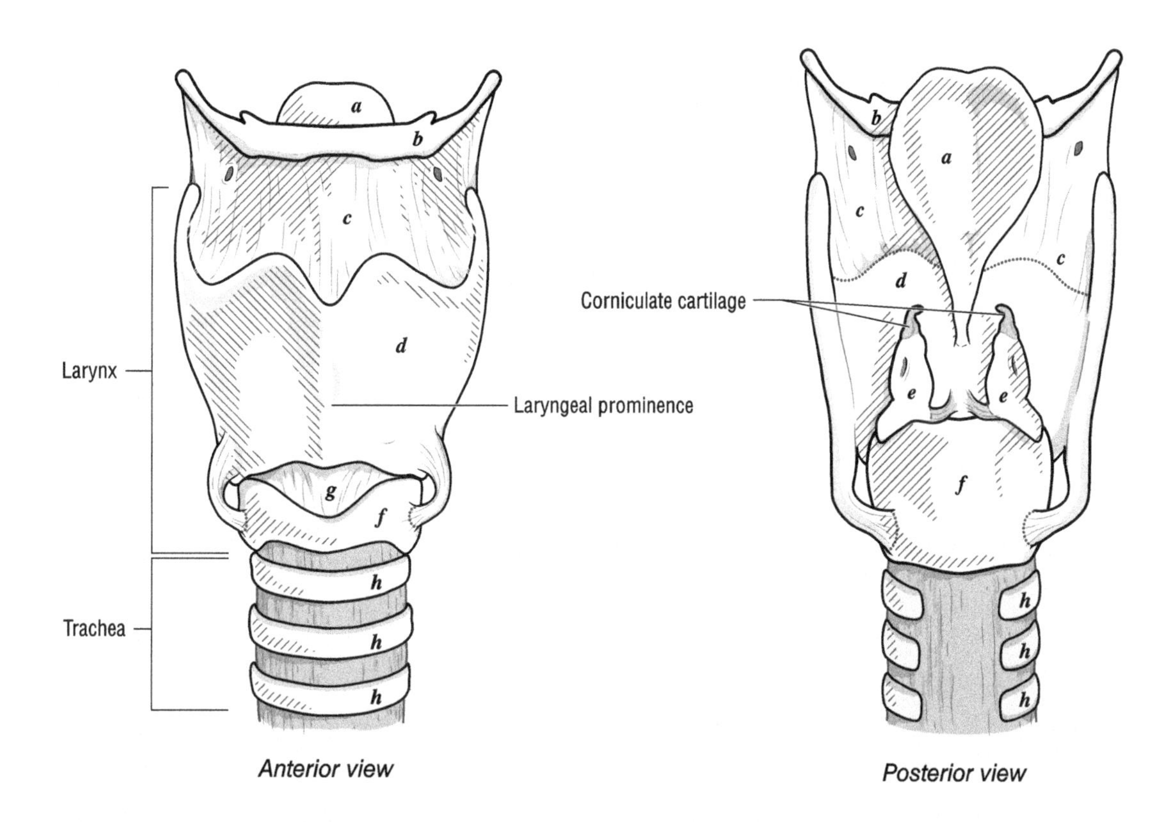

Anterior view

Posterior view

Pharynx and larynx

- (a) Epiglottis
- (b) Hyoid bone
- (c) Thyrohyoid membrane
- (d) Thyroid cartilage
- (e) Arytenoid cartilage
- (f) Cricoid cartilage
- (g) Median cricothyroid ligament
- (h) Tracheal cartilage
- (i) Laryngopharynx
- (j) Esophagus
- (k) Vestibular fold
- (l) Vocal fold

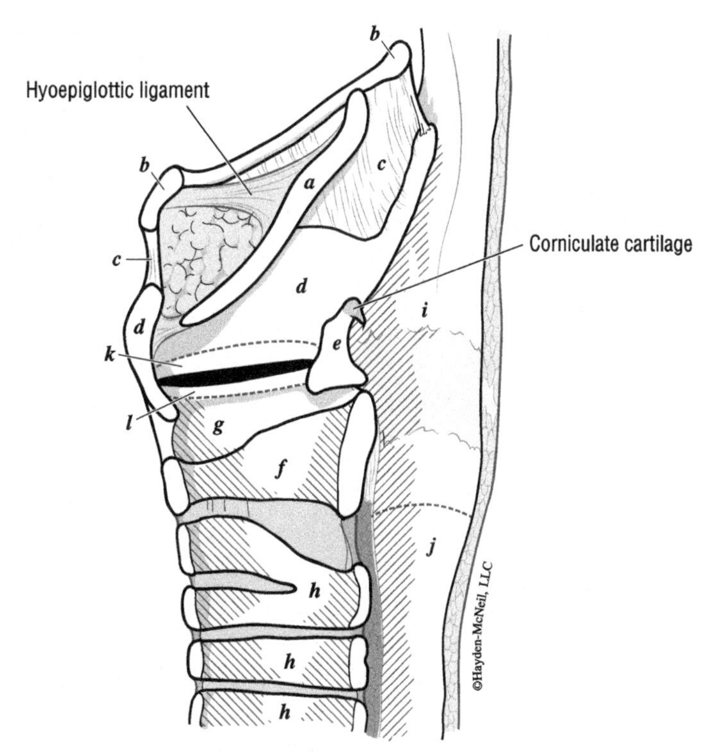

Sagittal section view

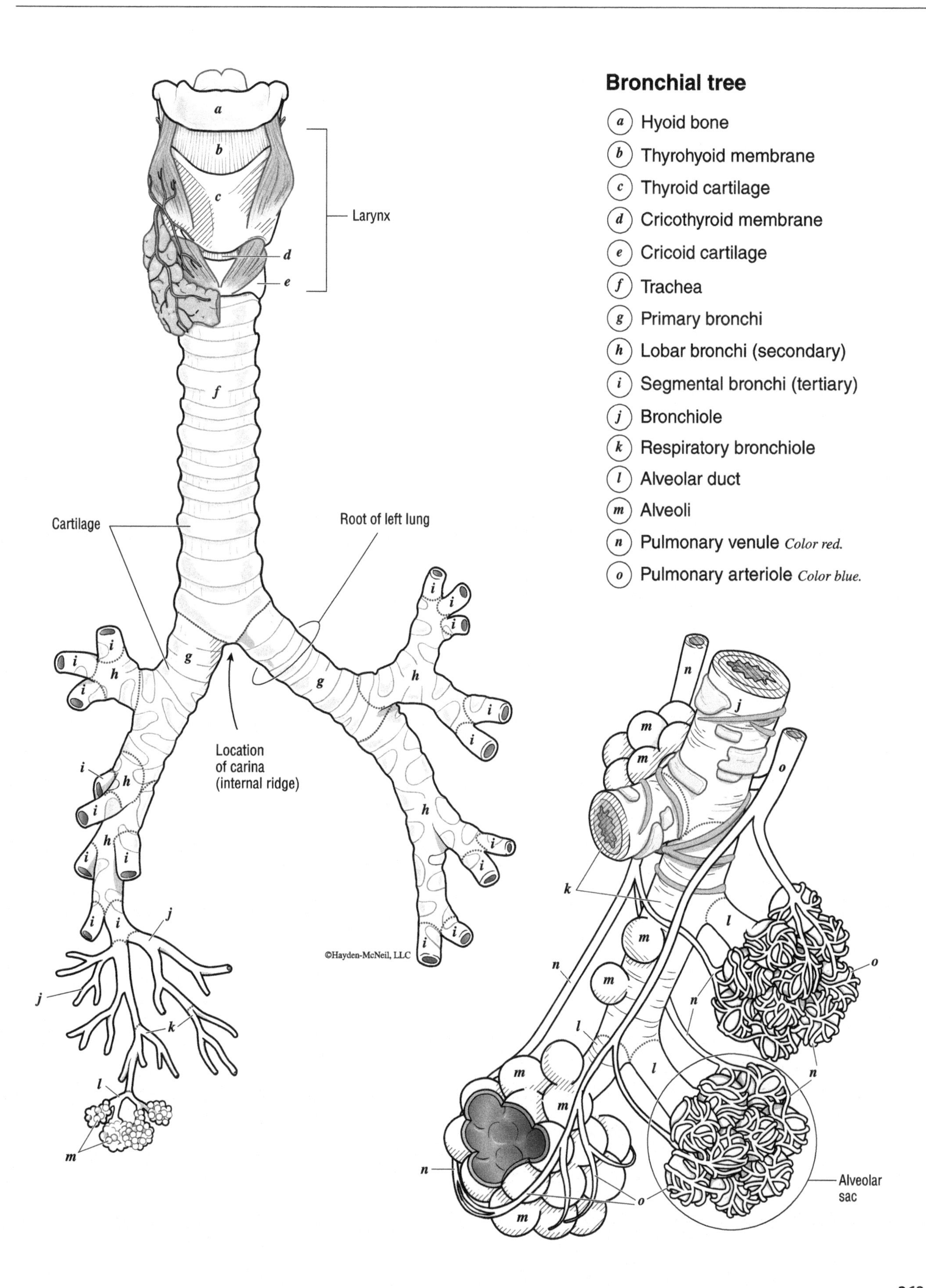

Bronchial tree
a Hyoid bone
b Thyrohyoid membrane
c Thyroid cartilage
d Cricothyroid membrane
e Cricoid cartilage
f Trachea
g Primary bronchi
h Lobar bronchi (secondary)
i Segmental bronchi (tertiary)
j Bronchiole
k Respiratory bronchiole
l Alveolar duct
m Alveoli
n Pulmonary venule Color red.
o Pulmonary arteriole Color blue.
Larynx
Cartilage
Root of left lung
Location of carina (internal ridge)
Alveolar sac
©Hayden-McNeil, LLC

Lobes and pleurae of the lungs

Be careful not to color the pleural cavity (*) between the parietal and visceral pleurae.

- (a) Larynx
- (b) Trachea
- (c) Right superior lobe
- (d) Right middle lobe
- (e) Right inferior lobe
- (f) Left superior lobe
- (g) Left inferior lobe
- (h) Parietal pleura
- (i) Visceral pleura
- (j) Primary bronchi

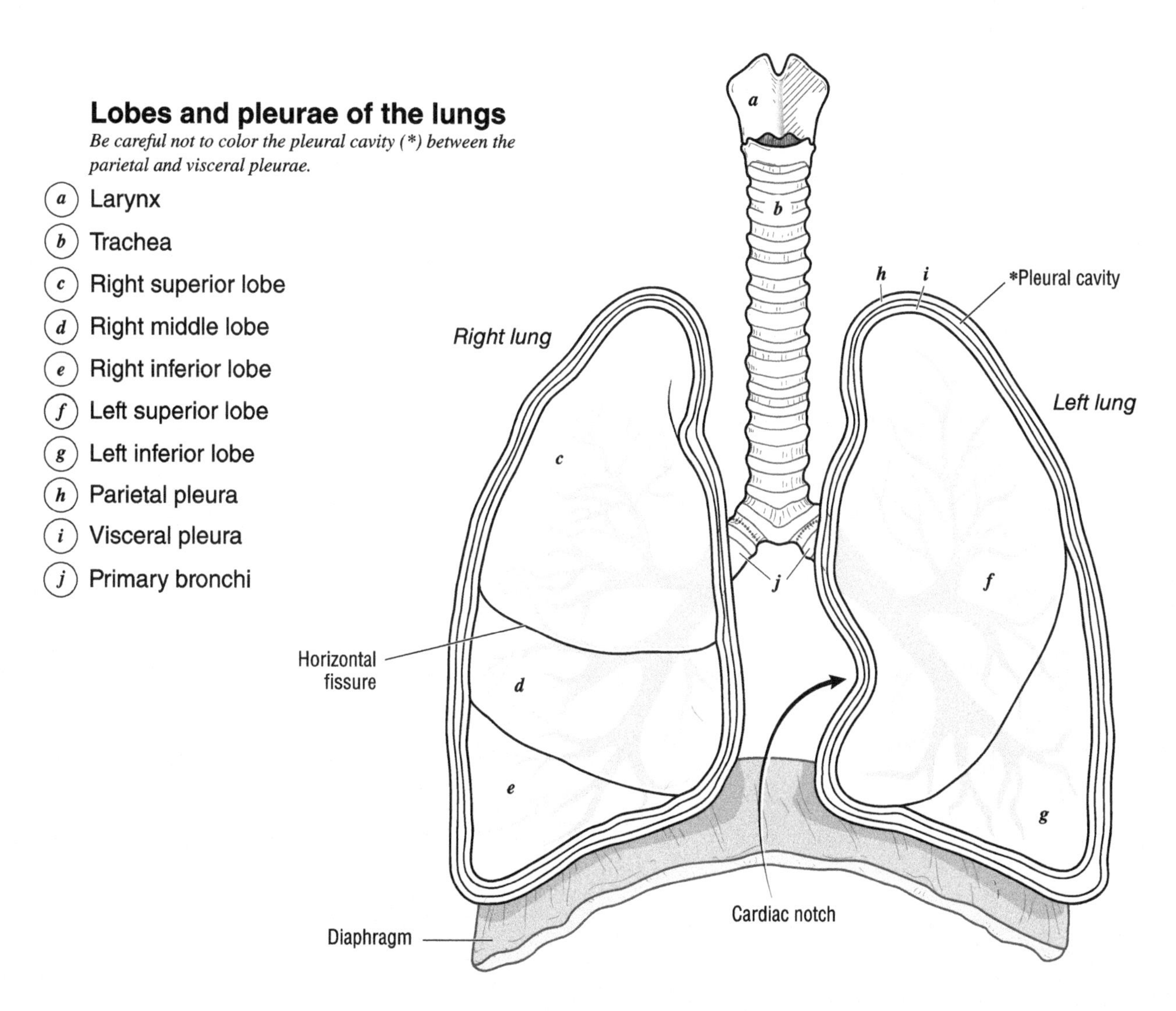

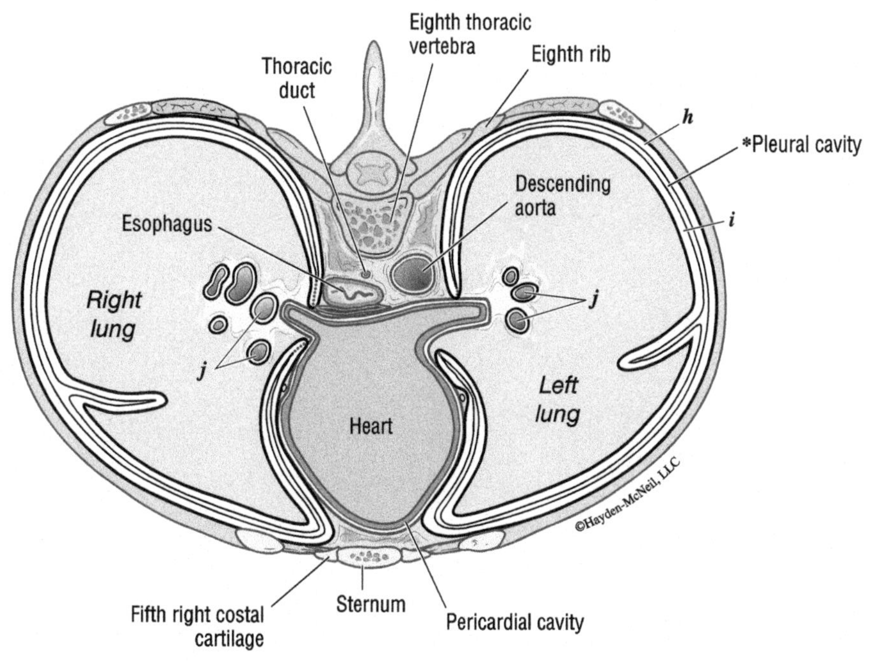

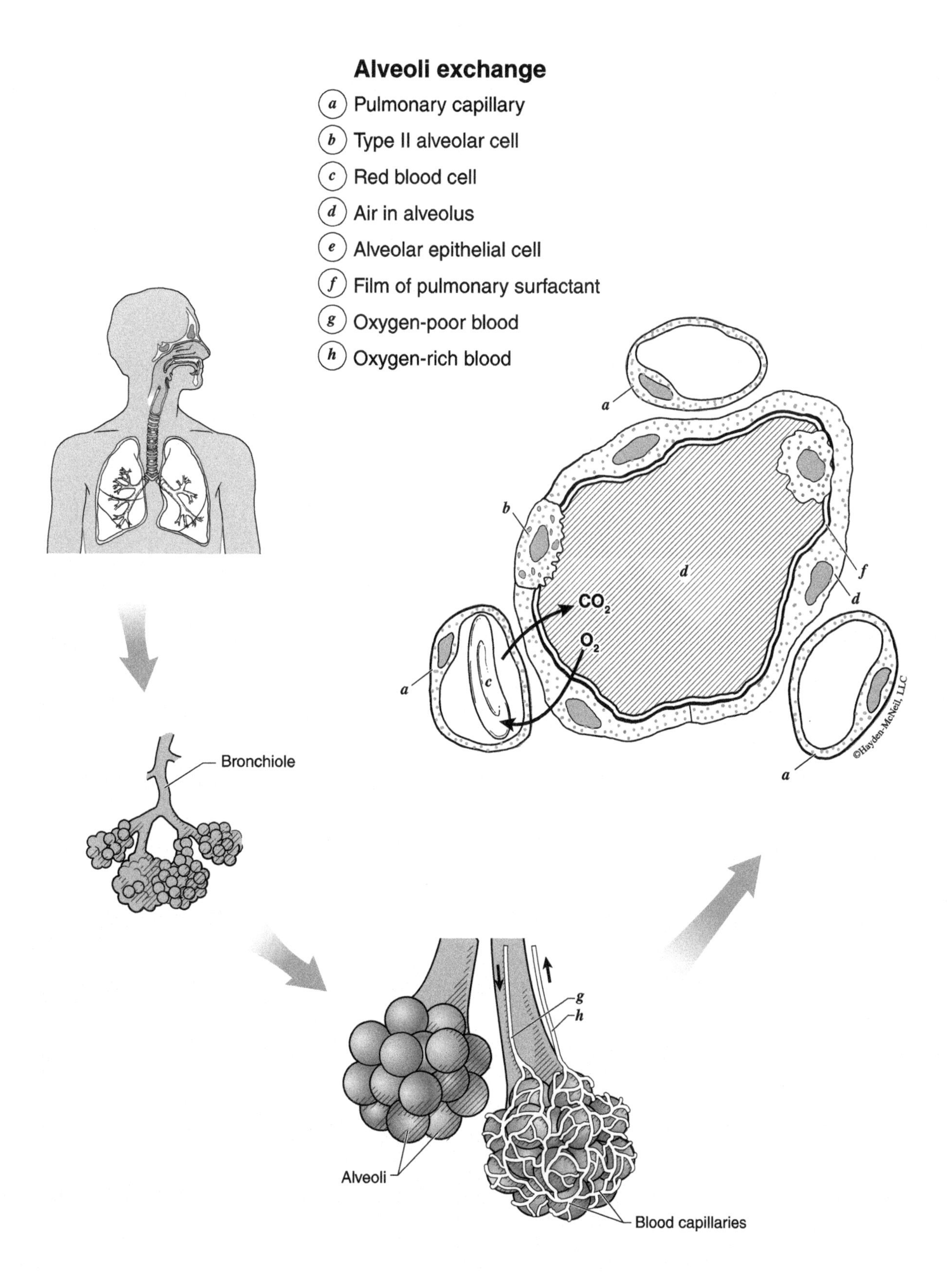
Alveoli exchange
a Pulmonary capillary
b Type II alveolar cell
c Red blood cell
d Air in alveolus
e Alveolar epithelial cell
f Film of pulmonary surfactant
g Oxygen-poor blood
h Oxygen-rich blood
a
b
d
f
d
CO_2
O_2
c
a
a
©Hayden-McNeil, LLC
Bronchiole
g
h
Alveoli
Blood capillaries

Respiratory muscles

Color muscles marked with (•) light gray. The internal intercostals and rectus abdominis are shown but not used in inhalation. The diaphragm, sternocleidomastoid, and scalenes are shown but not used for exhalation.

(a) Diaphragm
(b) Sternocleidomastoid m.
(c) Scalene muscles
(d) External intercostals
(e) Internal intercostals
(f) External and internal obliques
(g) Rectus abdominis

Inhalation

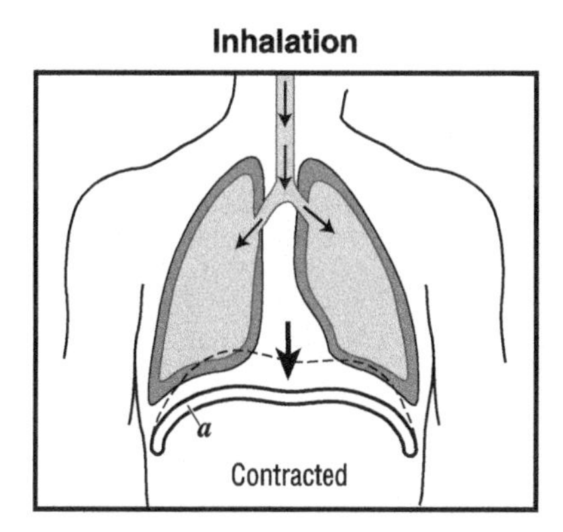

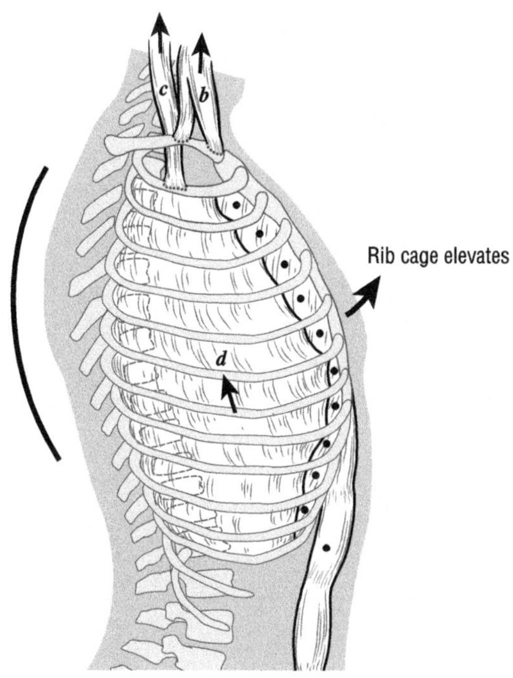

Exhalation

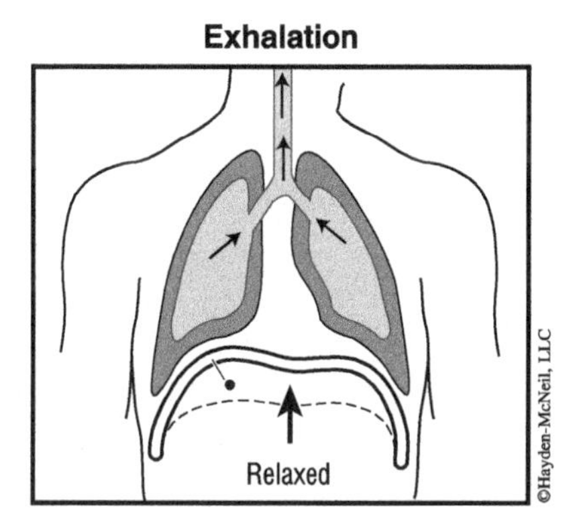

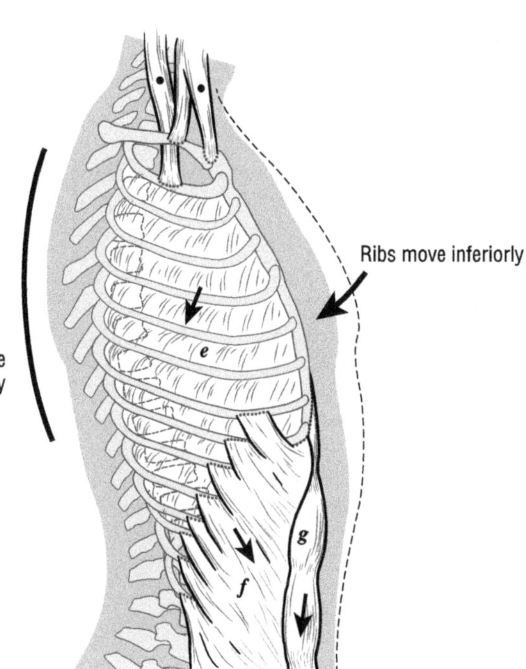

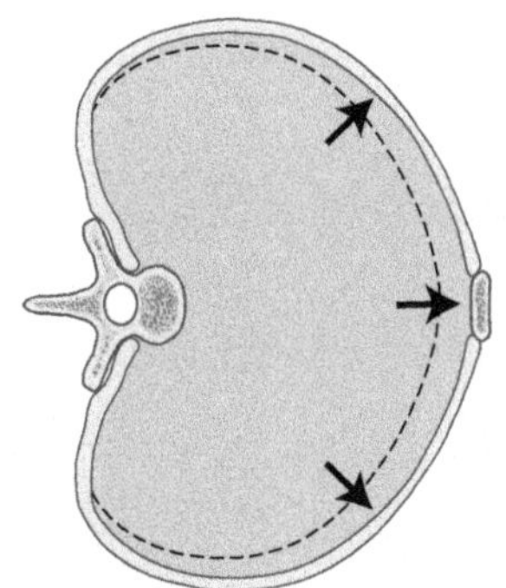

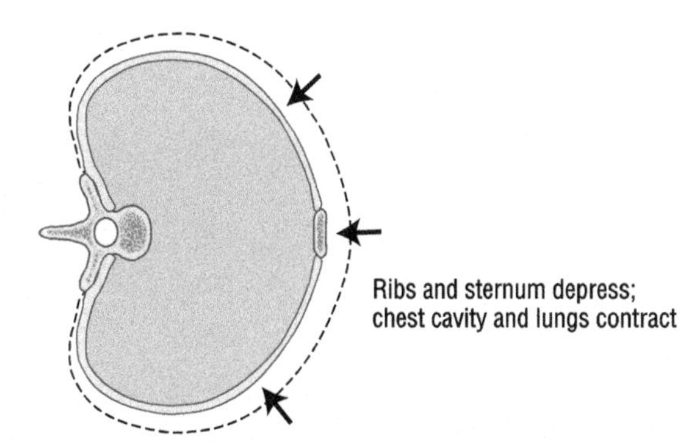

Ribs and sternum depress; chest cavity and lungs contract

14

The Digestive System

- *Digestive system overview*
- *Mouth*
- *Tooth anatomy*
- *Teeth*
- *Swallowing*
- *Mesentery and peritoneal spaces*
- *Stomach*
- *Small intestine*
- *Large intestine*
- *Layers of intestinal wall*
- *Rectum and anus*
- *Liver*
- *Accessory digestive organs*
- *Digestion*

Step inside the banquet hall as you partake in our next chapter on the **digestive system**. Here, you will plunge into the dynamics of nourishment, starting from the first bite, down the corridors of the esophagus, and into the vast dining hall of the stomach. As you continue, you will wander through the winding hallways of the digestive tract and see the intricate preparation of nutrients in the intestinal kitchen, where eager enzymes work tirelessly to convert food into fuel. You will witness the essential processes that ensure our body's overall health and well-being and marvel at the tenacity of the organs in charge. So, grab your favorite coloring utensil and join us as we journey through our digestive system!

Digestive system overview

- (a) Salivary glands
- (b) Tongue
- (c) Oral cavity
- (d) Pharynx
- (e) Esophagus
- (f) Stomach
- (g) Pancreas
- (h) Liver
- (i) Gallbladder
- (j) Small intestine
- (k) Large intestine
- (l) Appendix
- (m) Rectum
- (n) Anus

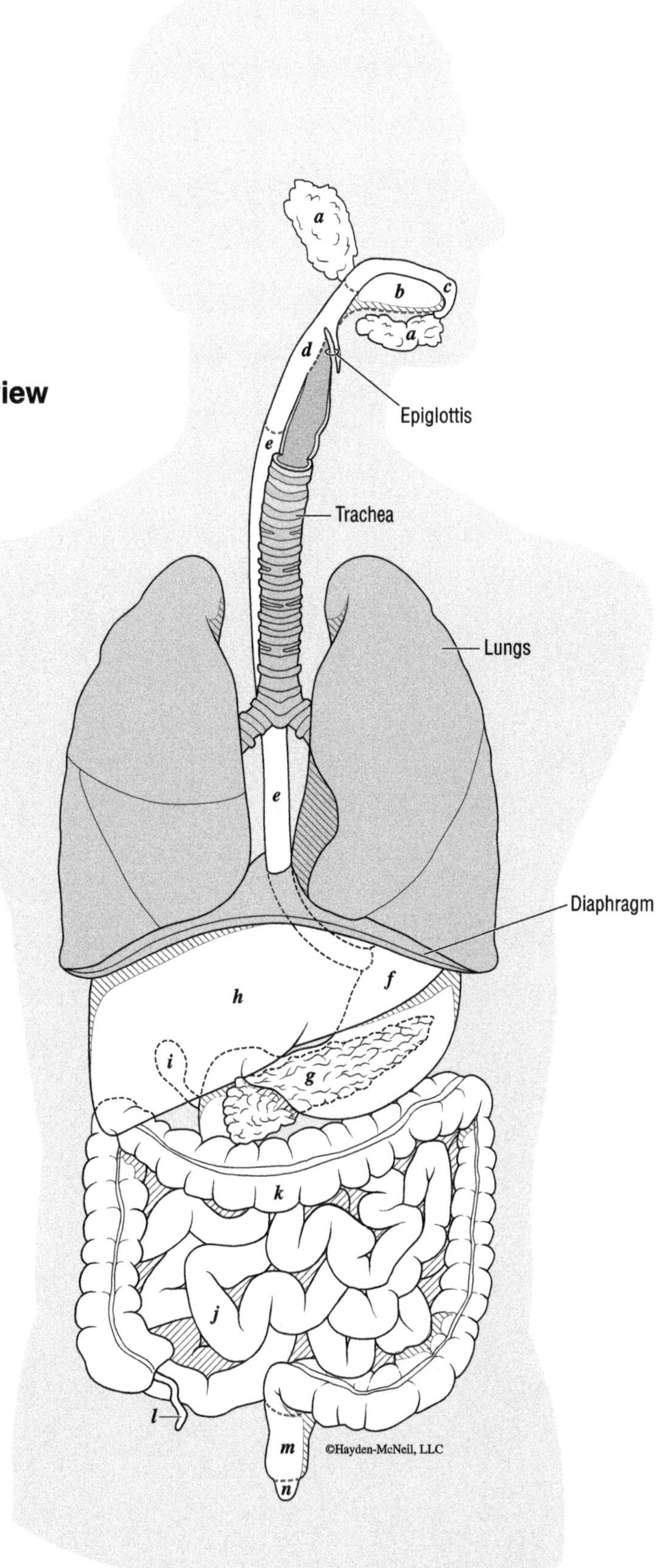

Mouth

- (a) Hard palate
- (b) Soft palate
- (c) Uvula
- (d) Palatine tonsils
- (e) Lips
- (f) Teeth
- (g) Tongue
- (h) Submandibular salivary gland
- (i) Sublingual salivary gland
- (j) Parotid salivary gland
- (k) Oral cavity
- (l) Pharynx
- (m) Esophagus

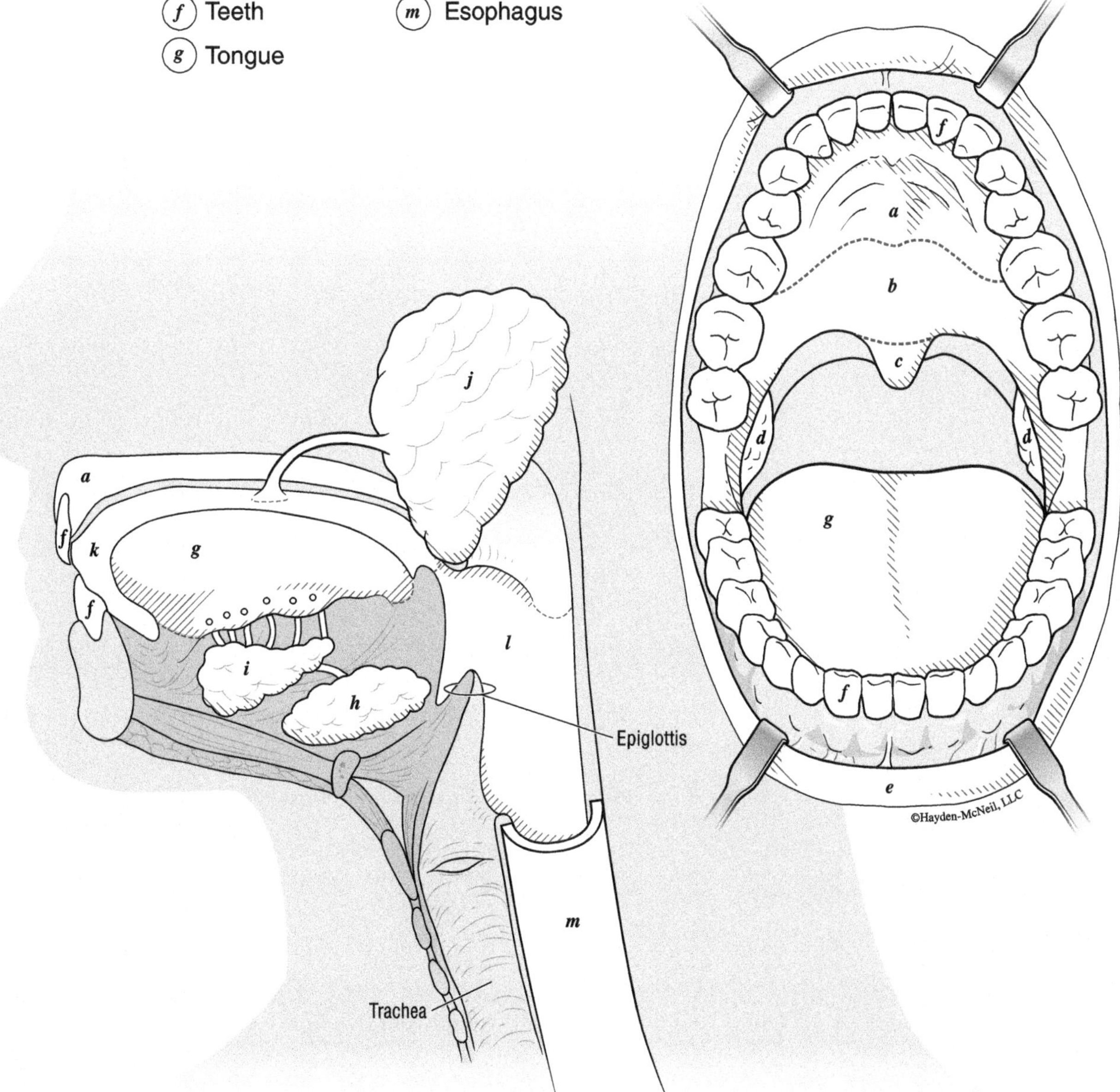

Tooth anatomy

- (a) Enamel
- (b) Dentin
- (c) Pulp cavity
- (d) Gingiva
- (e) Periodontal ligament
- (f) Cementum
- (g) Alveolar bone
- (h) Nerve
- (i) Blood vessels
- (j) Crown
- (k) Neck
- (l) Root

Teeth

- (m) Incisor
- (n) Canine (cuspid)
- (o) Premolar (bicuspid)
- (p) Molar

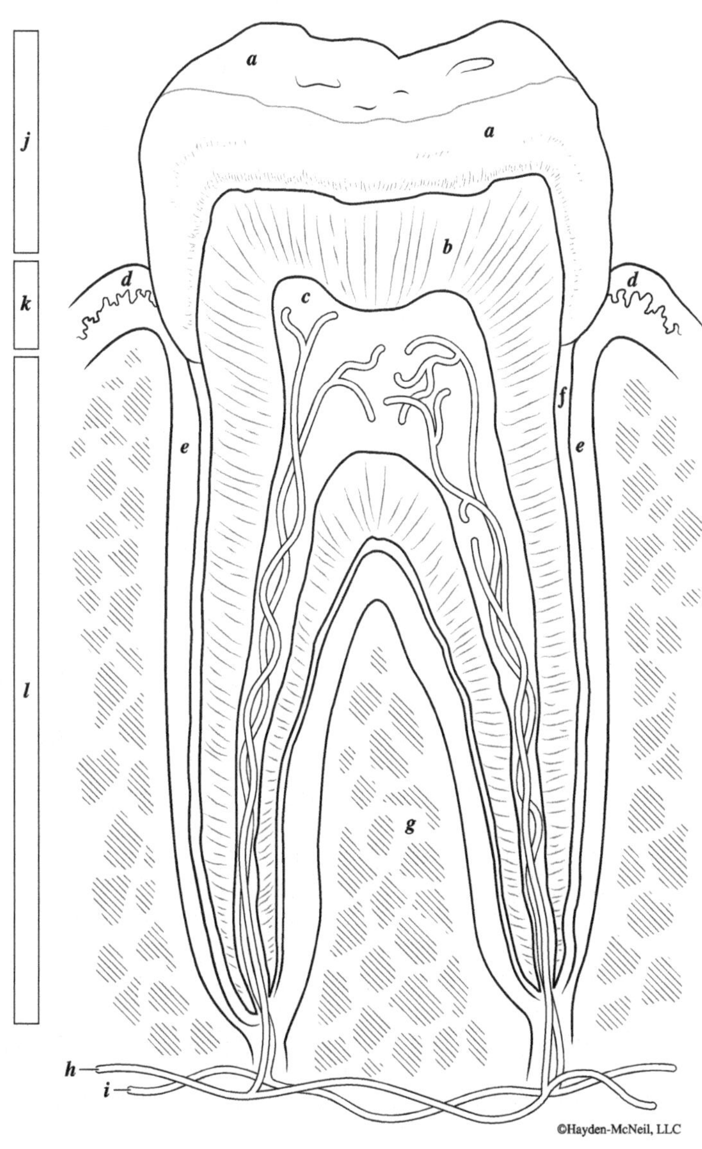

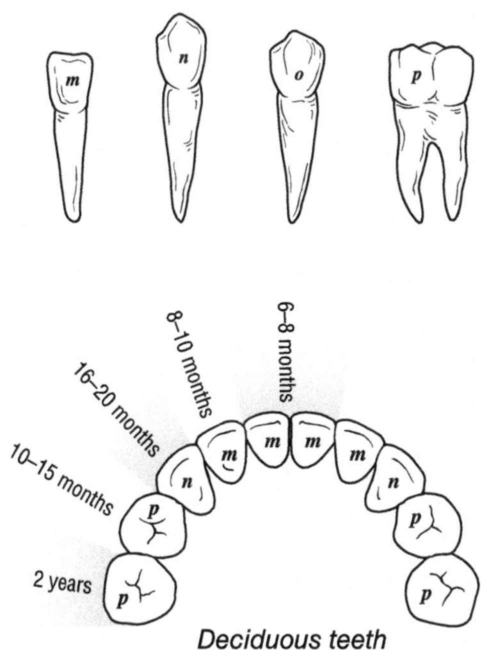

Deciduous teeth

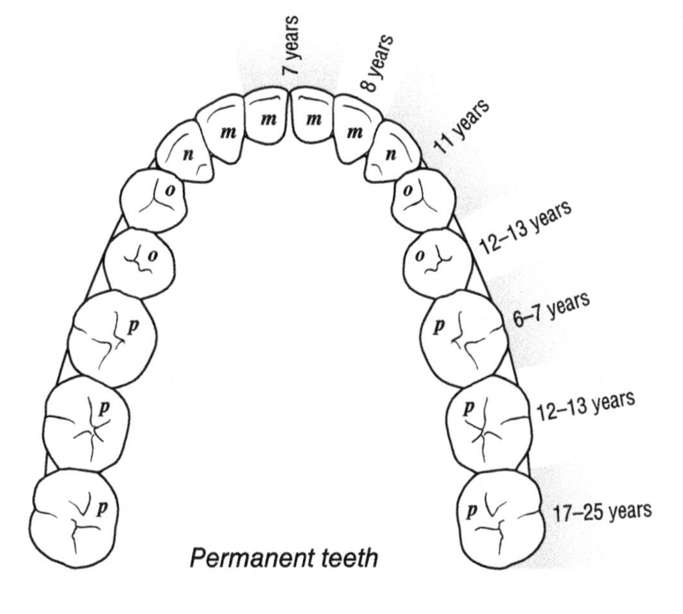

Permanent teeth

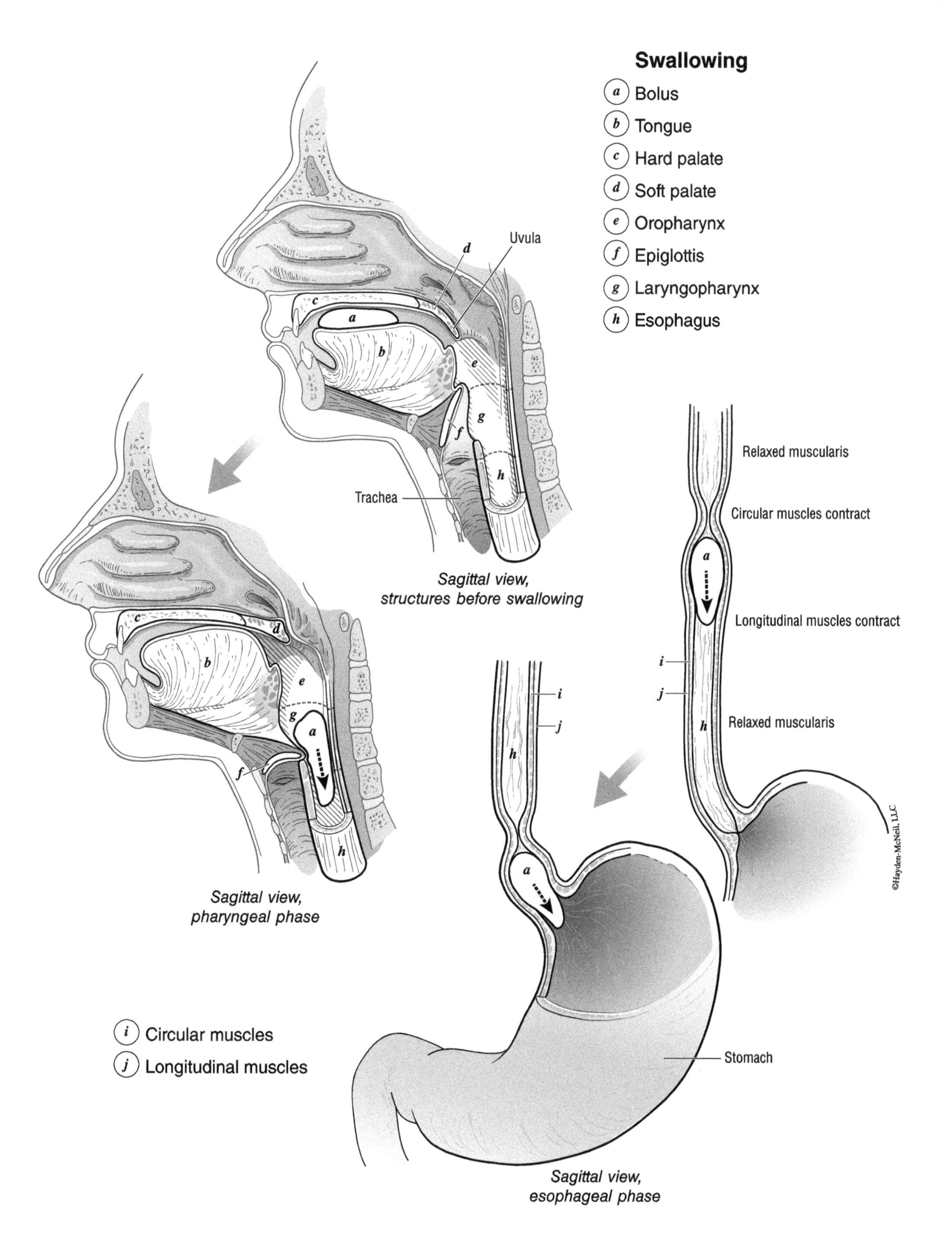

Sagittal view, structures before swallowing

Sagittal view, pharyngeal phase

Sagittal view, esophageal phase

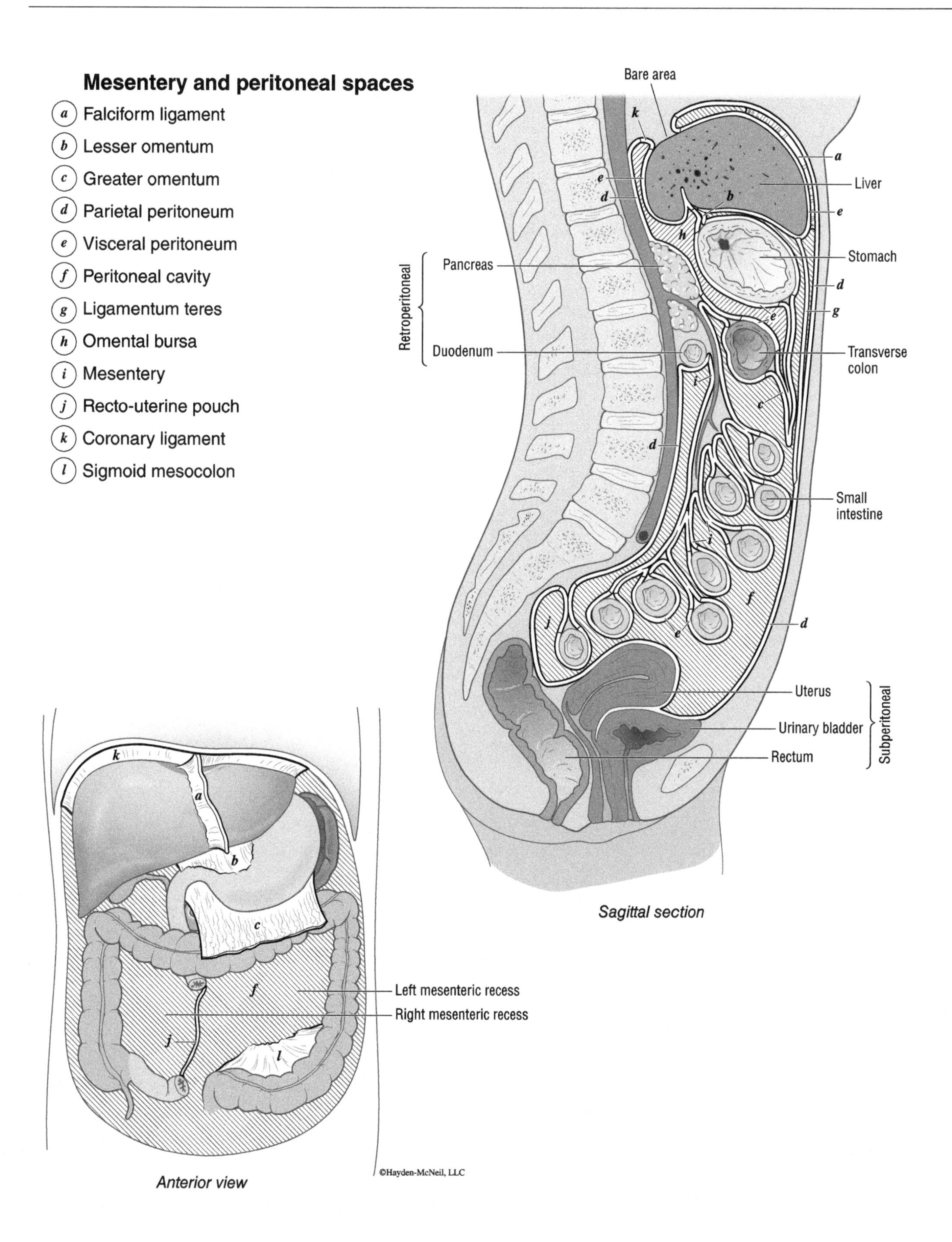

Sagittal section

Anterior view

Stomach

(a) Esophagus
(b) Fundus
(c) Body
(d) Pylorus
(e) Pyloric sphincter
(f) Duodenum
(g) Lesser curvature
(h) Greater curvature

(i) Serosa

(j) Longitudinal muscle
(k) Circular muscle

(l) Oblique muscle
(m) Submucosa
(n) Rugae/epithelium

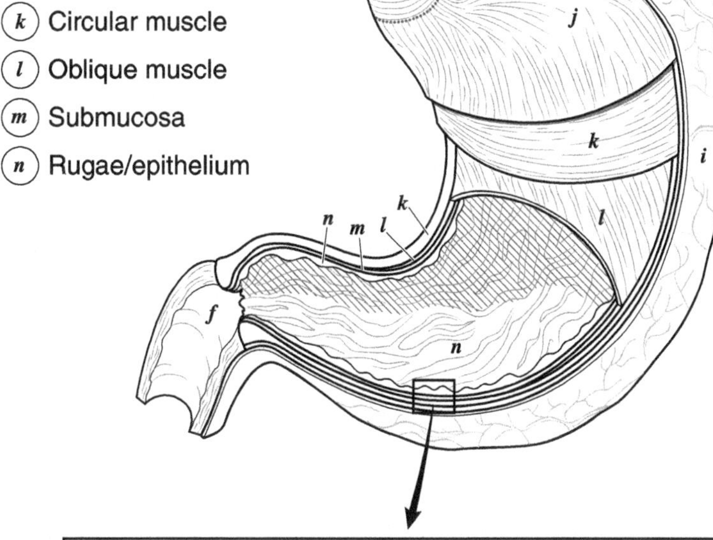

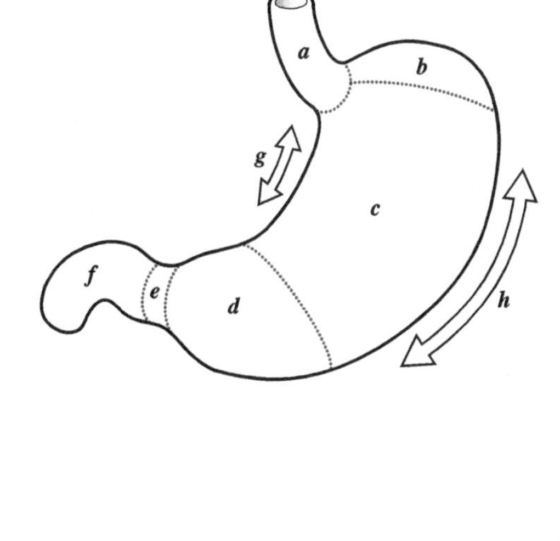

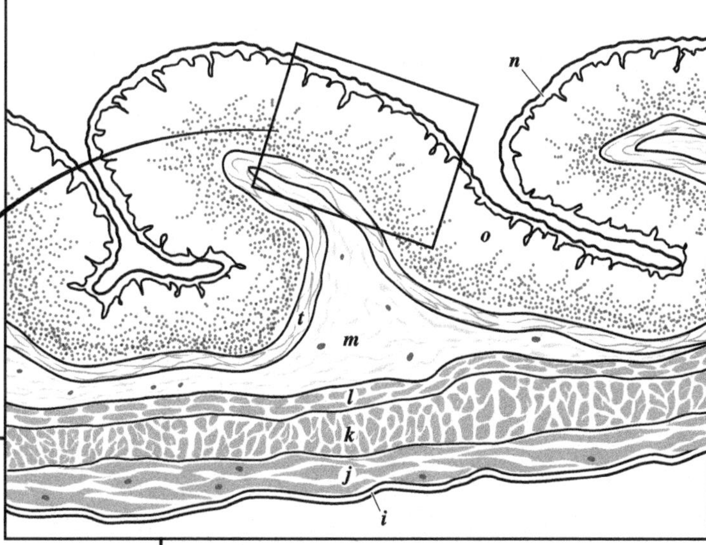

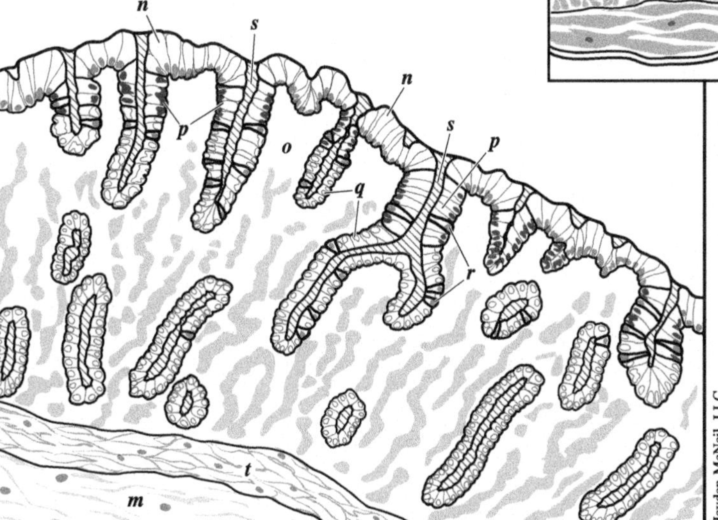

(o) Lamina propria
(p) Mucous cells
(q) Parietal cells
(r) Chief cells
(s) Gastric pit
(t) Muscularis mucosae

Small intestine

Use the same color for the circular folds with villi and the enterocytes.

(a) Mesentery/peritoneal membrane
(b) Longitudinal muscle
(c) Circular muscle
(d) Submucosa
(e) Muscularis mucosae
(f) Circular folds with villi

(g) Duodenum
(h) Jejunum
(i) Ileum

(f_1) Enterocytes (absorptive cells)
(j) Goblet cells
(k) Lamina propria
(l) Intestinal glands (crypts)

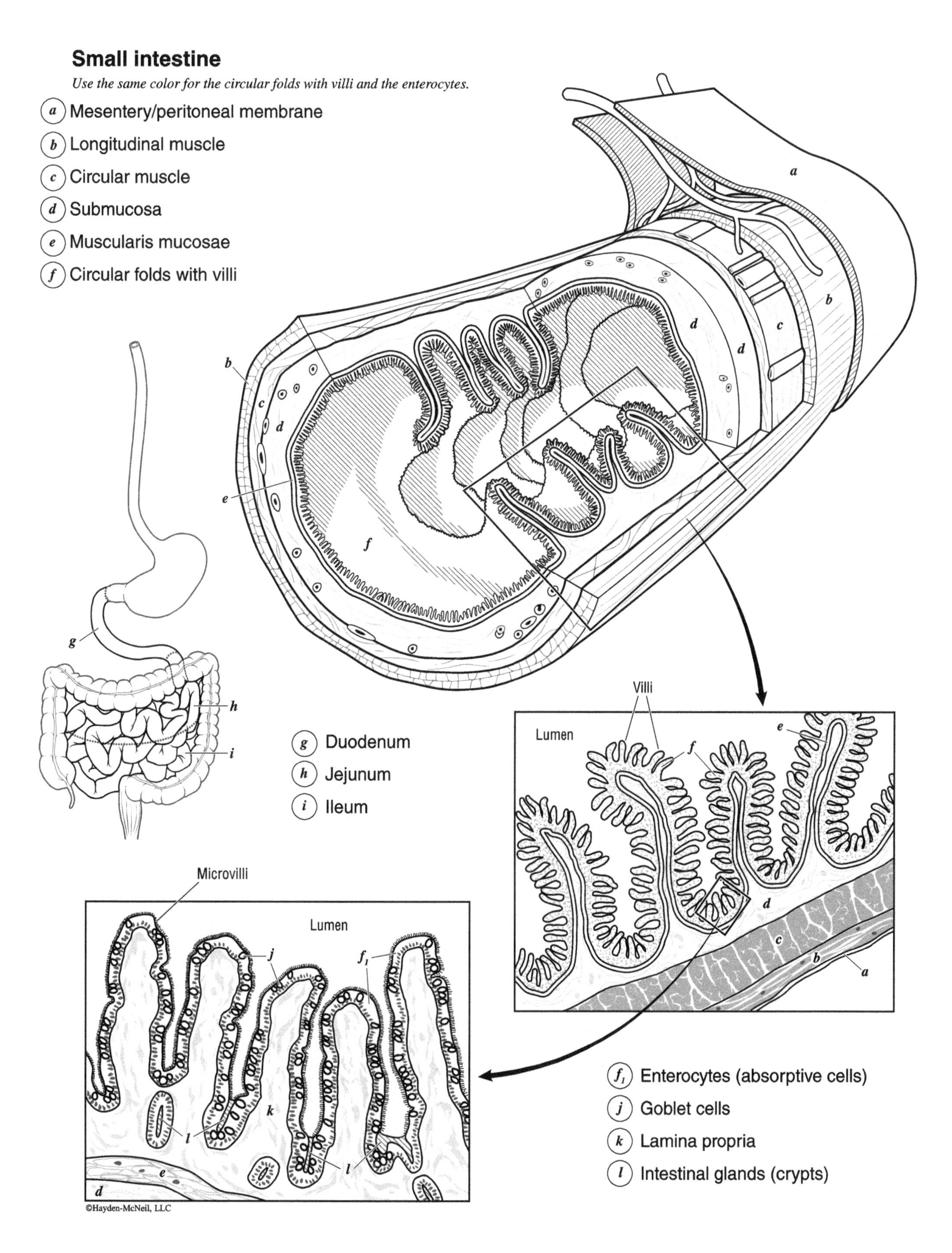

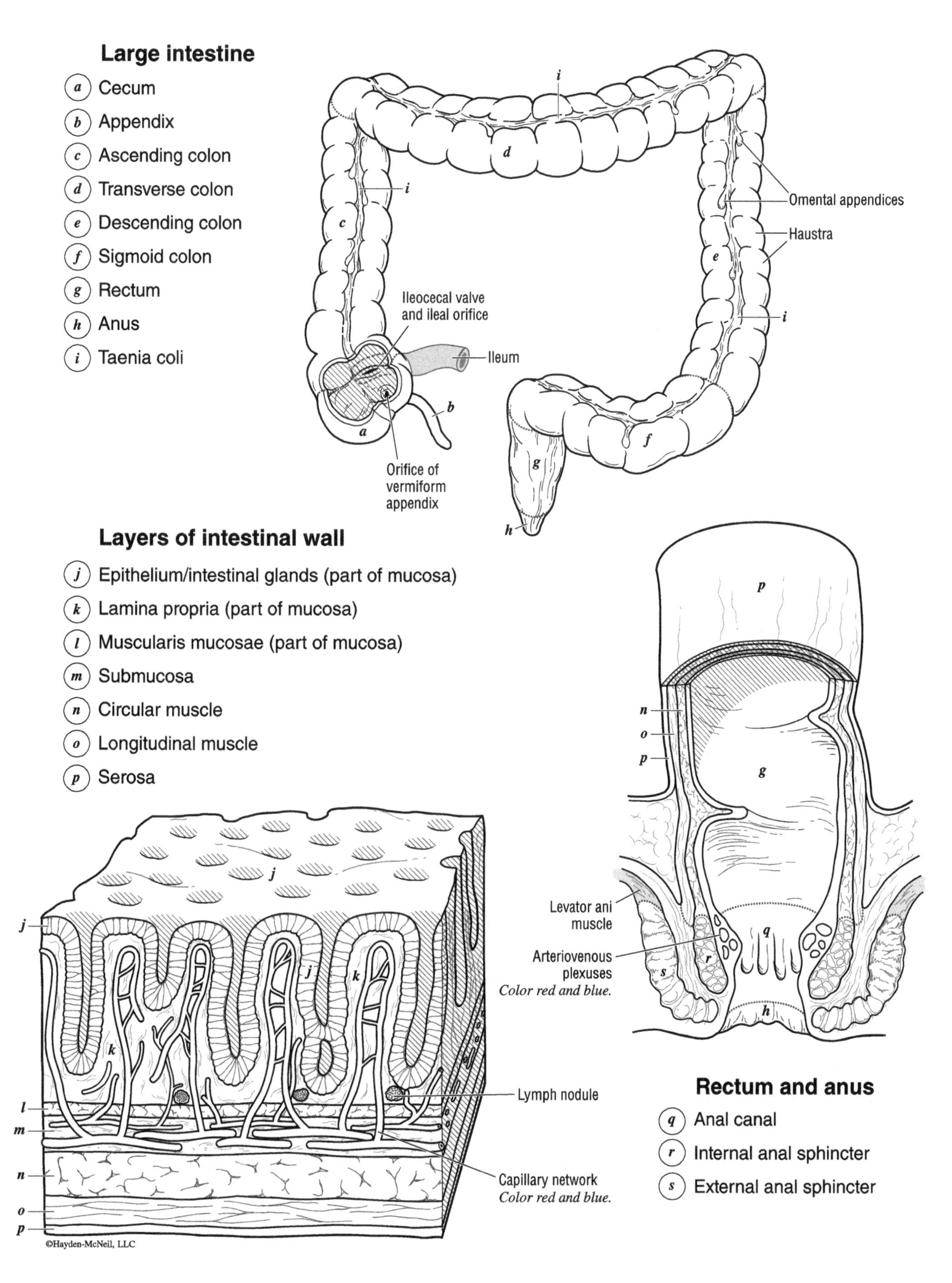
Large intestine
a Cecum
b Appendix
c Ascending colon
d Transverse colon
e Descending colon
f Sigmoid colon
g Rectum
h Anus
i Taenia coli
Omental appendices
Haustra
Ileocecal valve and ileal orifice
Ileum
Orifice of vermiform appendix
Layers of intestinal wall
j Epithelium/intestinal glands (part of mucosa)
k Lamina propria (part of mucosa)
l Muscularis mucosae (part of mucosa)
m Submucosa
n Circular muscle
o Longitudinal muscle
p Serosa
Levator ani muscle
Arteriovenous plexuses
Color red and blue.
Lymph nodule
Capillary network
Color red and blue.
Rectum and anus
q Anal canal
r Internal anal sphincter
s External anal sphincter
©Hayden-McNeil, LLC

Liver

(a) Ligaments
- a_1 Coronary lig.
- a_2 Falciform lig. and ligamentum teres (round lig.)
- a_3 Triangular lig.
- a_4 Lesser omentum and ligamentum venosum

(b) Right lobe
- b_1 Bare area

(c) Left lobe

(d) Quadrate lobe

(e) Caudate lobe

(f) Gallbladder

(g) Cystic duct

(h) Hepatic ducts

(i) Common bile duct

(j) Inferior vena cava *Color blue.*

(k) Portal vein *Color purple.*

(l) Hepatic artery *Color red.*

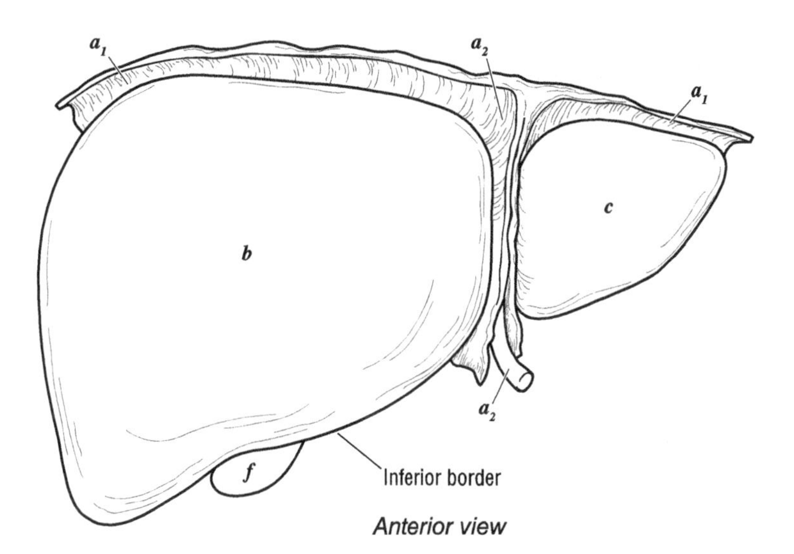

Anterior view

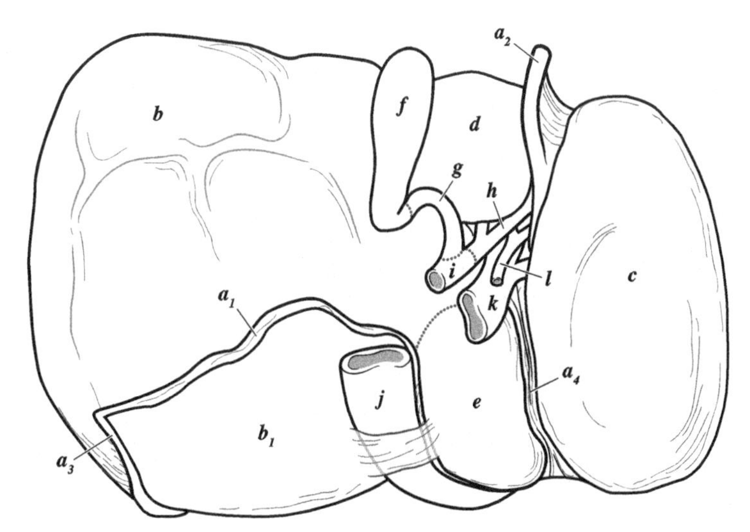

Inferior view

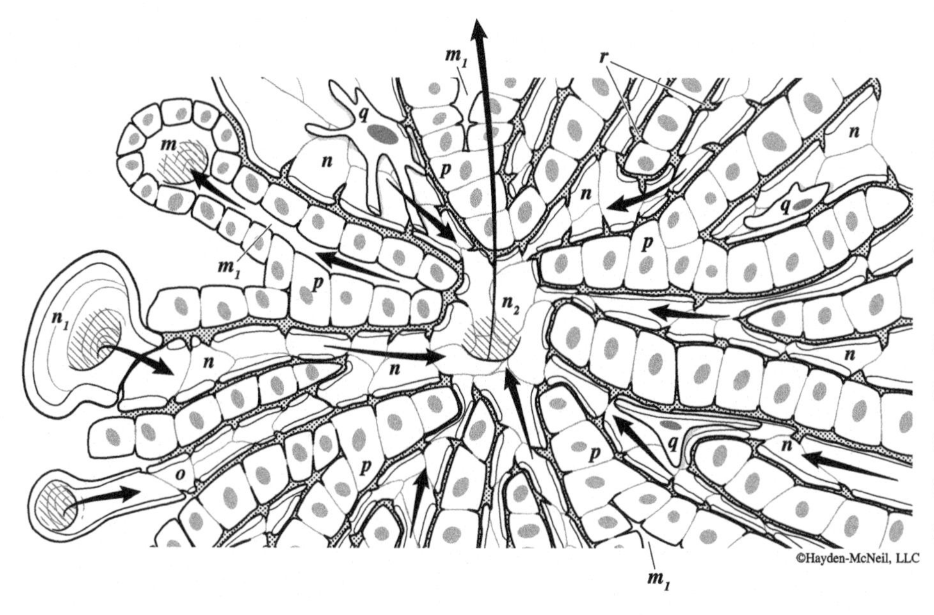

Liver lobule

Color the bile duct and bile canaliculus green. Color the sinusoids and veins blue and the arterioles red.

(m) Bile duct

(m_1) Bile canaliculus

(n) Hepatic venous sinusoids

(n_1) Branch of hepatic portal vein

(n_2) Central vein (to hepatic vein)

(o) Hepatic arteriole

(p) Hepatocytes

(q) Kupffer cell (macrophage)

(r) Perisinusoidal (Disse's) space

Accessory digestive organs

Color the aorta and its branches red. Color the portal vein and its branches purple.

- (*a*) Left and right hepatic ducts
- (*b*) Common hepatic duct
- (*c*) Cystic duct
- (*d*) Common bile duct
- (*e*) Gallbladder
- (*f*) Pancreas
- (*g*) Pancreatic duct
- (*h*) Accessory pancreatic duct
- (*i*) Major duodenal papilla
- (*j*) Minor duodenal papilla
- (*k*) Spleen
- (*l*) Aorta
  - l_1 Celiac trunk
    - l_2 Splenic a.
    - l_3 Common hepatic a.
  - l_4 Superior mesenteric a.
- (*m*) Portal vein
  - m_1 Superior mesenteric v.
  - m_2 Splenic v.

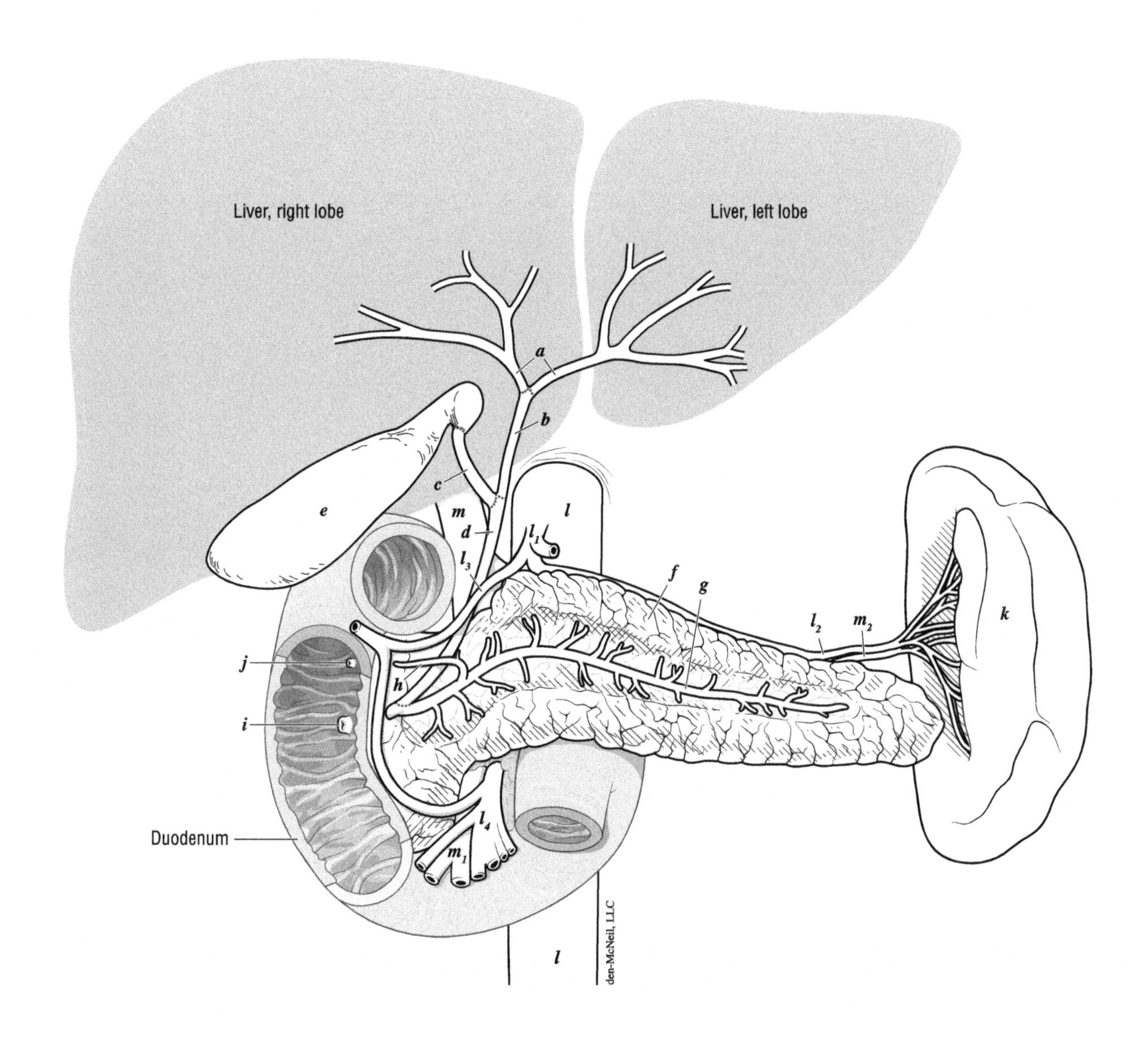

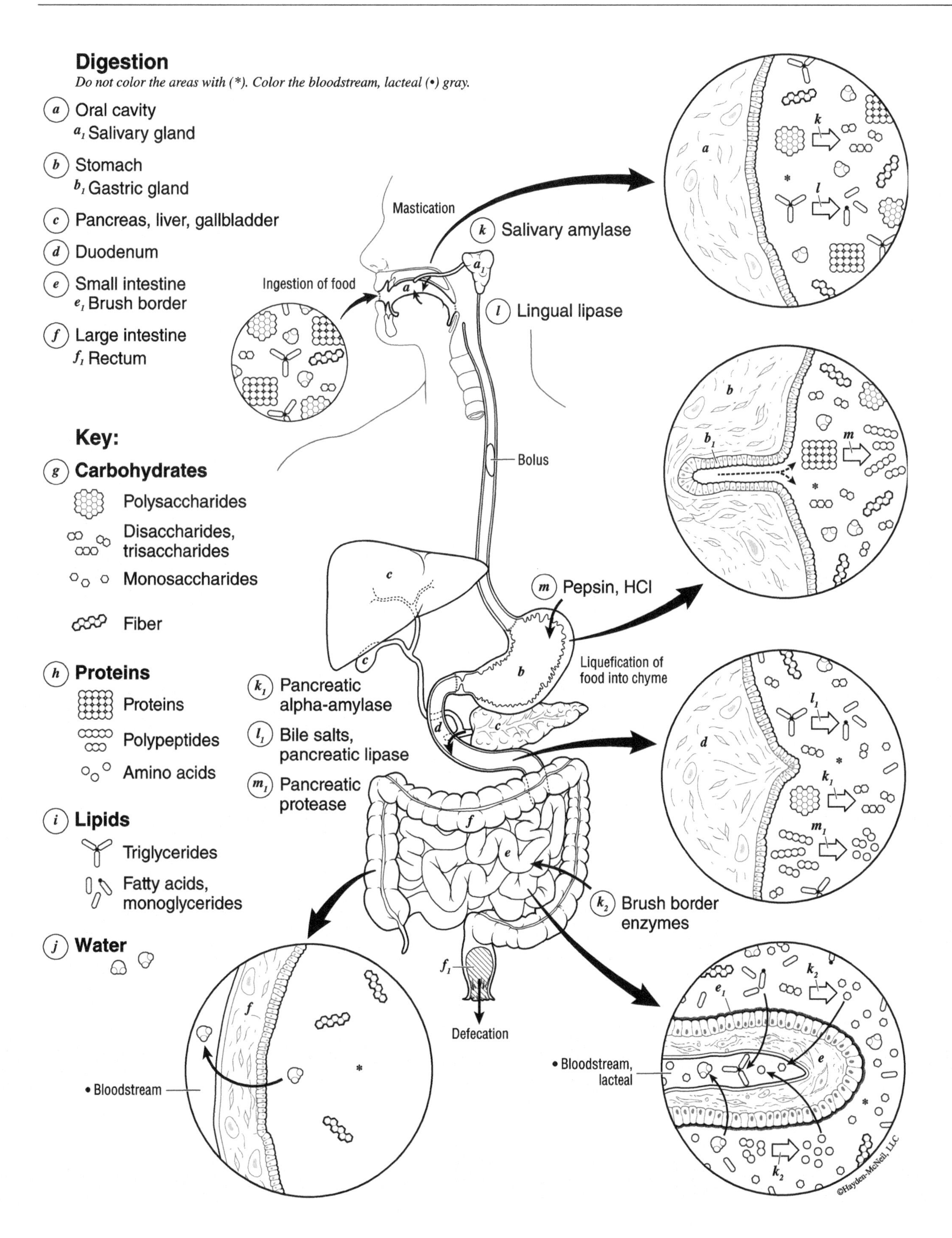
Digestion
Do not color the areas with (*). Color the bloodstream, lacteal (•) gray.
a Oral cavity
a1 Salivary gland
b Stomach
b1 Gastric gland
c Pancreas, liver, gallbladder
d Duodenum
e Small intestine
e1 Brush border
f Large intestine
f1 Rectum
Key:
g Carbohydrates
Polysaccharides
Disaccharides, trisaccharides
Monosaccharides
Fiber
h Proteins
Proteins
Polypeptides
Amino acids
i Lipids
Triglycerides
Fatty acids, monoglycerides
j Water
Mastication
Ingestion of food
k Salivary amylase
l Lingual lipase
Bolus
m Pepsin, HCl
Liquefication of food into chyme
k1 Pancreatic alpha-amylase
l1 Bile salts, pancreatic lipase
m1 Pancreatic protease
k2 Brush border enzymes
Defecation
• Bloodstream
• Bloodstream, lacteal
©Hayden-McNeil, LLC

15

The Urinary System

Urinary system

Kidney

Renal pyramid with juxtamedullary and cortical nephrons

Nephron function and formation of urine

In this chapter, you will be learning how the **urinary tract system** is like the body's internal janitor, keeping things clean and tidy by balancing fluids and kicking out waste. You will observe how hard the kidneys work to strain out unwanted bacteria and matter and keep our blood sparkling clean. This whole process isn't just about keeping things tidy, however; it's also the body's way of ensuring we stay balanced, hydrated, and feeling productive all day long. Get ready to take a dip into this dynamic system, where every drop counts!

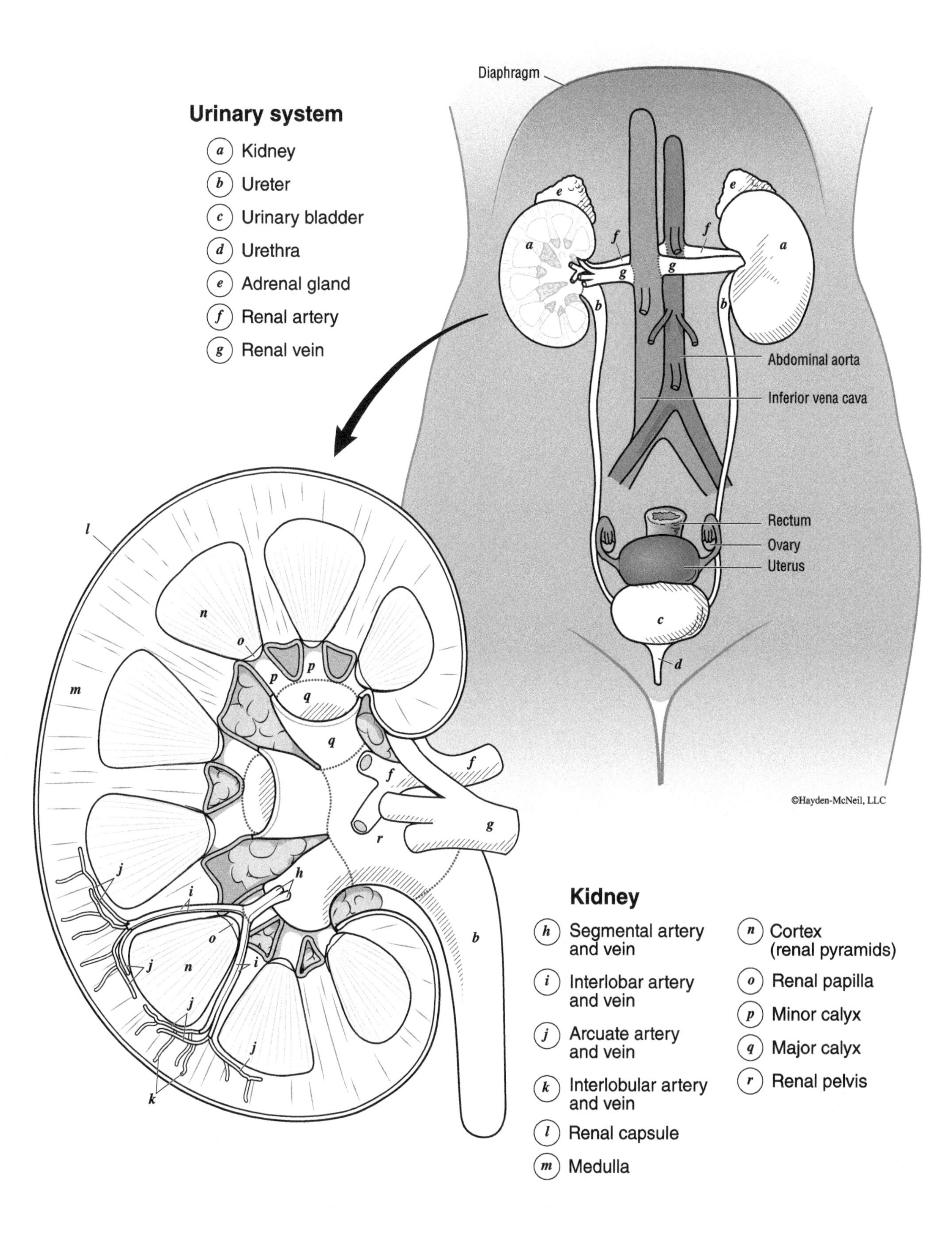
Urinary system
a Kidney
b Ureter
c Urinary bladder
d Urethra
e Adrenal gland
f Renal artery
g Renal vein
Diaphragm
Abdominal aorta
Inferior vena cava
Rectum
Ovary
Uterus
©Hayden-McNeil, LLC
Kidney
h Segmental artery and vein
i Interlobar artery and vein
j Arcuate artery and vein
k Interlobular artery and vein
l Renal capsule
m Medulla
n Cortex (renal pyramids)
o Renal papilla
p Minor calyx
q Major calyx
r Renal pelvis

Kidney

- (a) Renal capsule
- (b) Medulla
- (c) Cortex (renal pyramids)
- (d) Renal papilla
- (e) Minor calyx
- (f) Major calyx
- (g) Renal pelvis
- (h) Ureter
- (i) Renal artery
- (j) Renal vein
- (k) Segmental artery and vein
- (l) Interlobar artery and vein
- (m) Arcuate artery and vein
- (n) Interlobular artery and vein
- (o) Afferent arteriole
- (p) Glomerulus
- (q) Efferent arteriole
- (r) Nephron
  - r_1 Glomerular (Bowman's) capsule
- (s) Collecting duct

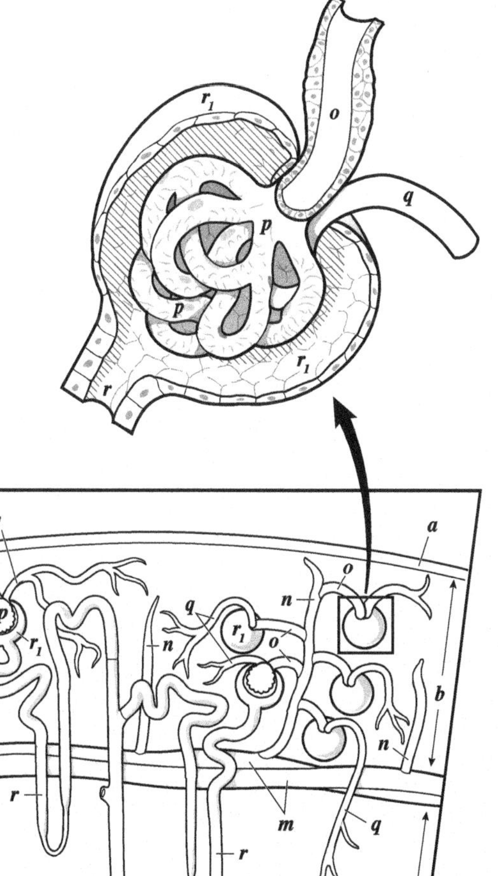

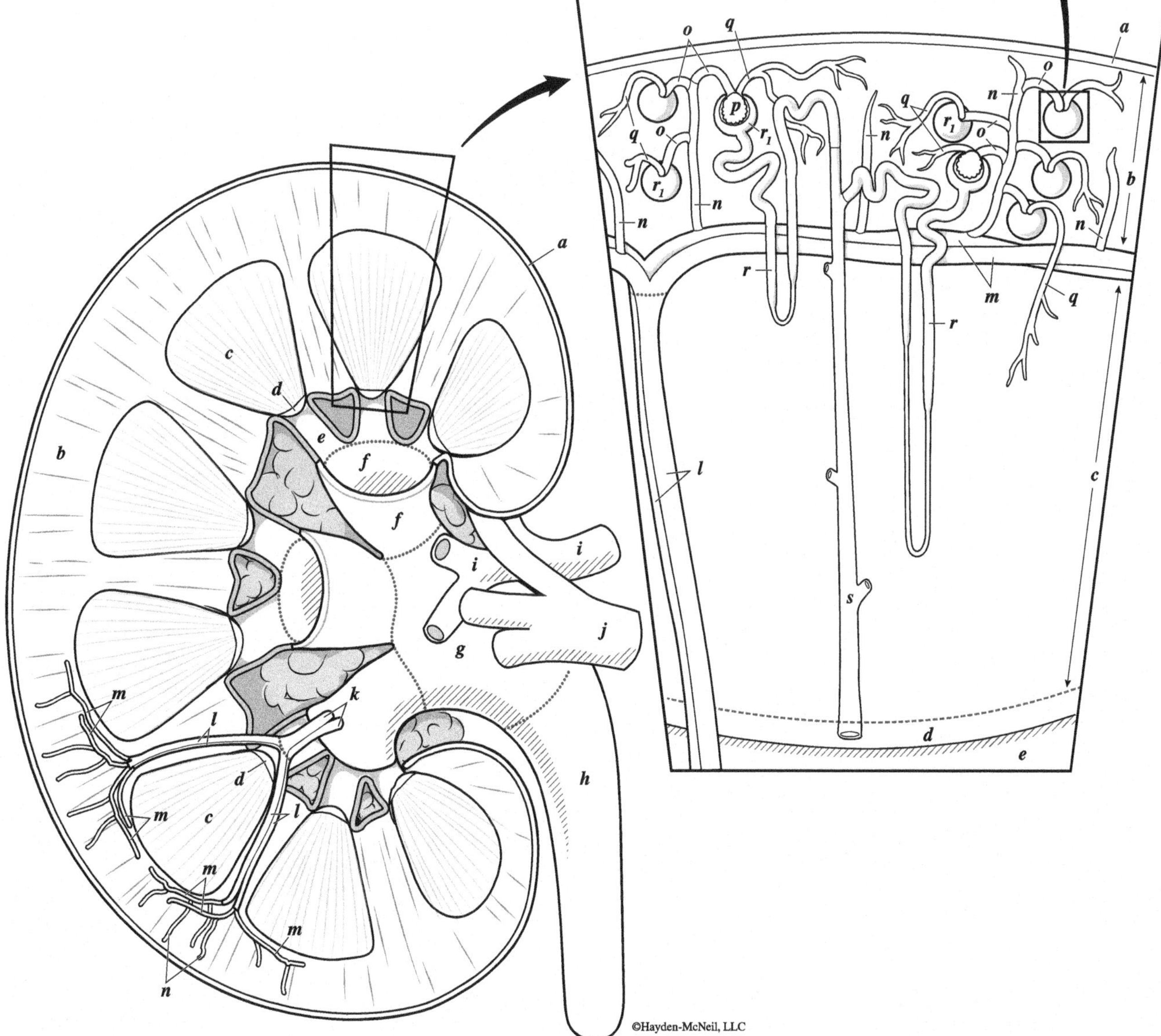

Renal pyramid with juxtamedullary and cortical nephrons

- (a) Glomerular (Bowman's) capsule
- (b) Proximal convoluted tubule
- (c) Loop of Henle
  - c_1 Descending limb
  - c_2 Ascending limb
- (d) Distal convoluted tubule
- (e) Collecting duct
- (f) Renal papilla
- (g) Arcuate artery
- (h) Interlobular artery
- (i) Afferent arteriole
- (j) Efferent arteriole
- (k) Peritubular capillaries
  - k_1 Vasa recta
- (l) Interlobular vein
- (m) Arcuate vein
- (n) Renal capsule
- (o) Renal cortex
- (p) Renal medulla

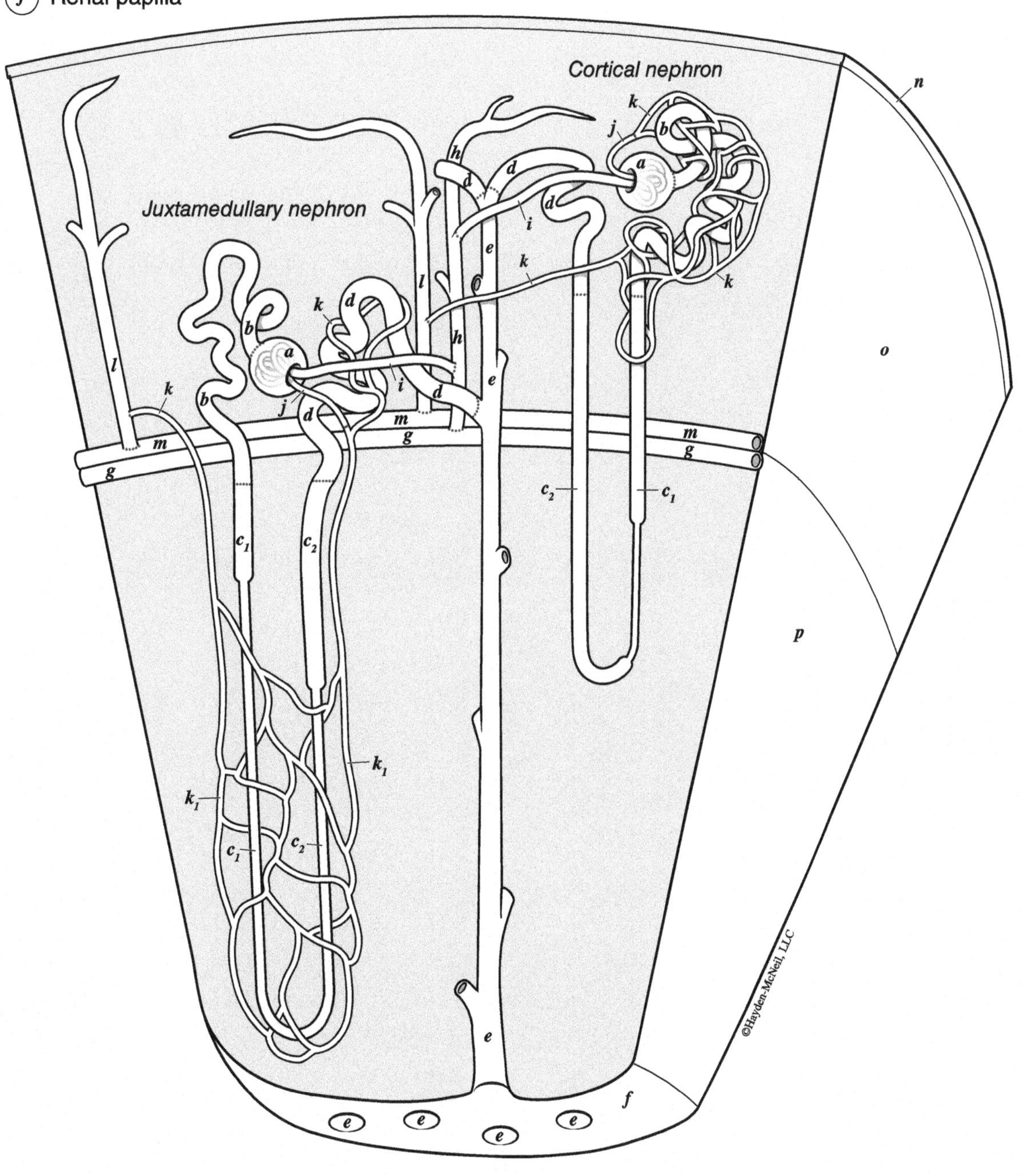

Nephron function and formation of urine

- (a) Glomerular (Bowman's) capsule
- (b) Proximal convoluted tubule
- (c) Loop of Henle
  - c_1 Descending branch
  - c_2 Ascending branch
- (d) Distal convoluted tubule
- (e) Collecting duct
- (f) Filtration
- (g) Reabsorption (from nephron into capillaries)
- (h) Secretion (from capillaries into nephron)
- (i) Excretion
- (j) Water
- (k) Potassium ions (K^+)
- (l) Creatinine
- (m) Urea
- (n) Sodium and chloride ions (Na^+, Cl^-)
- (o) Organic nutrients (glucose and amino acids)
- (p) Calcium ions (Ca^{2+})
- (q) Bicarbonate ions (HCO_3^-)
- (r) Uric acid
- (s) Hydrogen ions (H^+)
- (t) Magnesium ions (Mg^+)

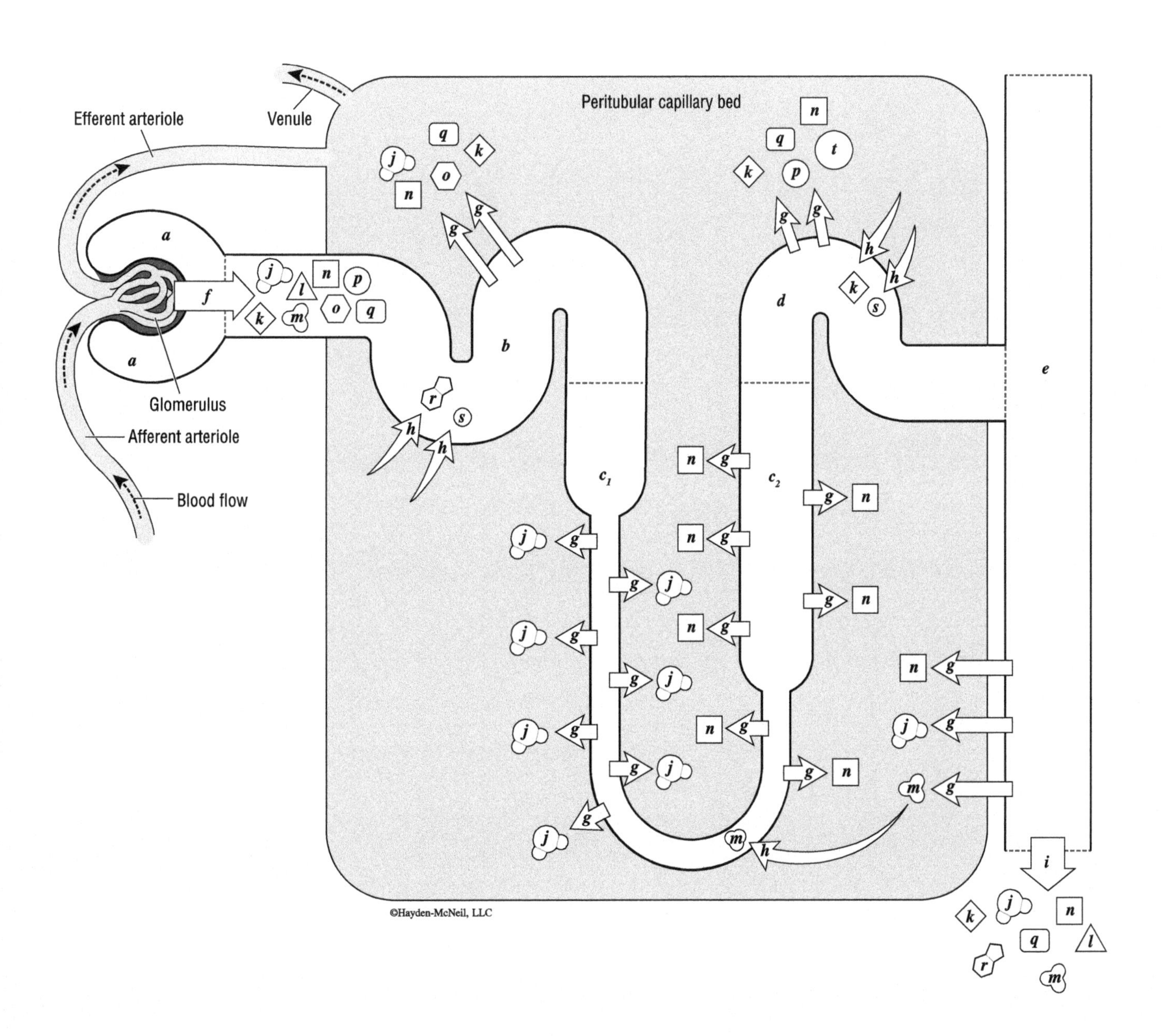

16

The Reproductive System

Male reproductive system
Spermatic cord
Penis and testes
Spermatogenesis
Female reproductive system
Uterus and ovaries
Follicle maturation
Reproductive cycle
Breast

In the final chapter of this journey, you will explore the complex structures of the male and female **reproductive systems**. You will discover the remarkable organs that enable the creation of new life. In males, the reproductive system includes the testes, which produce sperm, and other organs that assist in sperm delivery. In females, the reproductive system comprises the ovaries, which produce eggs, and the uterus, where fertilized eggs implant and grow into fetuses during pregnancy.

The upcoming sections will be a continuous reminder of the profound beauty and complexity inherent in the creation of life, inspiring us to reflect upon the true miracle of human existence.

Male reproductive system

- (*a*) Ureter
- (*b*) Urinary bladder
- (*c*) Seminal vesicle
- (*d*) Prostate gland
- (*e*) Bulbourethral gland
- (*f*) Urethra
  - f_1 Membranous urethra
- (*g*) Corpus spongiosum *(part of penis)*
- (*h*) Corpus cavernosum *(part of penis)*
- (*i*) Foreskin *(part of penis)*
- (*j*) Glans penis *(part of penis)*
- (*k*) Epididymis
- (*l*) Testis
- (*m*) Scrotum

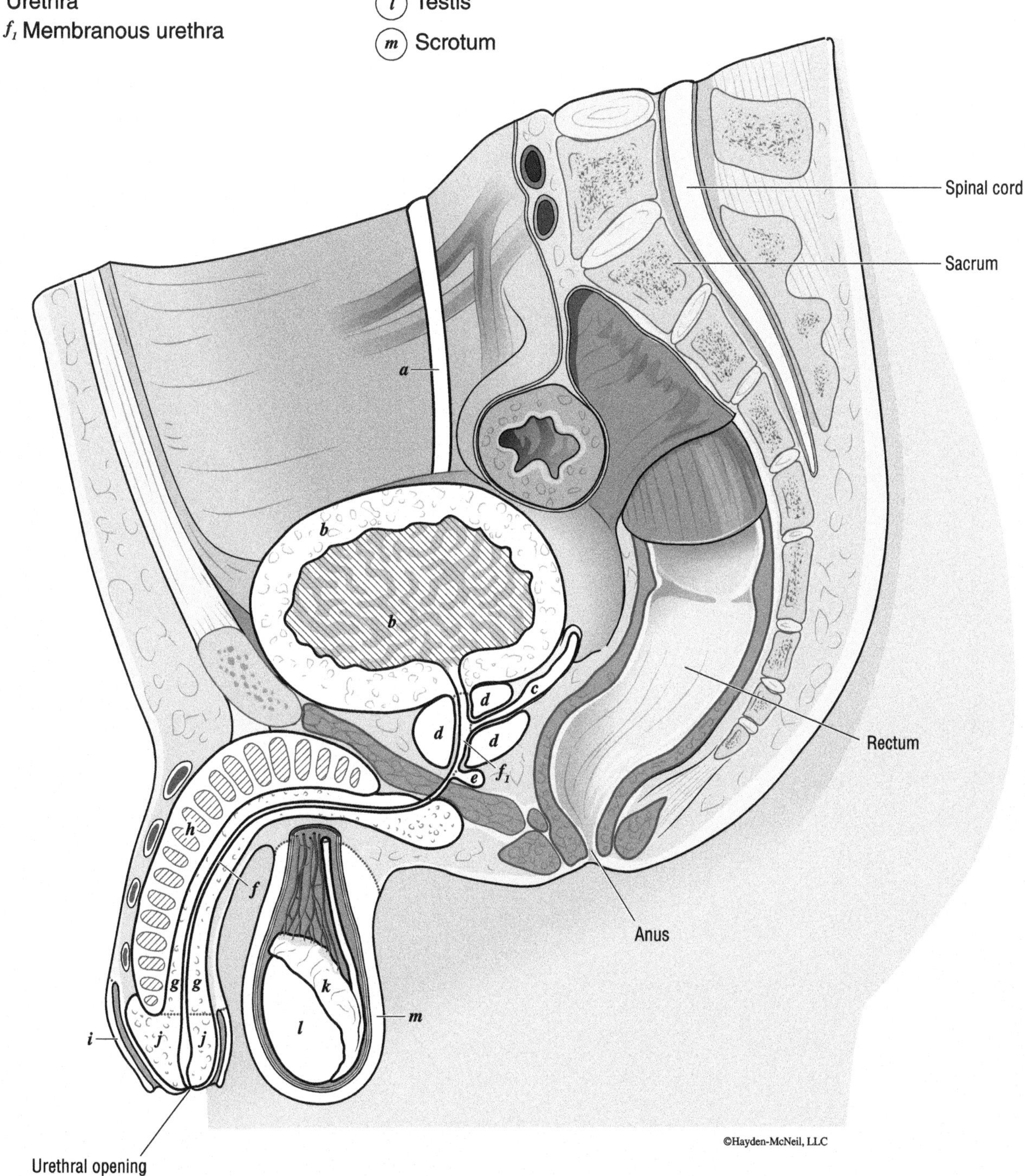

Spermatic cord

Color the femoral and testicular arteries red. Color the femoral vein and pampiniform plexus blue.

- (a) External spermatic fascia
- (b) Cremaster muscle
- (c) Tunica vaginalis (parietal layer)
- (d) Tunica dartos
- (e) Internal spermatic fascia
- (f) Testicular n. (plexus)
- (g) Testicular a.
- (h) Pampiniform plexus
- (i) Ductus deferens
- (j) Epididymis
- (k) Testis

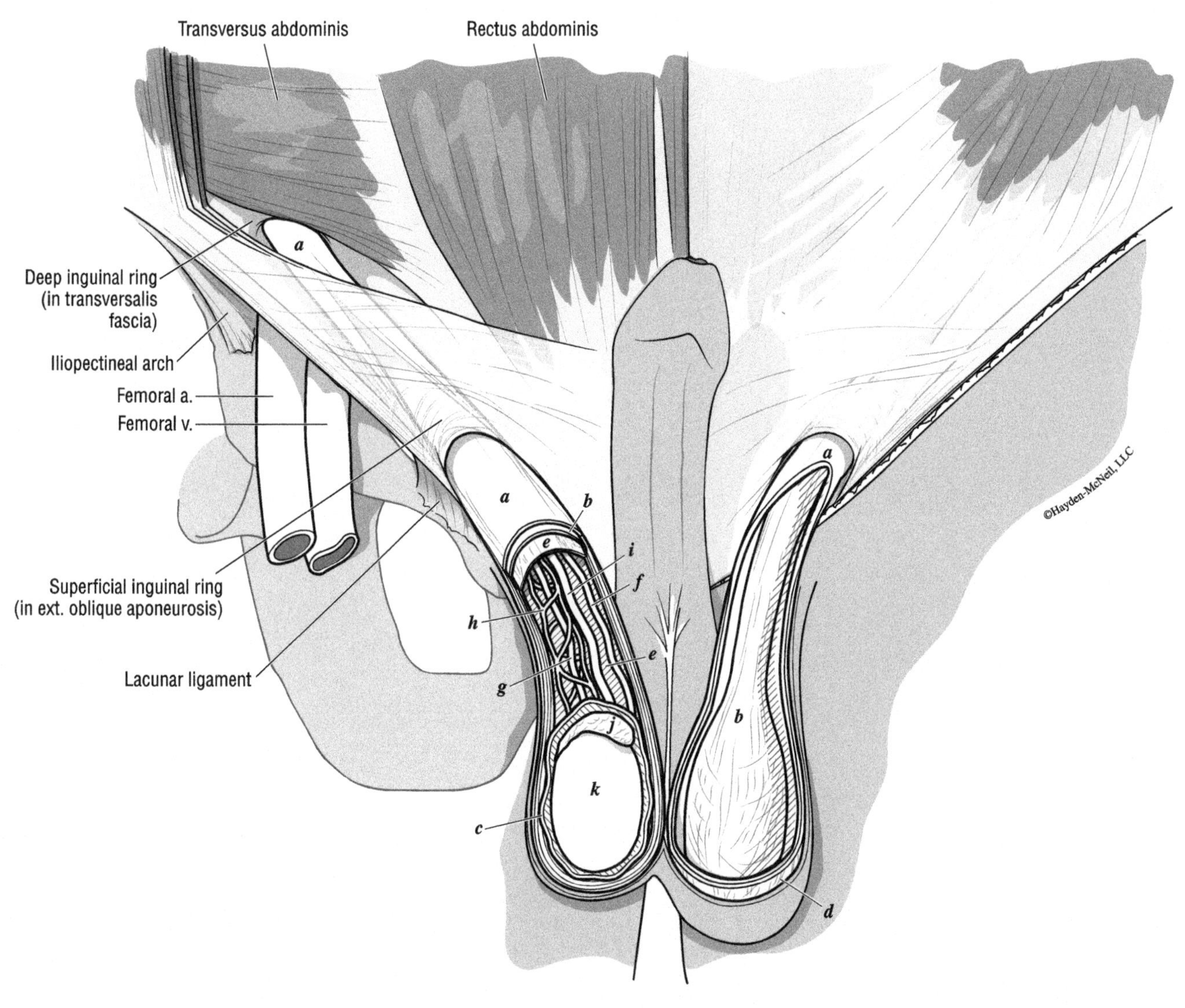

Anterior view

Penis and testes

- (a) Urinary bladder
- (b) Ureter
- (c) Seminal vesicle
- (d) Ejaculatory duct
- (e) Prostate gland
- (f) Bulbourethral gland
- (g) Corpus spongiosum *(part of penis)*
- (h) Corpus cavernosum *(part of penis)*
- (i) Urethra
- (j) Vas deferens
  - j_1 Ampulla
- (k) Epididymis
- (l) Seminiferous tubules
  - l_1 Rete testis
  - l_2 Efferent tubules
  - l_3 Duct of the epididymis
- (m) Testis

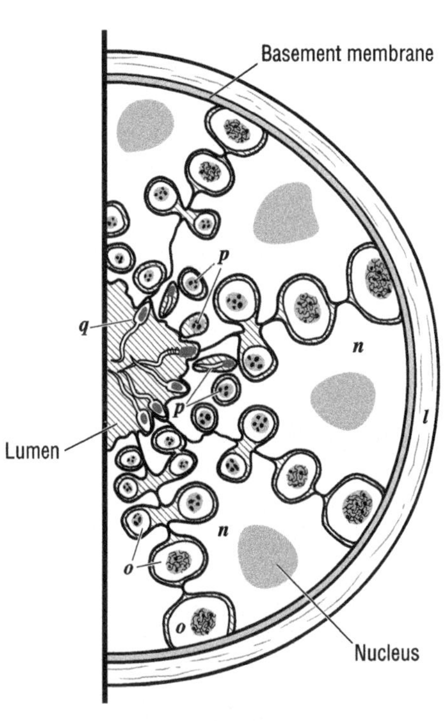

Cross section of seminiferous tubule

Spermatogenesis

- (n) Sertoli cells
- (o) Spermatocytes
- (p) Spermatids
- (q) Sperm cells
- (r) Head
- (s) Midpiece of tail (with mitochondria)
- (t) End piece of tail

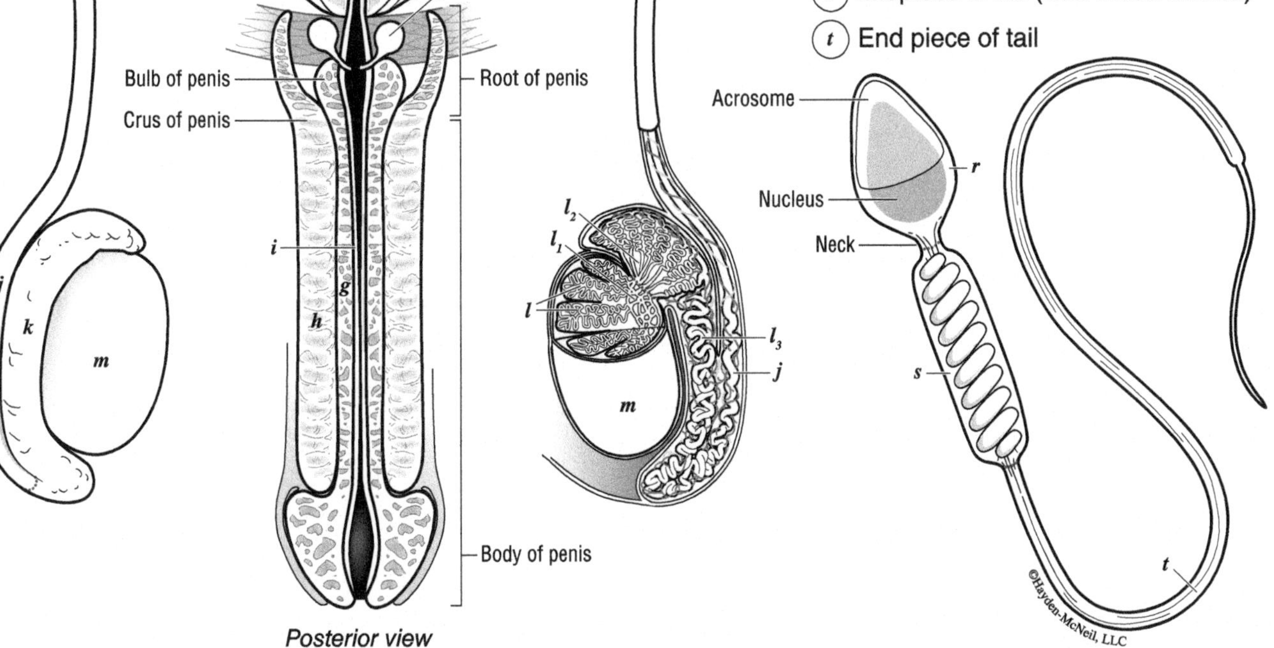

Posterior view

Female reproductive system

- (a) Ureter
- (b) Urinary bladder
- (c) Uterine (fallopian) tube
- (d) Ovary
- (e) Uterus
  - e_1 Fundus
  - e_2 Body
  - e_3 Cervix
- (f) Urethra
- (g) Vagina
- (h) Clitoris
- (i) Prepuce of clitoris
- (j) Bartoli glands
- (k) Labia minora
- (l) Labia majora
- (m) Vestibule

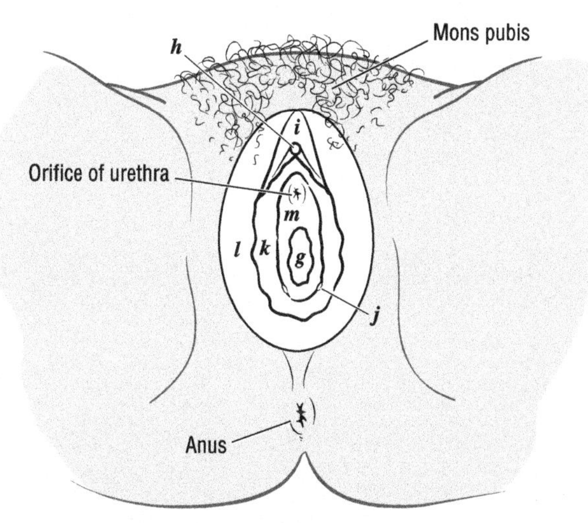

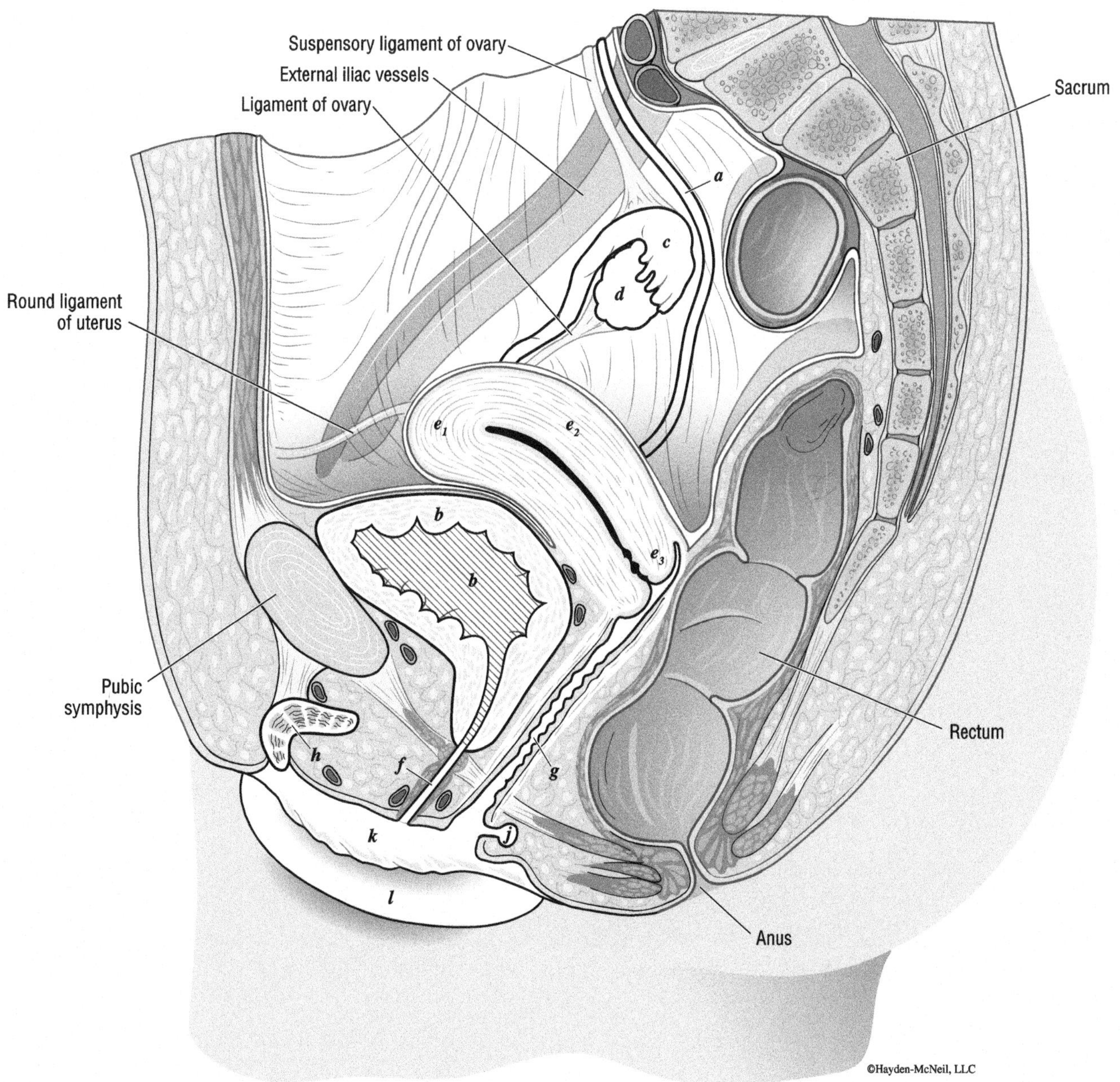

Uterus and ovary

- (a) Suspensory ligament of the ovary
- (b) Uterine (fallopian) tube
- (c) Ovary
- (d) Ovarian ligament
- (e) Vagina

Uterus

- (f) Perimetrium
- (g) Myometrium (smooth muscle)
- (h) Endometrium

a
Fundus
Mesosalpinx
b
g
h
f
b
c
d
h
m
n
o
l
Fimbriae
Body
h
g
Broad ligament
Cervix
e
Cervical os

Anterior view

Follicle maturation

- (i) Ovarian vessels *Color red and blue.*
- (j) Connective tissue *Use a light gray.*
- (k) Tunica albuginea
- (l) Oocytes
- (m) Growing follicles
- (n) Mature (Graafian) follicle
- (o) Ruptured follicle
- (p) Corpus luteum *Color yellow.*
- (q) Corpus albicans *Leave white.*

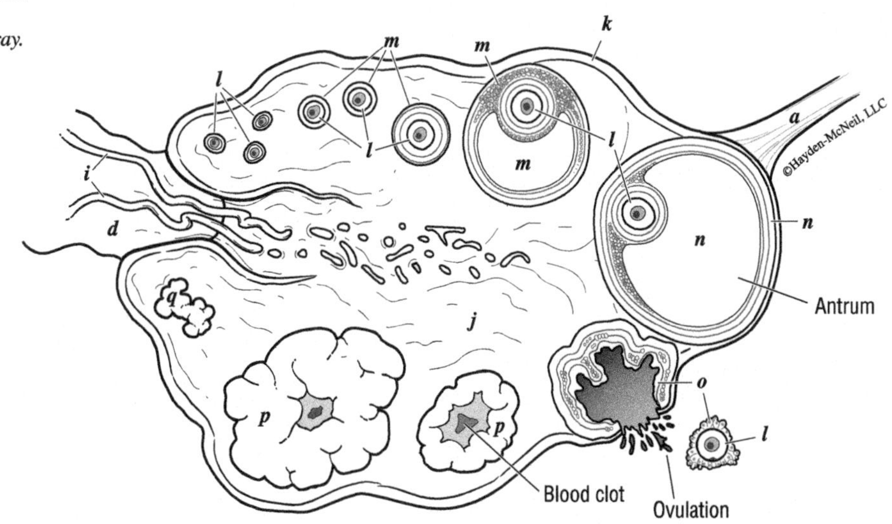

Transverse view of ovary

Reproductive cycle

Use yellow for the corpus luteum. Leave the corpus albicans white.

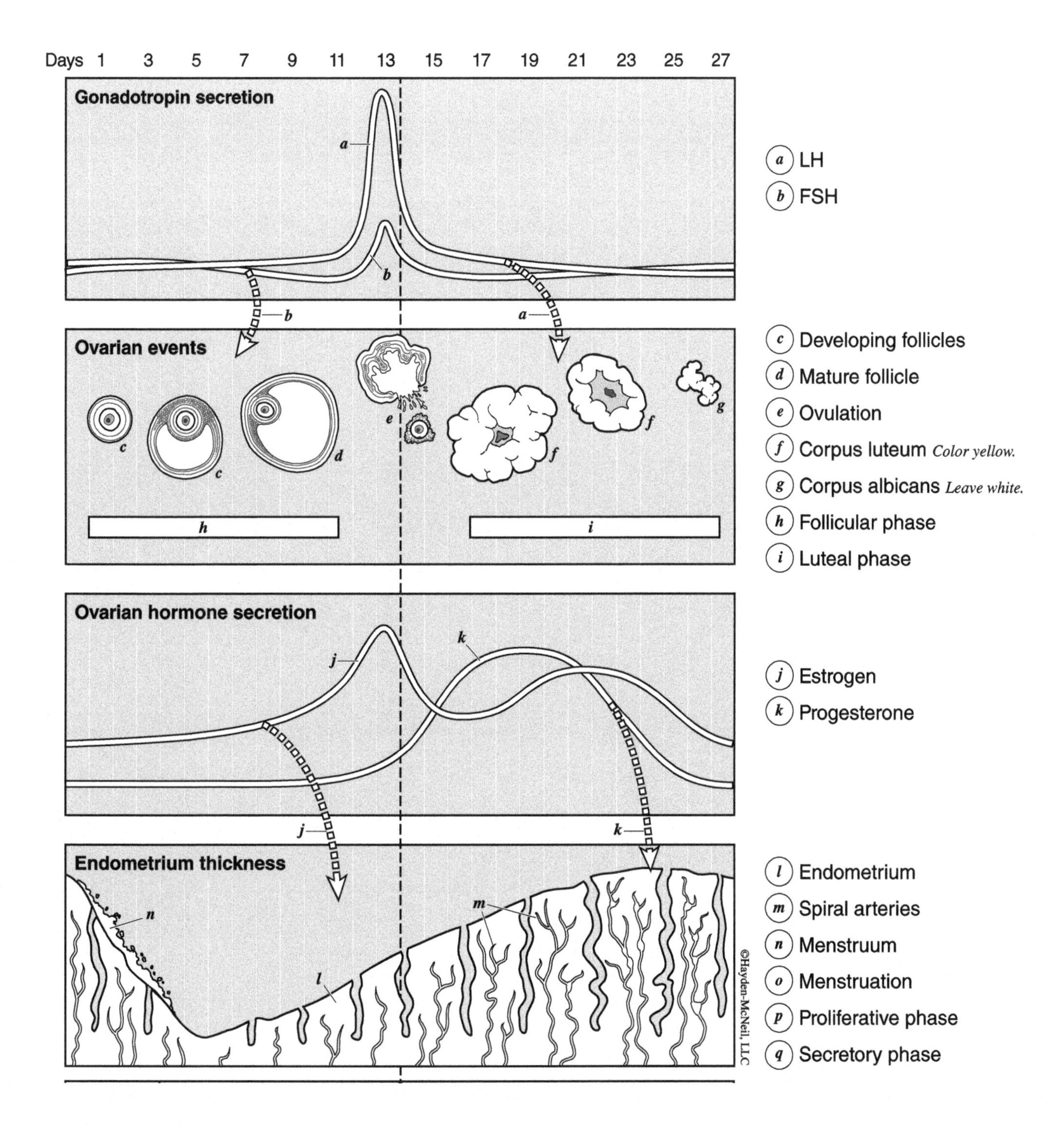

Breast

- (a) Pectoralis major m.
- (b) Intercostal muscles
- (c) Superficial fascia (fat)
- (d) Suspensory ligaments
- (e) Glandular lobe
- (f) Lactiferous duct
  f_1 Lactiferous sinus
- (g) Nipple
- (h) Areola

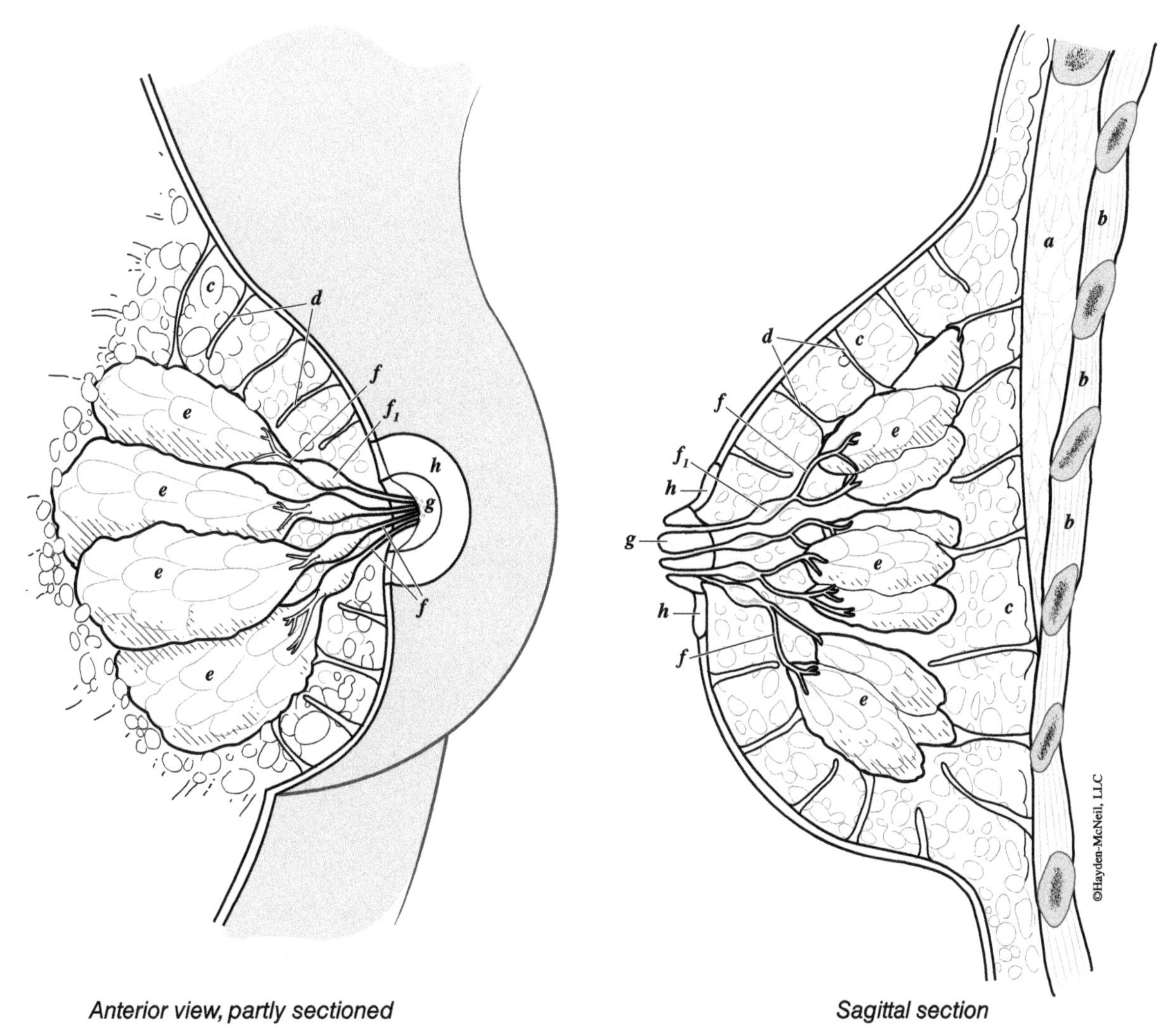

Anterior view, partly sectioned

Sagittal section

Index

A

abdomen 5, 115
abdominal cavity 7
abduction movement 9, 11
accessory digestive organs 291
acetylcholine 99
acetyl CoA 23
adduction movement 9, 11
adipocytes 31
adipose tissue 31
adrenal gland 201, 207, 297
aldosterone 207
alveolar epithelial cell 267
alveolus 267
amygdala 155
anatomical areas 5
ankle 95
anterior direction 3
anterior scalene muscle 103
anus 273, 287, 307
aorta 215, 239
appendicular skeleton 47
appendix 273, 287
arachnoid granulation 151
arachnoid mater 157, 165
areola 319
arms 9, 63, 119, 123, 125. *See also* upper limbs
arteries
  of brain 229
  of head and neck 223
  of lower limb 235
  of upper limb 231
  pelvic 243
  structure of 221
articular cartilage 45
ascending colon 287
astrocytes 143
atlas 57
atoms 17
auditory ossicles 49
autonomic nervous system 181
axial skeleton 47
axis 57
axon 143

B

back muscles 111
ball and socket joint 81
basal ganglia 153
basophils 213
biceps brachii muscle 119
bladder. *See* urinary bladder
blood cells 213
B lymphocyte (B cell) 251
body directions 3
body movements 9, 11
Body of Nail 41
body planes 3
bones. *See also* skeleton; *See also* individual bone names
  bone tissue 45
  long bones 45, 47, 67
  of the foot 77
  of the lower limb 73, 75
  of the pelvis 71
  of the skull 51, 53, 55
  of the spine 57
  of the upper limb 63, 67
  of wrist and hand 69
  shapes of 47
Bowman's capsule 299, 301, 303
brachialis muscle 119
brachial plexus 173
brain
  arteries of 229
  brainstem 149
  cerebellum 147, 149, 151, 161
  cerebral basal ganglia 153
  cerebrospinal fluid circulation 151
  cerebrum 147, 151, 167, 169
  circle of Willis 229
  cranial nerves 149, 161
  fiber tracts 153
  limbic system 155
  medulla 147, 149, 167, 169
  meninges 157
  midbrain 149, 167, 169
  precentral and postcentral regions 159
  ventricles 151
brainstem 149
breasts 319
bronchial tree 263
buccinator muscle 103
bulbospongiosus muscle 117
bulbourethral gland 307, 311

C

calcaneus bone 77, 95
cancellous bone 45
capillary structure 221
capitate bone 69
carbohydrate metabolism 23
cardiac muscle tissue 33
cardiac notch 255, 265
cardiovascular system
  aorta and branches 239
  arteries of the brain 229
  artery and vein structure 221
  blood cells 213
  cranial dural sinuses 227
  head and neck arteries 223
  head and neck veins 225
  heart 215, 217, 219, 239
  hepatic portal system 241
  lower limb arteries and veins 235, 237
  pelvic arteries 243
  upper limb arteries and veins 231, 233
carpals 47, 49, 63, 69
cartilage tissues 33
cartilaginous joint 81
caudal direction 3
cecum 287
cell body 143
cells
  atoms 17
  cell membrane 19
  cellular transport 21
  generalized animal cell 19
  mitochondrion 21
  multicellular organization 17
cellular respiration 23
central nervous system. *See also* brain
  autonomic nervous system 181
  brachial plexus 173
  cervical plexus 171
  dermatomes 179
  lumbosacral plexus 175
  nerves of lower limb 177
  nerves of upper limb 173
  neural pathways 167, 169
  neurons and neuron structure 143
  spinal cord 163, 165
  synapses and reflex arcs 145
centriole 19
cephalic area 5
cerebellar peduncles 149
cerebellum 147, 149, 151, 161
cerebral basal ganglia 153
cerebral cortex 157
cerebral peduncles 149, 169
cerebrospinal fluid 151
cerebrum 147, 151, 167, 169
cervical area 5
cervical plexus 171
cervical vertebrae 49, 57, 59
chest area 5
chondrocytes 33
choroid plexus 149, 151
cingulate gyrus 147
circle of Willis 229
circumduction movement 9, 11
circumvallate (vallate) papilla 197
clavicle 49, 63, 65, 85, 119
clitoris 117, 313
coccygeus muscle 117
coccyx 49, 57, 59, 71
cochlea 189, 191
collagen fibers 31, 33
compact bone 45
condylar joint 81
connective tissues 17, 31
coracobrachialis muscle 119
coronal plane 3
corpus callosum 147, 153, 155
corpus cavernosum 307, 311
corpus spongiosum 307, 311
corrugator supercilii muscle 103
cortisol 207
costal cartilage 61
cranial cavity 7
cranial direction 3
cranial dural sinuses 227
cranial nerves 149, 161
cranium 49
cricothyroid muscle 107
cuboid bone 77, 95
cytoplasm 19
cytoskeleton 19
cytotoxic T lymphocyte 251

D

deep transverse perineal muscle 117
deltoid muscle 119, 121
dendrites 143
dense irregular connective tissue 31
dense regular connective tissue 31
depression movement 11
depressor anguli oris muscle 103
depressor labii inferioris muscle 103
dermatomes 179
dermis 37
descending colon 287

diaphragm 113, 215, 255, 265, 269, 273
diaphysis 45
digastric muscle 107
digestive system
  accessory organs 291
  digestion 293
  large intestine 17, 273, 287
  liver 17, 273, 281, 289
  mesentery 281
  mouth 275
  multicellular organization 17
  overview 273
  small intestine 17, 273, 281, 285
  stomach 17, 273, 279, 281, 283
  swallowing 279
  teeth 275, 277
directions, body 3
distal direction 3
distal phalanx 69, 77
dorsal direction 3
dorsal nerve root 165
dorsiflexion 11
duodenum 209, 281, 283, 285
dura mater 151, 157, 165

E

ears 189, 191
ejaculatory duct 311
elastic cartilage tissue 33
elastic connective tissue 31
elastic fibers 31, 33
elbow joint 87
electrocardiogram 219
electrons 17
electron transport chain 23
elevation 11
endocrine system
  adrenal gland 201, 207, 297
  glands 201
  hormone secretions 207
  Islet of Langerhans 209
  pancreas 201, 209, 273, 281, 291
  parathyroid glands 201, 205
  pituitary gland 147, 201, 203, 227
  thyroid gland 201, 205
endomysium 101
eosinophils 213
epicranial aponeurosis muscle 103
epidermis 37, 39
epididymis 307, 309, 311
epiglottis 255, 261, 273, 275, 279
epimysium 101
epinephrine 207
epiphysis 45
epithelial tissues 17, 29
eponychium (cuticle) 41
erythrocytes 213
esophagus 17, 205, 215, 261, 265, 273, 275, 279, 283
ethmoid bone 51, 53, 55
ethmoid sinus 259
eversion 11
extension movement 9, 11
external abdominal oblique muscle 115
external anal sphincter muscle 117
external intercostals 269
eyes 185, 187

F

face 103
fallopian tubes 313, 315
false ribs 61
feet 11, 77, 95, 135, 139
female reproductive system 313–319
femur 49, 73, 75
fiber tracts 153
fibroblasts 31
fibrocartilage tissue 33
fibrous joint 81
fibula 49, 73, 75, 95
filliform papilla 197
film of pulmonary surfactant 267
fingers 89
flat bone 47
flexion movement 9, 11
floating ribs 61
folate papilla 197
follicle maturation 315
forearm 123, 125
foreskin 307
fornix 147, 149, 155
free edge of Nail 41
frontal bone 51, 53, 55, 257
frontalis muscle 103
frontal lobe 147
frontal plane 3
frontal sinus 255, 259
fungiorm papilla 197

G

gallbladder 273, 289, 291
generalized animal cell 19
genioglossus muscle 107
geniohyoid muscle 107
glands 201–208. *See also* endocrine system
glandular epithelium 29
glans penis 307

glial cells 33, 143
glomerular capsule 299, 301, 303
glycolysis 23
Golgi apparatus 19, 21
gray matter 165

H

hair 37
hamate bone 69
hands 11, 69, 127
hard palate 257, 275, 279
Haversian canal 45
head 5, 223, 225
heart 215, 217, 219, 239. *See also* cardiovascular system
Helper T lymphocyte 251
hepatic portal system 241
hinge joint 81
hip 91, 131
hippocampus 155
homeostasis 15
horizontal plane 3
hormone secretions 207
humerus 47, 49, 63, 67, 85, 87, 119, 121
hyaline cartilage tissue 33
hyoglossus muscle 107
hyoid bone 49, 103, 107, 261, 263
hyperextension movement 9, 11
hyponychium 41
hypophysis. *See* pituitary gland
hypothalamic nuclei 155
hypothalamus 147, 155, 203

I

ileum 285
iliac crest 71, 73
iliac fossa 71, 73
ilium 71
incus 49, 189
inferior direction 3
infundibulum 149
insula 149
insular lobe 147
integumentary system
  dermis 37
  epidermis 37, 39
  hair 37
intercostal muscles 319
interior nasal concha 51, 55
intermediate cuneiform bone 77
internal abdominal oblique muscle 115
internal intercostals 269
intervertebral disc 57
inversion 11
irregular bone 47
ischiocavernosus muscle 117
ischium 71, 73
Islet of Langerhans 209

J

jejunum 285
joints
  ankle and foot joints 95
  classification 81
  elbow joint 87
  hip joint 91
  knee joint 93
  movement 81
  shoulder joint 85
  wrist and finger joints 89

K

kidney 207, 297, 299
knee joint 93
Krebs cycle 23

L

lacrimal bone 51, 55
lacunae 33
Langerhans cells 39
large intestine 17, 273, 287
laryngopharynx 261, 279
larynx 255, 261, 263, 265
lateral cuneiform bone 77
lateral direction 3
lateral flexion movement 9, 11
lateral pterygoid muscle 105
lateral rotation movement 9
legs 9, 75, 135, 137. *See also* lower limb
leukocytes 213
levator anguli oris muscle 103
levator ani muscle 117
levator labii superioris alaeque nasi muscle 103
levator labii superioris muscle 103
ligaments
  ankle and foot 95
  elbow 87
  hip 91
  knee 93
  shoulder 85
  spinal 83
  wrist and finger 89
limbic system 155
lips 275

liver 17, 273, 281, 289
long bones 45, 47, 67
loop of Henle 301, 303
loose areolar connective tissue 31
lower limb
  area of 5
  arteries 235
  bones of 73, 75
  muscular regions of 129
  nerves of 177
  veins 237
lower respiratory tract 255
lumbar vertebrae 49, 57, 59
lumbosacral plexus 175
lunate bone 69
lungs 215, 217, 255, 265, 273
lunula 41
lymphatic system 247
lymph nodes 247, 249
lymphocytes 213
lymphoid stem cell 251
lymph vessels 247, 249
lysosome 19, 21

M

male reproductive system 307–311
malleus 49, 189
mammillary body 149, 155
mandible 49, 51, 53, 55, 105, 107
manubrium 61, 119
masseter muscle 103, 105
masticatory muscles 105
mastoid process 51, 53
matrix 33
mature B cell 251
mature NK cell 251
mature T (c) cell 251
mature T (H) cell 251
maxillary bone 51, 53, 55
maxillary sinus 259
medial cuneiform bone 77, 95
medial direction 3
medial pterygoid muscle 105
medial rotation movement 9
mediastinum 215
medulla 147, 149, 167, 169
medulla oblongata 149, 151
medullary cavity 45
mentalis muscle 103
mesentery 281
metabolic physiology 23
metacarpals 49, 63, 69
metatarsals 49, 73, 77
microglial cells 143
microscope 25
midbrain 149, 167, 169
middle phalanx 69, 77
mitochondrion 19, 21, 99, 101, 145
monocytes 213
monosynaptic reflex arc 145
motor end plate 99
motor neurons 99, 143, 145, 169
mouth 107, 275
movements, body 9, 11
multicellular organization 17
muscles. *See also* names of individual muscles
  masticatory muscles 105
  of abdominal wall 115
  of deep back 111
  of eye 187
  of face and neck 103, 109, 171
  of foot 135, 139
  of leg 135, 137
  of rib cage 113
  of the hip 131
  of the mouth floor 107
  of the pelvic diaphragm 117
  of thigh 133
  respiratory muscles 269
  shoulder and arm 85, 119
muscle tissues 33
myelin sheath 143
mylohyoid muscle 107
myofibrils 99, 101

N

nail bed 41
nail matrix 41
nasal bone 51, 53, 55
nasal cavity 193, 255, 257, 259
nasalis muscle 103
Natural Killer Cell (NK cell) 251
navicular bone 77, 95
neck 5, 11, 103, 109, 171, 223, 225
nephron function 303
nerves
  cervical plexus 171
  cranial nerves 149, 161
  lumbosacral plexus 175
  of lower limb 177
  of upper limb 173
nervous tissues 33
neucleolus 143
neural pathways 167, 169
neuromuscular junction 99
neurons 33, 143

neuron structure 143
neurotransmitters 145
neutrons 17
neutrophils 213
nipples 319
nodes of Ranvier 143
norepinephrine 207
nose 257
nucleus 17, 19, 21, 33, 143

O

oblique muscle 269
occipital bone 51, 53, 55
occipitalis muscle 103
occipital lobe 147
olfactory reception 193
oligodendrocytes 143
omohyoid muscle 103
optic nerve 185, 187
optic pathway 187
oral cavity 255, 273, 275
orbicularis oculi muscle 103
orbicularis oris muscle 103
oropharynx 279
osteocyte 45
ovaries 201, 313, 315
oxygen poor blood 267
oxygen rich blood 267

P

palatine tonsils 275
pancreas 201, 209, 273, 281, 291
paranasal air sinuses 259
parasympathetic nervous system 181
parathyroid glands 201, 205
parietal bone 51, 53, 55
parietal lobe 147
parotid salivary gland 275
patella 47, 49, 73, 75
pectoralis major muscles 119, 319
pectoralis minor muscles 119
pelvic arteries 243
pelvic cavity 7
pelvis 5, 49, 71, 243
penis 117, 311
pericardial cavity 7
pericardium 215
perimysium 101
periosteum 45
Peyer's patches 247
phalanges 49, 63, 69, 73, 77
pharynx 255, 261, 273, 275
pH balance 15
pia mater 157, 165
pineal body 149
pineal gland 149, 155, 201
piriformis muscle 117
pisiform bone 69
pituitary gland 147, 201, 203, 227
pivot joint 81
plane joint 81
planes, body 3
plantar flexion 11
plasma 213
plasma membrane 19, 21
platelets 213
platysma muscle 103
pleural cavity 7
polysynaptic reflex arc 145
pons 147, 149, 151
posterior direction 3
precentral and postcentral regions 159
primary bronchi 255
primary motor cortex 159
pronation movement 9
prostate gland 307, 311
protons 17
protraction 11
proximal direction 3
proximal phalanx 69, 77
pubic arch 71
pubis 71, 73
pulmonary arteries 215
pulmonary capillary 267
pulmonary veins 215

Q

quadrigeminal plate 149

R

radius 49, 63, 67, 119
rectum 273, 281, 287, 307
rectus abdominus muscle 115, 269
red blood cell 267
red blood cells 213
red bone marrow 251
reflex arcs 145
regions of taste 197
renal artery 297
renal capsule 301
renal cortex 301
renal medulla 301
renal papilla 301
renal pyramid 301
renal vein 297
reproductive system

female 313–319
male 307–311
reproductive cycle 317
respiratory system
bronchial tree 263
external nose 257
lungs 215, 217, 255, 265, 273
nasal cavity 193, 255, 257, 259
overview 255
paranasal air sinuses 259
pharynx and larynx 261
respiratory muscles 269
reticular connective tissue 31
retina 185
retraction 11
rib cage 49, 61, 113, 255, 269
ribosomes 19
risorius muscle 103
root of nail 41
rotation movement 9, 11
rough endoplasmic reticulum 19

S

sacroiliac joint 71
sacrum 49, 57, 59, 71, 117, 307
saddle joint 81
sagittal plane 3
salivary glands 273, 275
sarcolemma 99, 101
sarcoplasmic reticulum 99, 101
scalene muscles 269
scaphoid bone 69
scapula 47, 49, 63, 65, 85, 119, 121
scrotum 117, 307
seminal vesicle 307, 311
seminiferous tubules 311
senses
ears 189, 191
eyes 185, 187
olfactory reception 193
taste reception 195
sensory neurons 143
sensory receptors 143
serratus anterior muscle 119
sesamoid bone 47
short bones 47
shoulder 11, 85, 119
sigmoid colon 287
simple columnar epithelium 29
simple cuboidal epithelium 29
simple pseudostratified columnar epithelium 29
simple squamous epithelium 29
sinuses 259
skeletal muscle 101
skeleton 47, 49. *See also* bones
skin. *See* integumentary system
skull 51–55
sliding filament theory 101
small intestine 17, 273, 281, 285
smooth endoplasmic reticulum 19
smooth muscle tissue 17, 33
soft palate 257, 275, 279
somatosensory cortex 159
spermatic cord 309
spermatogenesis 311
sphenoid bone 51, 53, 55, 257
sphenoid sinus 255, 257, 259
spinal cord 147, 149, 163, 165, 167, 169, 307
spinal ligaments 83
spine 57, 59, 83
spleen 247, 291
stapes 49, 189
sternocleidomastoid muscle 103, 269
sternohyoid muscle 103
sternum 49, 61, 265
stomach 17, 273, 279, 281, 283
stratified columnar epithelium 29
stratified squamous epithelium 29
stratified transitional epithelium 29
stratum basale 37, 39
stratum corneum 37, 39
stratum granulosum 37, 39
stratum lucidum 37, 39
stratum spinosum 37, 39
striated muscle tissue 33
stylohyoid muscle 107
styloid process 51, 53, 55
subarachnoid space 151
subclavius muscle 119
sublingual salivary gland 275
submandibular salivary gland 275
subscapularis muscle 119
superficial transverse perineal muscle 117
superior direction 3
superior sagittal sinus 151, 157
supination movement 9
swallowing 279
sympathetic nervous system 181
synapses 145
synovial joint 81

T

talus bone 77, 95
tarsals 49, 73
taste buds 197
taste reception 195

teeth 275, 277
temporal bone 51, 53, 55
temporalis muscle 103, 105
temporal lobe 147
testes 201, 307, 309, 311
thalamus 147, 149, 153, 155
thigh 75, 133
thoracic area 5
thoracic cavity 7
thoracic vertebrae 49, 57, 59
thumb, opposition of 11
thymus gland 201, 247, 251
thyroid gland 201, 205
tibia 49, 73, 75, 95
tissues
  bone tissues 45
  cartilage tissues 33
  connective tissues 17, 31
  epithelial tissues 17, 29
  muscle tissues 33
  nervous tissues 33
tongue 273, 275, 279
tonsils 247, 275
trachea 205, 215, 255, 261, 263, 265, 273, 275, 279
transverse colon 281, 287
transverse plane 3
transverse tubules 99, 101
transversus abdominus muscle 115
trapezium bone 69
trapezius muscle 103, 121
trapezoid bone 69
triceps brachii muscle 121
triquetrum bone 69
true ribs 61
trunk movements 9
type II alveolar cell 267

U

ulna 49, 63, 67, 119
upper limb
  area of 5
  arteries 231
  bones of 63, 67
  nerves of 173
  posterior muscles of 121
  veins 233
upper respiratory tract 255
ureter 297, 299, 307, 311, 313
urethra 297, 307, 311, 313
urinary bladder 281, 297, 307, 311, 313
urinary system 297–303
urine formation 303
uterine tubes 313, 315
uterus 281, 313, 315
uvula 275, 279

V

vagina 313, 315
vas deferens 311
veins
  hepatic portal system 241
  of head and neck 225
  of lower limb 237
  of upper limb 233
  structure of 221
ventral direction 3
ventral nerve root 165
ventricles 151
vertebrae 47, 49, 57, 59
vertebral cavity 7
vestibular nerve 189
vomer bone 51, 53, 259

W

white blood cells 213
white matter 165
wrist 69, 89

X

xiphoid process 61

Z

zygomatic arch 105
zygomatic bone 51, 53, 55
zygomaticus major muscle 103
zygomaticus minor muscle 103